Orthodoxy and the Cold War

Orthodoxy and the Cold War

Religion and Political Power in Romania, 1947–65

Lucian N. Leustean
Lecturer in Politics and International Relations, Aston University, Birmingham, UK

palgrave
macmillan

First published 2009 by
PALGRAVE MACMILLAN

Palgrave Macmillan in the UK is an imprint of Macmillan Publishers Limited,
registered in England, company number 785998, of Houndmills, Basingstoke,
Hampshire RG21 6XS.

Palgrave Macmillan in the US is a division of St Martin's Press LLC,
175 Fifth Avenue, New York, NY 10010.

Palgrave Macmillan is the global academic imprint of the above companies
and has companies and representatives throughout the world.

Palgrave® and Macmillan® are registered trademarks in the United States,
the United Kingdom, Europe and other countries

ISBN-13: 978-0-230-21801-7   hardback
ISBN-10: 0-230-21801-6   hardback

This book is printed on paper suitable for recycling and made from fully
managed and sustained forest sources. Logging, pulping and manufacturing
processes are expected to conform to the environmental regulations of the
country of origin.

A catalogue record for this book is available from the British Library.

Library of Congress Cataloging-in-Publication Data
  Leustean, Lucian.
    Orthodoxy and the Cold War : religion and political power in
    Romania, 1947–65 / Lucian N. Leustean.
      p. cm.
    Includes bibliographical references and index.
    ISBN 978-0-230-21801-7 (alk. paper)
    1. Romanian Orthodox Church—History—20th century. 2. Christianity and
politics—Romanian Orthodox Church—History—20th century.   3. Communism
and Christianity—Romanian Orthodox Church—History—20th century.
4. Christianity and politics—Romania—History—20th century.
5. Communism and Christianity—Romania—History—20th century. I. Title.

    BX698.L48 2009
    322'.10949809045—dc22   2008029985

10   9   8   7   6   5   4   3   2   1
18 17 16 15 14 13 12 11 10 09

Printed and bound in Great Britain by
CPI Antony Rowe, Chippenham and Eastbourne

To Deborah, Clara, Lucia and Nicolae

Contents

Figures

Acknowledgements

In conducting research for this book I am indebted to many people. The manuscript began as a doctoral thesis in the Department of Government at the London School of Economics and Political Science (LSE) in 2003. I would like to thank my supervisors, John Madeley and Dr John Hutchinson, for their thorough comments, Professor Rodney Barker for guiding my research skills, and my examiners Professor Emeritus David Martin and Professor Emeritus George Schőpflin for their valuable criticism. At the LSE, the Association for the Study of Ethnicity and Nationalism, led by Professor Emeritus Anthony Smith, Professor John Breuilly and Dr John Hutchinson, provided the intellectual home for writing and discussing research. I have greatly benefited from the incisive debates and inspiring atmosphere of the seminars and annual conferences.

In 2007 I joined the School of Languages and Social Sciences at Aston University in Birmingham where I completed the book. I am grateful to my colleagues, in particular, Professor Michael Sutton and Professor Anne Stevens, for their support and discussions on religion and European politics.

Research for this book was conducted in a number of archives and libraries abroad and I am grateful for the assistance of their staff who shared their time and expertise: the Central National Historical Archives of Romania, Bucharest; the Archive of the Romanian Information Service, Bucharest; the BBC Written Archives Centre, Reading; the National Archives of the United Kingdom, London; Lambeth Palace Archives, London; Radio Free Europe Archives, Budapest; Keston Institute, Oxford; the Fellowship of Saint Alban and Saint Sergius, Oxford; the Archives of the World Council of Churches, Geneva; the Library of the Faculty of Orthodox Theology, University of Bucharest; the Library of the Romanian Academy, Bucharest; the Library of the National Archives of Romania, Bucharest; Gennadius Library and the National Library of Greece, Athens; the Library of the Getty Institute, Los Angeles; and the Library of the Central European University, Budapest.

I am grateful to the Right Reverend John Richard Satterthwaite for recollecting his journeys to Romania in the 1950s and 1960s. His comments at our meeting were extremely valuable for understanding how the Anglican hierarchy perceived their counterparts in Romania during this time. I would like to thank Michael Strang, my editor, Ruth Ireland, his assistant, and Palgrave's anonymous reviewer, for their guidance and support in publishing this book.

This research would not have been possible without the generous financial contributions of the following bodies for which I am very grateful: Department of Government, London School of Economics and Political Science; Open

Society Institute, New York; Anglo-Romanian Bank, London; Central Research Fund of the University of London; Institute of Historical Research, London; Aston University, Birmingham; and Rațiu Family Foundation – Romanian Cultural Centre, London. In particular, at the latter I would especially like to thank President Nicolae Rațiu, Director Ramona Mitrică and Mihai Rîșnoveanu.

Earlier versions of some chapters of this book were published in refereed journals. I am grateful for the comments of the editors and anonymous reviewers of these journals and for permission to reproduce copyright material: Taylor and Francis for Lucian N. Leustean, '"For the Glory of Romanians": Orthodoxy and Nationalism in Greater Romania, 1918–1945', *Nationalities Papers*, 35 (3), pp. 717–42; Lucian N. Leustean, 'Ethno-symbolic Nationalism, Orthodoxy and the Installation of Communism in Romania, 23 August 1944 to 31 December 1947', *Nationalities Papers*, 2005, 33 (4), pp. 439–58; Lucian N. Leustean, '"There is No Longer Spring in Romania, It is All Propaganda": Orthodoxy and Sovietization, 1950–1952', *Religion, State and Society*, 35 (1), pp. 43–68; Lucian N. Leustean, 'Constructing Communism in the Romanian People's Republic. Orthodoxy and State, 1948–1949', *Europe-Asia Studies*, 2007, 59 (2), pp. 303–29; Sage Publishers for Lucian N. Leustean, 'The Political Control of Orthodoxy in the Construction of the Romanian State, 1859–1918', *European History Quarterly*, 2007, 37 (1), pp. 60–81; Lucian N. Leustean, 'Religious Diplomacy and Socialism: The Romanian Orthodox Church and the Church of England, 1956–1959', *East European Politics and Society*, 2008, 22 (1), pp. 7–43; and Maney Publishing for the Modern Humanities Research Association and the School of Slavonic and East European Studies for Lucian N. Leustean, 'Between Moscow and London: Romanian Orthodoxy and National Communism, 1960–1965', *Slavonic and East European Review*, 2007, 85 (3), pp. 491–522. Together with the publishers, I wish to thank the Romanian National News Agency ROMPRES, who gave copyright permission to include their photographs.

This book is dedicated to Deborah and Clara, and my parents Lucia and Nicolae. To the above people and all those near to me in the last years, a great thank you.

Abbreviations

ANIC	*Arhivele Naţionale Istorice Centrale* [The Central National Historical Archives]
ARLUS	*Asociaţia română pentru legăturile de prietenie cu Uniunea Sovietică* [The Romanian Association for Friendly Relations with the Soviet Union]
ASRI	*Arhiva Serviciului Român de Informaţii* [Archive of the Romanian Information Service]
AFSASS	Archives of the Fellowship of Saint Alban and Saint Sergius
AKI	Archives of the Keston Institute
AWCC	Archive of the World Council of Churches
BBC WA	BBC Written Archives Centre
BOR	*Biserica Ortodoxă Română* [The Romanian Orthodox Church]
CEC	Conference of European Churches
CC of RCP	Central Committee of the Romanian Communist Party
CECFR	Church of England Council on Foreign Relations
CNSAS	*Consiliul Naţional pentru Studierea Arhivelor fostei Securităţi*, [National Council for the Study of the Archives of the former *Securitate*]
Comecon	Council for Mutual Economics Assistance
CPC	Christian Peace Conference
CPSU	Communist Party of the Soviet Union
FO	Foreign Office
GS	General Secretariat
GNA	Grand National Assembly
HU OSA	Open Society Archives – The Archives of Radio Free Europe
LNP	Liberal National Party
LPA	Lambeth Palace Archives
TNAUK	The National Archives of the United Kingdom
NKVD	Soviet Secret Police
PDF	Popular Democratic Front
PNP	Popular National Party
PRO	Public Record Office
ROC	Romanian Orthodox Church
ROEA	Romanian Orthodox Episcopate in America
RCP	Romanian Communist Party
RPR	Romanian People's Republic
RWP	Romanian Workers' Party
SDP	Social Democratic Party
WCC	World Council of Churches

Map 1 Romania, 1859–1965.

Introduction

In May 1965, in the middle of the Cold War, the Committee of the Judiciary United States Senate held an unusual hearing. Senators listened to the testimony of a Romanian reverend who had managed to escape from behind the Iron Curtain. His name was Richard Wurmbrand. Born in a Jewish family he converted to Christianity in 1936 and was twice convicted for his religious belief; he spent nine years in Romanian prisons, three of which were in solitary confinement in an underground cell. A few years before his escape from Romania, a group of Norwegian Christians raised $10,000 and, bribing communist officials, obtained a passport for the reverend and his wife. Wurmbrand spent a few years in Europe, but fearing the long arm of the Romanian Security Intelligence, the *Securitate*,[1] went to the US where he was now giving evidence of what he had experienced.[2] He shocked the audience when, as proof of his words, he unbuttoned his shirt and showed 18 deep wound marks inscribed on his body.[3] His testimony was a personal account of his inhumane experiences in prisons and of the general attitude of the atheist regime against religion.

At the end of his statement Wurmbrand recalled a touching story witnessed by himself and his wife while they were held in a slave labour camp constructing the Danube Canal. The canal was one of the most dreadful places in Romania. Among the inmates there was a special religious group of around 400 prisoners composed of priests, bishops and simple peasants who believed 'too much' in their religion and were perceived as dangerous to the political system. Wurmbrand remembered an event which strengthened his religious conviction:

And now a Sunday morning the political officer of the prison comes, the whole brigade is gathered, and just at random he sees a young man. He calls him, 'What is your name?' He says his name. 'What are you by profession?' He said 'A priest'. And then mocking the Communist said 'Do you still believe in God?' The priest knew that if he says yes, this is the last day of his life. We all looked to him. For a few seconds he was silent.

1

Then his face began to shine and then he opened his mouth and with a very humble but with a very decided voice he said 'Mr Lieutenant, when I became a priest I knew that during church history thousands of Christians and priests have been killed for their faith, and notwithstanding I became a Christian and then I became a priest. I knew what I became. And as often as I entered the altar clad in this beautiful ornate which priests wear I promised to God, 'If I wear the uniform of a prisoner, then also I will serve Him. Mr Lieutenant, prison is not an argument against religion. I love Christ from all my heart.'

I am sad that I can't give the intonation with which he said these words. I think that Juliet when she spoke about Romeo, she spoke like that. We were ashamed because we – we believed in Christ. This man loved Christ as a bride loves the bridegroom. This man has been beaten and tortured to death. But this is Romania. Romania is a country which is mocked, which is oppressed, but deep in the hearts of the people is a great esteem and a great praise for those who have suffered. The love to God, the love to Christ, the love to the fatherland has never ceased. My country will live.[4]

Wurmbrand's passion and eloquence make his account stand out from the thousands of testimonies from people who survived the horrors of communist prisons. Religion was denigrated and seen as the means through which people managed to escape indoctrination and were not completely controlled. Officially, political leaders stated that communism was the only regime under which people acquired complete religious freedom, but in practice any reference to religion was considered as a retrograde, bourgeois and subversive activity. Christianity was seen as the forerunner of the true ideology, communism, and its progress inevitably replaced any form of religious belief. The construction of the new man meant that religion should lose its meaning and people should embrace the communist ideal at the core of their beliefs. The history of Christian churches during Romanian communism is a sad one. As the leader of the party, Gheorghe Gheorghiu-Dej, officially recognised in 1961, from 1945 to 1952 more than 80,000 people were arrested of which 30,000 were imprisoned. Among them many priests and ordinary people were convicted for asserting their religious beliefs.

The installation of Romanian communism, under the military force of the Soviet Union, brought fears that the new authorities would follow a similar repressive attitude towards religion to that in Russia after the 1917 Revolution. Instead, Romania took a different trajectory and while the dominant religious institution, the Romanian Orthodox Church, suffered persecution, at times, its leadership benefited from its collaboration with the regime. Some church hierarchs praised the new leaders while others took a more ambivalent position. Those who showed dissatisfaction with the new regime suffered and were removed from their positions, while those who did not show interest in the political transformation of the country were closely monitored.

This book investigates the subtle and complex ways in which the Romanian Orthodox Church collaborated with and adapted to the communist regime in the Romanian People's Republic from 1947 to 1965. It considers the contradictions and ambivalent position of the church during the early Cold War period and addresses the following questions: How did the Romanian Orthodox Church, which openly opposed communism in the interwar period, survive an atheist regime? How did it adapt to the installation of communism? Did it have a strategy? What was the relationship between the Orthodox Church and other religions in Romania and what was its relationship with the Orthodox commonwealth? Did the regime use the church internally and, if so, how? Why did the regime allow the church to develop foreign ecclesiastical contacts? Moreover, to what extent were the party and its leadership atheist?

This study draws on unpublished material from recently opened archives, to examine the main political decisions which affected the life of the church. Access to these unpublished sources offers a better understanding of the place of religion in society and the evolution of Romanian politics. Central to the relationship between church and state was the role played by individuals at the top of the religious and political organisations. By presenting the trajectory of religious and political leaders, this book offers insights into the importance of individuals in shaping the history of religion and the Romanian state during this period.

This is neither a theoretical analysis of Orthodoxy and nationalism nor a study of Orthodoxy and the modern state.[5] It does not detail every historical event but focuses on those that influenced the evolution of church-state relations. It analyses the church as a political institution rather than Orthodoxy as a religious practice and community. By presenting the institutional dimension, it examines the means of survival of the church and the ways in which the church hierarchy sought collaboration with the regime. This study does not aim to accuse the church leaders of this period, some of whom are still alive and in hierarchical positions, but presents the ways in which the church survived one of the most contentious periods of Romanian history. This book offers a new understanding of this period in the light of new unpublished material.

The accommodation of the Romanian Orthodox Church within the communist regime sparked debates after the end of the Cold War period. On the one hand, some scholars argued that the Romanian Orthodox Church managed to survive by establishing an ideological pact with the communists. In this way, the attitude of the church during this period could be perceived as a form of Sergianism similar to the pact between the Moscow Patriarchate and the Soviet Comintern;[6] on the other hand, some scholars have pointed out that the church adapted to the communist regime, but this accommodation could be regarded as a form of resistance.[7] According to the latter view, Patriarch Justinian, whose position was approved by the party, did not completely obey the regime but, in fact, did everything in his position to oppose communism. This research will investigate both sides and assess the trajectory of church-state relations.

There are only a few studies of early Romanian communism.[8] They analyse the trajectory of the communist party and argue for various dates when the country embarked on a 'semi-autonomous' form of politics in the Eastern bloc. Stephen Fischer-Galaţi suggests that the Romanian road to communism could be perceived in the mid-1950s, in particular around 1955 when the country began to mediate the Sino-Soviet ideological conflict,[9] while David Floyd claims that it began in 1958–9 when the party changed its economic policy.[10] Alexandru Bârlădeanu, a communist leader during this period, suggested in an interview in 1998 that Gheorghiu-Dej and the Romanian government began approaching the West in 1956 in an attempt to foster economic relations which would allow the country a more flexible and independent politics in the Eastern bloc.[11]

This study of church-state relations in the Romanian People's Republic supports Fischer-Galaţi's and Bârlădeanu's views that in the 1950s the regime was interested in asserting a more autonomous politics in the Eastern bloc. However, a systematised politics of national communism only began to develop in the early 1960s and would reach its climax under Nicolae Ceauşescu's dictatorship. Romania developed a special position behind the Iron Curtain, being the only country which did not participate in the military invasion of Czechoslovakia in 1968, and, initially, national communism during the Ceauşescu period contributed towards improving relations with Western European governments.

Analysis of Romania as a case study is extremely valuable for comprehending the transformation of Orthodoxy in countries under communism. The regime used the institutional religious channels of the church in order to present a false image of religious freedom abroad and to foster the country's political interests. This book investigates the internal means through which the regime employed the church and the perceptions and activities of Western European religious leaders of the religious and political situation in Romania.

There are no detailed studies on Romanian Orthodoxy and communism in Romanian or English. The best-known study is Olivier Gillet's *Religion et Nationalisme: L'Ideologie de l'Eglise Orthodoxe Roumaine sous le Regime Communiste*. Gillet offers a useful examination of how the church adapted to the regime by appropriating themes of the country's national past. However, Gillet does not carry out any significant archival research but instead refers only to the main journals of the church which were published in English. This book takes Gillet's argument further and analyses the collaboration between church and regime highlighting those factors which helped the church maintain a strong position in society.

An important contribution to the field and somewhat unpublicised study is Kaisamari Hintikka's *The Romanian Orthodox Church and the World Council of Churches, 1961–1977*.[12] Hintikka deciphers many unknown details of the life of the church in the 1960s and 1970s but focuses only on relations between the Romanian Orthodox Church and the World Council of Churches without exploring Romanian communism. In addition, although she conducted archival research on the Orthodox Church in Romania, Finland, Switzerland and England, her study presents official letters between institutions without analysing the ways in which the church was affected by political decisions. This book presents not only documents written by religious leaders but also reports of Radio Free Europe and the *Securitate*. While these reports need to be read cautiously, they articulate the perceptions of those who were close to top-level religious and political leaders and who witnessed the evolution of church-state relations inside the country.

In 2005, two studies were published in Romanian which were based on archival research: Cristian Vasile's *Biserica Ortodoxă Română în primul deceniu communist* (The Romanian Orthodox Church in the First Communist Decade) and George Enache's *Ortodoxie şi putere politică în România contemporană* (Orthodoxy and Political Power in Contemporary Romania). Each book offers a fascinating introduction to the study of the Orthodox Church during communism, focusing especially on the personal histories of various church leaders, compiled from unpublished documents of the Romanian archives. However, neither author examines how church actions were connected to the political field and how the West perceived the Romanian Orthodox Church.

This book offers a detailed analysis of the role of the church during this period by drawing on newly accessible archival resources. In Bucharest, I have conducted research in the *Securitate*'s Archives and the Central National Historical Archives. Outside Romania I have conducted research in the Archives of Radio Free Europe, Budapest; the BBC Historical Written Archives, Reading; the Lambeth Palace Archives and the National Archives of the United Kingdom, London; the Archives of Keston Institute[13] and the Archives of the Fellowship of Saint Alban and Saint Sergius, Oxford; and the Archives of the World Council of Churches, Geneva.[14]

I contacted the Archive of the Holy Synod of the Romanian Orthodox Church; however, I have not been granted permission to see its documents and most scholars who have approached it have experienced a similar response. Hintikka is the only non-Romanian scholar who has been allowed access to documents. However, her research primarily focuses on the relations between the Romanian Orthodox Church and the World Council of Churches and, as she mentions, 'it is obvious that some documents are not included in the files but have either been destroyed or have disappeared. This seems to be the state of affairs especially in the case of the Patriarch's correspondence'.[15] In addition, in 2005, Father Iustin Marchiş, a well known Romanian theologian, claimed he saw the vicar bishop of the Patriarchate destroying documents from this Archive.[16] While the current leadership of the church continues to have deep connections with the communist past, it seems that, for the meantime, it will be difficult to have open access to the remaining archival material of the church. In addition, even when the archive is made completely available, researchers will face difficulties as most decisions during the early period of communism were transmitted by telephone between the Romanian Patriarchate and the communist authorities.[17] For this reason it is difficult to have a clear image of exactly who took the initiative on various political and religious issues as they have not always been recorded.

The opening of communist archives in Romania in recent years has offered access to a broad range of reports from the *Securitate* which reveal how the communists sought to influence religious life and provide evocative descriptions of the private lives of public religious figures. While one has to bear in mind who wrote these notes and for whom, they provide valuable material for the contemporary researcher. Small gestures by members of the church hierarchy and by the communists might seem insignificant. However, combined with the communists' general attitude towards religion in the Eastern bloc and detailed investigation of particular events, they offer a greater understanding of the dynamics between leading public figures and of the atmosphere 'behind the scenes' at the time. Analysis of the close timing of a number of events indicates the ambivalent position of the church hierarchy, state control of the church, the nationalist resistance of the church and the church's relationship with the Soviet Union.

Three years after my first application, in the summer of 2007, I was granted access to the *Securitate* archives which contain 24,000 metres of dossiers.[18] Although officially open to the public, the archive remains highly censored especially on those dossiers which contain sensitive information regarding the national security of Romania. In particular, material on the former Service of Foreign Affairs of the *Securitate* has been made only partially public. Top hierarchs travelled extensively outside Romania and it is highly likely that they had files in this archive, either due to being followed by *Securitate* agents or because they were working for it. This study does not aim to expose the clergy within the *Securitate*. Instead, it focuses on the ways in which the church collaborated and survived this period despite the personal trajectory of some of its members. The *Securitate* archives provide valuable material on the private lives of top hierarchs, seen through a political perspective.

In 2001 Cristina Păiuşan and Radu Ciuceanu of the Institute for the Study of Totalitarianism in Bucharest published a book in Romanian of *Securitate* documents spanning from 1945 to 1958 which throws light on aspects of the church's position during this period.[19] The main task of the *Securitate* was to ensure that the regime controlled almost every aspect of social life. Some documents were fabricated and it is hard to confirm the authenticity of some information. The same approach has to be applied to documents which were written for Radio Free Europe. While inside Romania, informers sometimes created documents to further their own positions; outside the country reports were written in order to show how bad the communist regime was, presenting obviously invented facts or a distorted image of Romanian communism.

Most church activities have been recorded in ecclesiastical journals and they offer useful material on church policy, showing the official actions and speeches of the church hierarchy.[20] I have found many copies of these journal articles in *Securitate* files, showing that its agents read them carefully.

In addition to these sources, particular documents in the Lambeth Palace Archives offer unique perspectives on this period. The Romanian Orthodox Church had enjoyed good relations with the Church of England during the interwar period. The dossiers in these archives contain reports of official and private meetings between Western religious leaders and Romanian Orthodox hierarchs. Analysis of these documents fleshes out both the personalities of the Romanian hierarchs seen through the eyes of their Anglican counterparts and confirms the impact of political leaders on religious life in Romania. The detailed conversations of those who travelled to Romania and met clergymen, from mere priests to members of the leadership, offer valuable material on the place of religion in Romanian society.

Analysis of the Romanian Orthodox Church during the early Cold War period has to take into account its social and political role in previous periods, dating back to the establishment of the state in 1859. Close relations between the religious and political realms influenced the political evolution of the country. For this reason, the study is structured in chronological order, discussing the most relevant events concerning relations between the church and successive political regimes.

The first chapter analyses the theoretical framework of the relationship between church and state in Orthodoxy. It presents the concept of *symphonia* as the basis of collaboration between the Orthodox hierarchy and political regimes in a tradition which dates back to the Byzantine period. It examines the place of Romanian Orthodoxy within the Orthodox commonwealth, highlighting the similarities and differences with other countries in the Eastern bloc.

The second chapter examines the relationship between the Orthodox Church and the state from the establishment of the Romanian state in 1859 to the end of the Second World War. In the Principalities of Wallachia and Moldavia the Orthodox Church became a state institution during the reign of Prince Alexandru Ioan Cuza who founded a Synod which reflected his political interests. In addition, with the declaration of independence of the

Romanian state from the Ottoman Empire in 1877, the Orthodox Church sought its own independence from the religious jurisdiction of the Ecumenical Patriarchate and declared its autocephaly. After the establishment of Greater Romania in 1918, the Orthodox Church asserted a nationalist discourse more vigorously and proclaimed the status of Patriarchate in 1923. Early church legislation, from the election of a hierarchy to the composition of local clerical assemblies, would set the conditions for political dominance of the church during the Cold War period. This historical background is important for analysing the evolution of church-state relations.

The third chapter investigates the installation of communism in Romania and the attitude of the church towards the new regime from 1944 to 1947. The government of Petru Groza, the son of a priest who was a lay member of the church, imposed control over the church hierarchy. Furthermore, the death of Patriarch Nicodim in 1948 led to the appointment of Patriarch Justinian, whose accession was supported by the communists.

The fourth chapter analyses the position of the Orthodox Church from the establishment of the Romanian People's Republic in 1947 to the climax of Sovietization as represented by the endorsement of the 1952 Constitution. Despite Orthodox churches in other communist countries facing public trials and mass persecutions, the Romanian church hierarchy officially enjoyed good relations with the authorities. The church had an important position in society mainly because the communists were able to use it in promoting their own political agenda. By strengthening its relations with Moscow and confirming its distance from capitalist Catholic and Protestant Western Europe, the church dominated the Romanian religious scene. Thus, the Greek Catholic Uniate Church was forcibly incorporated into the Orthodox Church and official relations with the Vatican were interrupted, while the Orthodox hierarchy praised the communist leaders for promoting 'religious freedom'. This study argues that during the 1950s the church reintroduced a nationalist discourse which would become more visible with the development of the regime's stance against Moscow. It investigates the impact of the new law of religious confessions and the propagandistic employment of the church in the battle for peace. In addition, it presents the personal experiences of an Arab priest who was a guest in the Patriarchal Palace for five months from 1952 to 1953, offering an eyewitness account of the church hierarchy and of the place of religion in Romania.

The fifth chapter focuses on the role of the Orthodox Church at the beginning of the Romanian road to communism from 1953 to 1955. As part of the communists' intention to distance the country from the Soviet Union and to create a more cohesive national agenda, the Orthodox Church was extensively used in propaganda. For the first time in church history, Romanian saints were canonised in October 1955.

The sixth chapter examines one of the most strained periods in church history from the Second Congress of the Romanian Workers' Party in 1955 to the issuing of the decree for the regulation of monasteries in 1959. The party leadership intensified its control of the church, culminating in restrictions on monastic life. If in the first years of communism Patriarch Justinian showed subservience towards the regime, after 1956 he faced difficulties. On the one hand, the regime nominated new members of the hierarchy as his ecclesiastical competitors; on the other hand, it began a campaign aimed at reducing the influence of the church in society. The ambivalent position of Justinian towards the regime led to rumours in the West in 1958 that he was under house arrest. Facing political pressure from the regime, the patriarch turned towards Western religious leaders as a means of protecting the church. The most important relations of the Romanian Orthodox Church were with the Church of England, in a tradition which dated back to the interwar period when a delegation of the Romanian hierarchy visited London in 1936. This research argues that fostering closer relations with the Anglican hierarchy became one of the most important issues on the patriarch's agenda. At the same time, Justinian's connections with the West were encouraged by the regime which wanted to expand its influence abroad and saw the value of engaging the church in its political mission.

The last chapter presents the connections between the rise of Romanian national communism and the activity of the church from 1960 to 1965. At the Party Plenum in 1964, Gheorghiu-Dej officially claimed the right of every communist party 'to elaborate, choose or change the forms and methods of socialist construction',[21] excluding the idea of a 'parent' and 'son' party and asserting national communism as state policy.[22] While this was a political decision, the church followed the same line and strengthened its relations with other Orthodox countries which were also seen as promoting independent politics, especially Yugoslavia and Greece, and with Western countries, especially England. Increased contacts with the West reached a climax with the visit of the archbishop of Canterbury, Arthur Michael Ramsey, to Romania in 1965. The unexpected death of Gheorghiu-Dej in 1965 led to the rise of Nicolae Ceauşescu as his successor, who maintained the same attitude towards the church. The new position of Romania within international communism and the transformation of the party leadership set the country on a new political path, renamed as the Socialist Republic of Romania.

1
Orthodoxy, *Symphonia* and Political Power in East European Communism

Orthodoxy and *symphonia*

Orthodoxy is the dominant religion in south-eastern Europe and Russia. Alongside Roman Catholicism and Protestantism, it is one of the three major branches of Christianity. It numbers around 200 million faithful and largely spans from the Far East to Eastern Europe, while important diaspora communities are present in Western Europe, North America, Africa and Australia. *The Oxford Dictionary of the Christian Church* defines Orthodoxy as

> a family of Churches, situated mainly in Eastern Europe: each member Church is independent in its internal administration, but all share the same faith and are in communion with one another, acknowledging the honorary primacy of the Patriarch of Constantinople.[1]

The 'Orthodox' in Orthodox Christianity means 'correct belief' or 'right thinking', while the sometimes-used adjective 'Eastern' in the construction of the term 'Eastern Orthodoxy Christianity' refers to the eastern part of the Roman Empire and the Christian conversion of Eastern Europe under the political domination of Constantinople. In contrast with the Roman Catholic Church and Protestant churches, Orthodox churches lack a systematised doctrine regarding their relationship with the state.

In Orthodoxy, the relationship between church and state is characterised by the concept of *symphonia* (συμφωνία) or the 'system of co-reciprocity' (Σύστημα συναλληλίας, lat. *consonantia*), a doctrine which developed in Byzantium. As John Meyendorff states, 'the great dream of Byzantine civilisation was a universal Christian society, administered by the emperor and spiritually guided by the Church'.[2] According to this vision, the empire was considered the kingdom which would last forever, whose political and religious domination would be without competitor on earth, as it was the reflection of Christ's kingdom. Both the church and the state should collaborate towards 'achieving a sublime destiny'[3] of the people under their jurisdiction and there is no conflict between the means employed by the church or the state in promoting the welfare of their subjects. While there is a separation between the completely laic character of the state and the religious status of the church, *symphonia* promotes equality and an intimate relationship between these institutions, however, with different priorities and methods of operating. The state is interested in its survival within a system of states and the projection of its power in international politics. The church operates with religious methods leading the community towards the best way of achieving spiritual progress and salvation.

According to *symphonia*, both church and state use their own laws in order to promote their purposes and there is no confusion between them. The state does not rule itself according to church law or vice versa. There is no interdependency nor is there a complete separation. In fact, the major problem of the concept of *symphonia* is that the demarcation line between church and state remains unclear. For this reason, religious leaders could achieve strong political roles in society and political leaders could influence the church's position. From this perspective, the ruler and the priest are the major political and, at the same time, religious figures on earth.

Both the church hierarchy and the emperor had special status in the Byzantine Empire. The emperor was considered the thirteenth apostle or equal with the apostles (*isapostolos*) who fought for the 'right' faith, while the patriarch was in charge of ensuring that the community was following the spiritual path towards salvation.[4] The church–state relationship was influenced by the mutual cooperation between the emperor and the patriarch on their respective paths to achieving their individual and their subjects' salvation. According to Orthodoxy, the emperor had a special place in the material and spiritual worlds as the chosen leader, 'similar to God, who is over all, for he does not have anyone higher than himself anywhere on earth'.[5] As Steven Runciman stated,

> the Church was a democratic institution. It was possible for any Orthodox Christian, however humble his origin, to attain to the Patriarchal throne; merit was in theory the sole criterion. And in practice, except when an Emperor deliberately appointed a nonentity – an action that was always unpopular – the Patriarchs were of a very high level of ability.[6]

The patriarch's main attributions were related to the spiritual progress of his faithful while the emperor regulated the life of the clergy, appointing the highest hierarchical positions and mediating conflicts between its clergy. The history of Byzantium also offers examples of patriarchs who acted against imperial policy. Thus, Patriarchs Photius, Germanus and Arsenius were deposed for defying their emperors who attempted to interfere too much in ecclesiastical policy while John Chrysostom lost his patriarchal throne for censuring the morals of the Court.[7] From the fifth century onwards, religious heresy was considered a state crime. Byzantium supported only one religion, Orthodoxy, and condemned its opponents, especially when they were politically dangerous for the stability of the empire, such as in the case of the Bogomils sect which had its own concept of Christianity and preached disobedience to state order.[8]

The concept of *symphonia* acquired a stronger dimension during the reign of Emperor Justinian, who, in the systematisation of civil law, set out some aspects of the relationship between church and empire. Referring to the importance of Christianity, Justinian stated in his *Edict to the People of Constantinople Concerning the Faith* in 554:

> We believe that the first and greatest blessing for all mankind is the con-fession of the Christian faith, true and beyond reproach, to the end that it may be universally established and that all the most holy priests of the whole globe may be joined together in unity and with one voice may confess and preach the ORTHODOX Christian faith, and that every plea devised by heretics may be rendered null and void.[9]

Furthermore, the classical text which indicates the boundaries of the priest-hood and imperial offices is Justinian's *Sixth Novel* in which he writes:

> There are two major gifts which God has given unto men of His supernal clemency, the priesthood and the imperial authority – *hierosyne* and *basileia; sacerdotium* and *imperium*. Of these, the former is concerned with things divine; the latter presides over the human affairs and takes care of them. Proceeding from the same source, both adorn human life. Nothing is of greater concern for the emperors as the dignity of the priesthood, so that priests may in their turn pray to God for them.[10]

Justinian's model was followed by his successors and became the tenet of the relationship between Orthodoxy and politics. His *Novels* were further devel-oped in the ninth century in a document titled *Epanagoge*, most probably written by Patriarch Photius. Even if this document was only a draft and was not officially adopted by the state, it was widely circulated in Orthodox territories and influenced the development of further legislation outside Byzantium, through the Middle Ages, to the creation of nation-states. *Epanagoge* states that 'the temporal power and the priesthood relate to each other as body and soul; they are necessary for state order just as body and soul are necessary in a living man. It is in their linkage and harmony that the well-being of a state lies.'[11]

In Orthodoxy the body of the church is understood differently from that in Western Europe, where it is perceived as merely an ecclesiastical organi-sation within the state. According to Orthodoxy, the church includes 'the whole body of the faithful, the "holy catholic Church" of the Creed, or at least the faithful of his own persuasion'[12] and for this reason, the emperor was obliged to interfere in church life, as he was responsible for the salvation of his own subjects before God.

Even if the Byzantine Empire disappeared from Europe's map with the fall of Constantinople on 26 May 1453, its religious and political legacy for Orthodox regions has remained.[13] As Henri Grégoire states, 'The Byzantine Church is the most important of Byzantine survivals. The Empire has disappeared, but the Church remains.'[14] Through its religious ceremonies and jurisdictional organisation, the church has continued to remind the faithful of the Byzantium model regarding the relationship between church and state. Myths, religious and political symbols of liturgical ceremonials continued to bring together the Orthodox people even if they were now subjects of other regimes. The commonwealth survived through the transformation of Orthodox churches into national churches which shared the same religious faith.[15]

The construction of nation-states in the Balkans was closely connected with the support of Orthodox churches for their countries' political regimes. The historical evolution of most Balkan states shows that politicians used Orthodoxy because of its nationalist message in order to induce national cohesion and gain support for their political programmes.[16] Looking back in history, Orthodoxy influenced the nation-building process as its hierarchy saw the possibility of reviving the Byzantine dream of a Christian state while political leaders sought to transform the religious identity of their communities into a national identity.[17]

The evolution of Orthodox nation-states in the Balkans showed direct connection between religious and political spheres. Thus, four years after obtaining independence from Turkish rule in 1829, Greece ruptured spiritually from the Ecumenical Patriarchate and proclaimed the Greek Autocephalous Church. Comparably, the proclamation of independence of the Romanian Principalities led to autocephaly for the Romanian Orthodox Church in 1885. The establishment of the Kingdom of Yugoslavia in 1918 was followed by the rise of the Serbian Patriarchate in 1920, while that of Greater Romania in 1918 led to the Romanian Patriarchate in 1923.

Orthodoxy and East European communism

Orthodox churches faced a new challenge with the emergence of communist ideology. Political leaders not only had to take into account the religious background of their countries but also sought to use the church's influence in order to benefit their regimes. Because Orthodoxy was moulded on political power, the church hierarchy generally remained passive to the threat of atheism and acted on the belief that 'every regime is the will of God'. Communists sought to control Orthodox churches because religious symbols remained deep rooted in society and offered legitimacy to traditional institutions. Collaboration between church and state remained on a public level while, at the same time, the regime implemented its atheist doctrine by systematically persecuting religion.

Political employment of religion remained the norm throughout the communist period. After the October Revolution in 1917, the Russian Orthodox Church was raised to the rank of Patriarchate and the regime combined international ecclesiastical prestige with internal mass religious persecution. From 1917 to 1922, 28 Russian bishops and 1215 Orthodox priests were executed, while the clergy were deprived of civil rights.[18] Stalin's intention to transform the church into a national museum failed mainly because it retained its authority in society. Despite persecutions, the beginning of the Second World War led to a higher level of cooperation between the Soviet communists and the church. The 1937 census in the USSR showed that 50 per cent of the urban and 70 per cent of the rural population considered themselves religious.[19] The regime employed the church as a propaganda tool both internally in motivating the people to participate in the war and externally in propagating the superiority of communist ideology in other predominantly Orthodox countries.[20]

Metropolitan Sergii, who was responsible for the Moscow See, sent a Pastoral Letter to the faithful on 22 June 1941 urging the people to fight against the enemies of the Soviet Union.[21] For his subservience the regime allowed more than 17,000 churches to be reopened and the church began to recover rapidly.[22] The main claim of the Russian Patriarchate was that it was the heir of the third Rome and that the other Orthodox countries should recognise the dominance of Moscow over Rome and Constantinople. This religious point went hand in hand with the suggestion that the other Orthodox countries might follow communism as the leading political ideology of the time. With the regime's support, the Russian Orthodox Church started a campaign sending delegations to Bulgaria and Yugoslavia in April 1945, Romania in May 1945 and attending a Slavonic conference in Belgrade in 1946. Patriarch Alexius went even further and visited the ancient Orthodox Patriarchates (Antioch, Jerusalem and Alexandria) in Syria, Palestine and Egypt in 1945 attempting to bring Oriental Orthodoxy under Russian influence. Furthermore, Moscow sought to dominate the Russian diaspora; its delegation, led by Metropolitan Nikolai of Krutic, went to London in June, to Paris in August and to the US in September 1945 failing, however, to bring these communities under its control.[23] Soviet interest in employing the church in its foreign policy was evident through the fact that the Department of External Church Affairs, which was established in 1946, had the largest number of employees of all the Patriarchate's departments.[24]

Another political method of Moscow was directed against the independent autocephalous Orthodox churches in its vicinity. The Latvian, Lithuanian and Georgian Orthodox churches were incorporated under Russian jurisdiction, while the small Orthodox Church in Poland was forced to declare illegal its autocephaly offered by Constantinople on 13 November 1924, and to ask the Russian Patriarchate, as the third Rome, to issue a new *Thomos* of recognition.[25] A pattern of offering autocephaly to the Orthodox churches in the region extended after the Second World War. Church leaders who had not previously held prominent positions but who visited Moscow were suddenly elected as their countries' leading hierarchs.[26] Metropolitan Timothy of the Polish Orthodox Church visited Moscow in July 1948 and was made head of his church in November 1948. Metropolitan Justinian of the Romanian Orthodox Church had a similar visit in October 1946 and became patriarch in May 1948. Bishop Paisi from Albania went to Moscow in January 1948 and became head of the Albanian Orthodox Church in August 1949. Thus Moscow extended its religious authority to the Orthodox commonwealth and this was to be followed by the political ideology of the Soviet Union. The prime position of the Ecumenical Patriarchate in Orthodoxy was perceived as a thing of the past as the new political situation led to the rise of Moscow's religious and political authority.[27]

The ecclesiastical domination of Moscow was not only over broad Orthodoxy but also at the expense of other religions. In the confessional clash between the Orthodox Church and the Vatican, the Greek Catholic Uniate churches were forced to 'reunite' with the Orthodox churches and major religious confessions were forcibly abolished by the communists.[28] The climax of these combined political and religious methods was visible on 17 July 1948 when delegates of the most important Orthodox churches, which were already under Soviet influence, sent an 'appeal to all Christians' urging them to fight against the terror of the West while Stalin was portrayed as the 'genius peace-maker'.[29]

The installation of communism in Eastern Europe revealed comparable religious patterns. In those countries where religion was perceived as indissolubly tied to national identity, political leaders had to ensure that their authority was drawn from both the political and the religious spheres. Across the region, communists appeared at mass rallies together with religious hierarchs mainly because their association with the church's representatives strengthened their own political legitimacy. This had a direct effect as by combining religion with politics the regime prevented opposition and ensured stronger control of the population. The communists took advantage of the ecclesiastical organisation of the church, which followed not only the civil law of the state but also its own canon law, parallel to state structures. Political leaders interfered in church matters and imposed their verdicts while the church presented itself as independent from politics, giving the false notion of religious freedom. Those members of the hierarchy who were seen as undesirable for the regime were tried by the church and expelled by their hierarchical fellows rather than by communist authorities. Religious trials were followed by civil trials suggesting that the church was active in supporting the establishment of a new society.

Moreover, the church became engaged in communist discourse and its prayers were invaded by communist slogans. By combining religious and communist language, communist regimes increased their authority. The portrayal of Stalin as a 'Saviour of the people' and the propagandistic construction of 'a new man and society' were slogans adopted by both the party and the church. This type of language influenced the ways in which the position of the church and the authority of the communist leadership were regarded, and ultimately the evolution of their political regimes.

Communists sought control not only through the subservience of church hierarchs but also through a systematic method of controlling opposition. Following the Soviet model of the People's Commissariat for Internal Affairs, communist countries developed their own apparatus of surveillance and state security. By censoring every pastoral visit and sermon of the church hierarchy the regime ensured that religion was a method of controlling the people, and especially those who preached or showed public dissatisfaction. Even the space of the church confessional was no longer private and, in many cases, words uttered there led directly to political and religious persecution.

The church was effectively transformed into a state department, a process with a history as previous regimes had also used the church in their political designs. The employment of religion by the communists had a direct impact on the political evolution of the country and on the place of religion in society. Officially, in all communist countries, people were free to profess their faiths, but in practice the state sought to erase and destroy any form of religious belief. The regime's stance on religion was especially evident in education from primary to university levels where special courses on atheism were taught. Furthermore, atheism was an academic course even in church seminaries in Bulgaria and Czechoslovakia.[30] However, despite anti-religious propaganda, in some cases the regimes reserved a special attitude towards religion. It is still uncertain how deeply anti-religious some members of the party were during the Cold War period. For example, in Romania, Prime Minister Petru Groza was the son of a priest and a layman within the church's structures. Many Romanian communists continued to bring up their children in the Orthodox faith attending religious ceremonies outside their cities where the party could not easily control them.

Analysing religious factors which influenced the evolution of church–state relations in East European communism, Pedro Ramet proposed six main elements:

1   the size of a religious organisation,
2   its amenability to infiltration and control by the secret police,
3   its allegiance to any foreign authority,
4   its behaviour during World War Two,
5   the ethnic configuration of the country in question, and
6   the dominant political culture of the country.[31]

A large religious organisation represented a direct challenge to communist authorities. Orthodox churches were not only controlled through political infiltration of the secret police but also through the church's own perception of the concept of *symphonia* which argued for cooperation with political power. In addition, because Orthodox churches were not under the jurisdiction of any external ecclesiastical authority, the actions of regimes were seen as internal affairs. Most Orthodox churches supported the invasion of the Soviet Union during the Second World War, representing another reason for the communists to replace hostile hierarchs. Moreover, the ethnic structure of the state led to competition between churches which wanted to expand their influence at the expense of destroying others; for this reason, Orthodox churches were eager to 'unite' with the Greek Catholic churches. The communists took into account the previous political culture of the state, and their attitudes towards religion followed existing forms of collaboration between church and state.

In addition to these factors, a fundamental point in analysing Orthodoxy in post-war Europe had been the employment of religious channels for international purposes. Orthodox churches proved to be important vehicles through which the image of a regime could be perceived and often improved abroad. The Orthodox commonwealth provided the framework for political cooperation between countries, while interwar relations between Orthodox churches and clergymen in the West were reopened in an attempt to infiltrate the diaspora and propagate communism abroad. By preserving relations within the Orthodox commonwealth and with the West, Orthodox hierarchs asserted not only their religion but also reinforced the national identity of their countries. In Orthodoxy, international contacts led to an awareness of similarities and divisions between religions and political regimes and encouraged religious revival after the fall of communism. This book will explore the religious and political contacts between East and West and how Western religious leaders perceived the place of church–state relations during this period.[32]

The religious and political authority of Romanian Orthodoxy

Although the first historical records mention the presence of Romanians as the dominant ethnic group in the fourteenth and fifteenth centuries, their evolution as a nation has been inextricably linked to Orthodoxy which was instrumental in preserving their cultural identity. Romanians spoke a language derived from Latin which distinguished them from the mass of Slavic neighbouring countries. Language and religion became major elements which unified the populations in these territories even when they were subject to different empires. In the Millet system, Turkish control of the region assigned the church the task of providing education while the first manuscripts in Romanian came from the Orthodox monasteries.[33]

In Transylvania, Romanians had a comparable historical trajectory. During Habsburg rule, despite forming the major ethnic group, Romanians lacked political and religious rights being considered a 'tolerated' nation. At the end of the seventeenth century the Habsburg authorities sought to diminish the position of the Orthodox Church and established a rival church, the Greek Catholic Uniate Church, which accepted jurisdiction of Rome while maintaining most of its doctrinal corpus with Orthodoxy. The Uniate Church took further the task of protecting the Romanians in Transylvania, and Uniate leaders were the first to use a nationalist discourse in asserting political rights for their communities.[34]

The evolution of Romanian nationalism was one of the most complex issues in the history of the Balkans. This was especially so since the construction of the Romanian state was itself the political product of various factors, such as the political interests of the European empires in the region, the political opportunism of the Romanian elite and historical chance. Unification of the Principalities of Moldavia and Wallachia in 1859 is considered to mark the start of the Romanian state-building process which reached its climax in 1918 when Transylvania, Bessarabia and Bukovina united to form Greater Romania.

In building the political architecture of the Romanian state, the Orthodox Church played a vital role as one of the major elements of political attention: politicians believed that controlling the church hierarchy would lead to control of the masses and to the development of a Romanian national identity. Following the unity of the principalities, the church contributed to the construction of the Romanian state by reinforcing the mythologisation of political figures from the Romanian past, thus making a connection between the newly-established state and previous rulers from the Middle Ages who fought for the sovereignty of their states.[35]

The Orthodox Church acquired a new place in the institutional design of the state and the church's evolution ran parallel to the political trajectory of Romania. The state benefited by including religion in its national policy while Orthodoxy was used as a political tool because it confirmed the historical legitimacy of the state and enabled the construction of national myths. The Orthodox Church also benefited by collaborating with the state, aspiring to identify Orthodoxy with the Romanian nation. Close relations between Orthodoxy and Romanian nationalism reached an extreme in the interwar period when right-wing parties claimed the superiority of Orthodoxy over other religions and promoted the idea of an indissoluble relationship between Orthodoxy and nationality. The unification of the state and close relations between the monarchy and top religious hierarchs revealed a thin line between religion and politics. At times of political crises the monarchy looked to the church as the only stable institution. The patriarch became a member of the regency and led the government three times. The church supported the invasion of the Soviet Union, blessing its troops to fight against 'those without God'.

The gradual installation of communism after 6 March 1945 had to take into account both religious and political positions of the church and, for this reason, the regime spared it from violent religious purges and mass persecutions. The official attitude of the Romanian communist leadership towards religion followed the same principles as their Soviet counterparts in the 1940s indicating that collaboration was possible between the church and the regime. After 1947, having gained complete control of political power, the communists saw the church as part of the Orthodox commonwealth; it could foster the country's relationship with the Soviet Union and could help in imposing its domination over the people.

Romania remained a unique case in the Orthodox bloc during the communist period as the only state which did not have mass religious persecution on the scale of its neighbours. Being a dominantly agrarian society, the authority of the Orthodox Church remained strong and posed a serious threat to a sudden political change. For this reason, in the first years of communism, the regime was careful in organising anti-religious campaigns. By controlling the church hierarchy and employing the church in the propagandistic battle for the construction of a new society, the regime promoted the false idea of religious freedom. The general passive attitude of the church towards the regime did not lead to mass protests; on the contrary anti-clericalism remained low, as the main discourse of the church was that every regime is the will of God and people should not follow political disputes but rather focus on spiritual progress. While this type of discourse did not lead to public opposition, it indirectly retained the authority of the church throughout communism.

The security police in Romania was one of the harshest in the Eastern bloc and religion became one of its methods of eliminating political opposition.[36] The main charges against those condemned were that they propagated mysticism and were dangerous to society. In many cases the church hierarchs acted on the borderline between religious and police realms. The effectiveness of religious control was evident through the fact that the number of religious dissidents was reduced and not capable of mobilising the masses; in some cases they only denounced the regime and attracted the attention of mass media when in exile. The 'Lord's Army', a religious movement founded in the interwar period in Transylvania, which posed a challenge to traditional Orthodoxy, was banned by the communists in 1947. However, while the movement clandestinely continued its activity on a smaller scale, some of its members joined Protestant churches rather than becoming involved in political opposition to the collaboration between the Orthodox hierarchy and the regime.[37]

The religious situation in Romania could be compared with that in Poland, where the Roman Catholic Church did not suffer mass religious persecutions. This was especially due to the prime role of Catholicism for Polish identity and the fact that one-fourth of its clergy was killed in the Second World War. Poland was the only country in which Catholic bishops were sent to concentration camps and after the war the communists had to take into account the influence of religion.[38] Some Catholics were even members of parliament and supported the establishment of a Polish construction of socialism. The main difference between Romania and Poland lies at the core of church–state relations. Owing to the concept of *symphonia*, the Romanian Orthodox hierarchy did not hesitate to fully engage the church according to the regime policy. By contrast, in Poland the church remained independent and was not completely under the party control.[39] This fact later fostered the rise of the Solidarity Movement in the 1980s and public opposition to the regime.[40]

Another comparable case was provided by the Bulgarian Orthodox Church. After the fall of the Coburg regime on 9 September 1944, the Orthodox Church was incorporated into state structures, becoming highly dependent on its financial support. At the end of the Second World War, Bulgaria was one of the least religious countries in Europe and there was limited church opposition to the communist regime.[41] Georgi Dimitrov, who fled to Moscow in 1923 and returned in 1946 as president of the Central Committee of the Fatherland Front, appeared at mass rallies with church hierarchs. He publicly presented his own vision of the church's place in Bulgarian society in a speech at the millennial commemoration of the death of Saint Ivan Rilsi on 26 May 1946.[42] In his opinion, the church preserved the national sentiments of the people and had a significant contribution in Bulgaria's history. He followed the Soviet view that collaboration between church and state was possible for the reconstruction of the country and was beneficial to the regime.

Despite official good relations between communists and hierarchs, in order to impose its domination, the regime launched extensive campaigns of religious persecution, to the extent that Metropolitan Kiril, the future patriarch, was imprisoned and tortured. Many high-ranking clergy were executed and the church was transformed into a propaganda tool. The 1947 Constitution stated the freedom of religion and separation between church and state while the 1949 Law of Confessions named the Bulgarian Orthodox Church as 'the traditional Church of the people'.[43] The church had been in dispute since 1870 with the Ecumenical Patriarchate regarding its autocephaly. However, in 1953 with the direct support of the Russian Orthodox Church, it re-established its Patriarchate, a sign of Soviet influence abroad.

The religious situation in Bulgaria was similar to that in Romania. The Orthodox clergy were incorporated into the Priests' Union and the church's educational and social settlements were confiscated. Metropolitan Kiril enjoyed good relations with top Bulgarian communists and other foreign clergymen. He attended the enthronement of Patriarch Justinian in Bucharest in 1948 and developed a close relationship with him. They visited each other and travelled together to Moscow; in 1953 Kiril was raised to the rank of patriarch (1953–71) and Justinian was the first to congratulate him during his enthronement. However, the church witnessed a constant decrease in the number of clergy from 2446 in 1948 to 2263 in 1951 and around 1700 in 1977.[44] After the international détente between the great powers, the church supported the Bulgarian version of national communism, manifested by the bicentenary celebration of the writing of the *Slavic-Bulgarian History* by monk Paisii of Khilendar in 1962, the centenary of the Bulgarian Exarchate in 1970 and the 130th anniversary celebration of the Bulgarian state in 1981.

The trajectory of Orthodoxy in Romania differed from that in Bulgaria in the ways in which the church hierarchy retained an autonomous religious discourse. Bulgaria suffered significant religious persecutions and the church was more dependent on their regime, while Romanian hierarchs were more eager to follow the party line. The Bulgarian church had a limited voice in political and social affairs and remained a weak institution without large popular support. This difference had a direct impact on the evolution of their churches after the fall of communism. In 1992 the Bulgarian Orthodox Church was split when some members of the hierarchy accused the patriarch for collaborating with the communists. In Romania, there was a brief crisis when the patriarch resigned in December 1989, however, he returned to his position in April 1990 and the church preserved its structure intact.

The Romanian Orthodox Church could also be compared with that of its Serbian counterpart. The Serbian Orthodox Church suffered extreme losses during the Second World War. Many hierarchs were imprisoned in German concentration camps, around 25 per cent of its buildings were destroyed and a fifth of the clergy were killed. The communist regime imposed its authority over the church by accusing many clergy of collaborating with the fascist regime, confiscating its properties in the 1945 Law on Agrarian Reform and Colonization and removing education from religious affairs.

The church's influence in society remained strong and the communists used its legitimacy in order to consolidate their position. Rather than appoint a new spiritual leader, Marshall Josip Broz Tito ensured that Patriarch Gavrilo returned to his hierarchical position, following his liberation from Dachau. He attended the 1948 Moscow meeting of religious leaders in the newly established communist countries. However, while other Orthodox hierarchs praised the new international situation, Gavrilo developed tense relations with the Russian Patriarch. In particular, he was critical of extending Russian jurisdiction over the Serbian communities in Hungary.[45] Serbian relations with Moscow remained difficult and followed Yugoslavia's political position in the Eastern bloc.

The Serbian Orthodox Church was challenged by the establishment of new religious structures such as the Croatian Orthodox Church in 1942, the Czechoslovak Orthodox Church in 1951, the autonomous American–Canadian diocese in 1963 and the autocephalous Macedonian Orthodox

Church in 1967. These divisions had a direct impact on the position of the church in Yugoslavia as, for example, the Macedonian clergy largely cooperated with the communists in order to obtain independence.

The main concern of the communist regime was that the use of nationalism by the Serbian Orthodox Church would diminish its influence. In order to control the church, the regime carried out extensive religious persecutions and, as in other countries in the region, set up priests' associations. Church–state relations improved under Patriarch Vikentije (1954–8). After Tito established independence within the Comintern, the hierarchs were able to restore relations with other churches behind the Iron Curtain.

In 1958 Tito imposed German, his own candidate, in the position of patriarch (1958–90), who would be compared to Justinian and also labelled as a 'Red Patriarch'. The Serbian Orthodox Church retained relative independence from political interference, focusing on the reconstruction of its destroyed churches and increasing its involvement in social issues.[46]

In the 1980s the Serbian Orthodox Church began to extend its authority by extensively employing a nationalist discourse.[47] The position of Serbian Orthodoxy during this period differed from that in Romania as it combined Orthodoxy and extreme nationalism while maintaining autonomy from direct political interference. In 1988 the church exhumed Prince Lazar's remains and toured them around the country promoting the idea of a heavenly Serbia. The combination of extreme nationalism and Serbian Orthodoxy would have an impact on Yugoslavia's future shape. The church's discourse developed alongside that of Slobodan Milošević's regime which used references to Orthodoxy in its political actions.[48]

After the fall of communism, despite half a century of atheist rule, the Romanian Orthodox Church retained its dominant position and authority in society. The country remains one of the most religious in Europe having one of the highest figures of believers in the existence of God.[49] In recent years more than 80 per cent of Romanians consider the church the most trusted institution, followed closely by the Army with 70 per cent while democratic institutions such as the Parliament and the Presidency gained less than 40 per cent.[50] The law of religious confessions of 1948 was in use until 2006 and many of the religious hierarchs of the late communist period remained in their ecclesiastical offices.[51]

At a time of increasing secularisation in Western Europe, shiny new churches were erected in Romania in contrast with derelict and bankrupt factories.[52] National communism began as official state policy in the early 1960s and was perpetuated during the Ceauşescu period. After the fall of communism in 1989, the nationalist elements of this policy were reiterated in a new light as the church and the democratic regime continued to make strong references to the national past. The current religious revival has to take into account the complexity of Romanian nationalism which considers the church as a major political institution. During the first years of communism the church showed the ways in which it could adapt to the new political regime. It is the task of this study to examine this transformation.

Conclusion

The relationship between church and state in Orthodoxy is characterised by the concept of *symphonia*. With the emergence of nation-states in the Balkans, political leaders used the powerful force of religion in order to gain authority and legitimacy. Orthodoxy, as the traditional institution which fostered religious identity in the region, was incorporated by political leaders in their institutional designs.

The spread of communism from the Soviet Union to Eastern Europe revealed a comparable pattern in which churches were employed for the regimes' benefits. Maintaining close relations between church and state, the regimes took advantage of the authority of the church and sought their own legitimacy. Analysis of the relationship between the Romanian Orthodox Church and the communist regime has to take into account both the status of religion in the Orthodox world and the role of the church in fostering national cohesion. For this reason, the historical trajectory of Romanian Orthodoxy, from the establishment of the state in 1859 to the first government with a communist presence in 1944, is essential for comprehending the development of Orthodoxy during communism and in contemporary Romanian politics. The concept of *symphonia* would become clearly visible in the attitude of religious leaders who stated that 'every regime is a will of God'. The church would continue to collaborate with the political powers, thereby imposing its authority in both the material and spiritual worlds.

2
The Political Control of Orthodoxy and Romanian Nationalism, 1859–1944

Orthodoxy and state during the reign of prince Cuza

The unification of the Principalities of Moldavia and Wallachia resulted from the political actions of indigenous leaders with the support of the great European powers, in particular France, which saw the new state as a barrier against the perceived Russian intention to dominate the Balkans.[1] Colonel Alexandru Ioan Cuza, Police Commander in Moldavia and a well known supporter of the Unionist cause, was elected prince of Moldavia on 5 January 1859 and shortly after, he was also elected ruler of Wallachia on 24 January 1859. Wallachia had been commonly known as 'The Romanian Country' (*Ţara Românească*) or 'Romania' and when Prince Cuza suggested that the newly unified state would unite the population into a single nation he faced international opposition from Austria, Russia and the Ottoman Empire, all of which had political interests in the region.[2] The main task of Romanian politics during Cuza's regime was to convince Europe that Moldavians and Wallachians were in fact the same nation, and that their new state integrated the political aspirations of the dominant ethnic group, the Romanians.[3] The deadlock of Cuza's election was resolved on 20 November 1861 when the Ottoman Empire finally accepted the union by issuing 'The Firman of Moldavia and Wallachia, concerning the Administrative Organisation', but on condition that the unification would cease at the end of Cuza's reign.

After Cuza managed to obtain international recognition for his regime, he sought to strengthen the internal unity of his state. He founded common institutions for both principalities, declared Bucharest as the capital of the country, organised a unified army, and in his speech on 23 December 1861 he was able to state that from that moment 'Romanian nationality was founded'.[4] A few days later, on 22 January 1862, the first Romanian government was set up by Conservative Barbu Catargiu who became prime minister and the term 'Romania' began to be used in official documents.[5]

The population of the newly established state was, according to the census of 1859, composed of 3,864,848 people. The data of this census did not indicate the ethnic structure of the population but rather the citizenship of its inhabitants, showing that Romanians were the majority, while the remainder belonged to other European states: 28,136 Austrians, 9545 Greeks, 3658 Prussians, 2823 British, 2706 Russians, 2631 Turks, 1142 French, 167 Italians and 569 citizens of other states.[6] The religious structure revealed that the population was mainly Orthodox with small Catholic and Jewish communities. Figures from the 1860s show that there was: one Orthodox priest for 100 families; one Orthodox monk for 1000 people; one Orthodox nun for 1000 people; one Orthodox church for 612 people; one Catholic church for 716 people; one Protestant church for 2408 people; one Armenian Gregorian church for 746 people; one Lipovan church for 1182 people; one synagogue for 486 people and one temple for 441 people of other religions.[7] The Orthodox Church had 9702 priests for its 6858 churches and 4672 monks and 4078 nuns lived in 173 monasteries while the other religions had the following number of buildings in the country: 63 Catholic churches, 12 Protestant churches, 11 Armenian Gregorian churches, 7 Lipovan churches, 176 synagogues and 3 temples of other religions.[8]

As the above figures suggest, in the Principalities of Wallachia and Moldavia before 1859, Orthodoxy occupied a central place in the life of the state. From a political perspective, the church endorsed and blessed the rulers in front of the faithful and its cooperation with the state was economically rewarded by enjoying the benefits of being one of the largest landowners. The Orthodox Church was in possession of almost one-third of the entire land in Moldavia and a quarter in Wallachia.[9] Possession of land accumulated due to the religious tradition of principalities and the generosity of princes and boyars. A significant proportion of this land was under the administration, of so-called 'dedicated monasteries',[10] which were under the protection of monasteries from holy places in Levant, Mount Athos, Jerusalem or Sinai.[11] The Orthodox Church's financial profits were impressive. Most of its revenue was controlled by Greek monks and directed to monasteries abroad.

The law which regulated political life in Moldavia and Wallachia before 1859, the Organic Statutes (*Regulamentele Organice*), had established two 'Central Houses' of the Orthodox Church (*Casa Centrală*) which collected all financial resources. These assets were in direct competition with state finances, constituting a separate financial department under the exclusive control of the church. Furthermore, it was usual for the ruling princes to borrow money from the Central Houses to have sufficient funds to travel to Constantinople and to begin their reigns, indebting the state to such an extent that on 1 January 1858, the state owed 8,136,000 lei to the church.[12]

Prince Cuza took steady political decisions aimed at imposing better control of the church and at using Orthodoxy to consolidate his regime in both principalities. The church hierarchs were perceived by the people as serving Greek Phanariot interests rather than those of Romanians, and for this reason Cuza faced little opposition from within the country. His first gesture was to trans-form the Central Houses of the church in Moldavia and Wallachia into one Financial Department of the state on 19 October 1860. In this way the state would now collect all financial resources from Romanian territory. On 17 December 1863, Cuza confiscated the land of most Orthodox monasteries claiming that the revenue belonged to the state and not to a foreign juris-diction.[13] In addition, Cuza issued a law on 30 November 1864 (*Decretul organic pentru reglementarea schimei monahiceşti*) aimed at reducing the num-ber of monks in monasteries by making it more difficult to become one: they now needed special approval from the Ministry of Religious Confessions. By restricting the number of monks, Cuza increased his control over those who aspired to become part of the church hierarchy, as according to Orthodox tradition the hierarchy was chosen only from monastic clergy.

Cuza's gestures had specific political aims. He made it possible to dispos-sess the church hierarchy of its vast economic revenues and to control those who wanted to become part of the hierarchical clergy. His laws were offi-cially directed against 'dedicated monasteries' and Greek monks who moved to the Romanian principalities, but in practice also had an effect on the very core of Orthodoxy's organisation. The most important effect was that, from then on, officially the Orthodox religious ceremonial could only be held in the Romanian language while the Greek monks were allowed only in a few churches.[14] This change indicated that the future role of the Orthodox Church was to promote its mission in society only in the dominant lan-guage of the state.

Cuza's next step in asserting political authority over the church hierarchy was to unify the church from both principalities into one national body. He gave a prime role to the metropolitan of Bucharest offering him the title of Primate Metropolitan in spring 1865, and on 11 May 1865 introduced a law which broke with tradition concerning the appointment of bishops and metropolitans in Romania (*Legea pentru numirea de mitropoliţi şi episcopi eparhioţi în România*) enabling him, rather than the church, to make new appointments. Some bishops' positions were empty, and Cuza took advan-tage of the situation to personally appoint new bishops who supported his political regime and the union of the principalities. Cuza's decisions found support in the Romanian political elite which saw his actions as a way of consolidating the nation against Greek influence.

Cuza organised the Orthodox Church into two metropolitanates with headquarters in Bucharest and Iaşi, and six bishoprics: in Buzău, Argeş and Râmnic in Wallachia and in Roman, Huşi and Ismail (the Lower Danube) in Moldavia. With his own people in the church and in order to increase his control of the hierarchy, on 3 December 1864, Cuza introduced a law to establish a Synod (*Decretul organic pentru înfiinţarea unei autorităţi sinodale centrale pentru afacerile religiei române*), the first article of which stated that the Romanian Orthodox Church was independent of the Patriarchate of

Constantinople, thus requiring all church decisions to need his approval. Cuza's main reason for creating the Synod was political and it would have authority as the official institution dealing with ecclesiastical matters. The first Synod of the united Romanian Orthodox Church was held on 1 December 1865 in Bucharest, composed of both metropolitans from Bucharest and Iaşi, six eparchial bishops (*episcopi eparhioţi*), eight vice-bishops (*arhierei*), three deputies from each of the six bishoprics elected from clergy or lay members with theological studies and two deans of theological faculties which would be founded in Bucharest and Iaşi.

An important element at the opening ceremony of the first Synod was the oath which each member had to take in front of the other members. Primate Metropolitan Nifon read the Orthodox Symbol of Faith and after every paragraph all members had to declare together 'I believe and truly confess', while after the ceremonial of kissing the cross and the Gospel, each member had to repeat the same formula again.[15] The oath introduced at the first meeting of the Synod was purely religious, but it would later acquire other dimensions. After Cuza's regime, at the Synods of the Orthodox Church and at the enthronement of new members of church hierarchy, the required oath contained both religious and political declarations, representing the allegiance of the church to the Romanian state.

The Synods of 1865 and 1867 proposed various changes regarding the discipline and the organisation of the Orthodox Church in Romania. However, most of these remained only proposals as Cuza endorsed only those which accorded to the state's political interests. The Synod's authority was thus reduced, and the bishops refused to meet; the last Synod was held in 1869 and was suspended as only four members took part. A clash between the old church hierarchy and the newly appointed bishops supported by the regime arose as two vice-bishops, Neofit Scriban and Ioanichie Evantias, and a priest, Father Păunescu, claimed that the Synod was in fact a pseudo-Synod. In other accusations, these vice-bishops suggested that the Synod was completely controlled by Cuza and had nothing to do with real Orthodoxy, that being composed of bishops appointed by Cuza the church was not accurately represented, that the president of the Synod was Cuza, that the Synod intended to change Orthodoxy and that ordinary people could not participate in the Synod's debates.[16] Most of their claims were published in the press but did not have a strong impact on the church as it remained intact and did not split. The vice-bishops and the priest were marginalized by the church, lost their positions and the conflict did not develop further. This lack of wider resistance indicated that the church could be brought under political control.

Cuza wanted to impose new reforms, such as land and electoral reforms, but his proposals faced the opposition of Conservatives and were rejected in Parliament. For this reason, and following the model of his political hero, Napoleon III, on 2 May 1864 Cuza established a coup d'état which dissolved

Parliament and asked for a plebiscite which would endorse his political decisions. The plebiscite was held from 22 to 26 May 1864 and was passed with
682,621 votes for, 1307 against and 70,220 abstentions.[17] According to the
new statute resulting from this plebiscite, the prince became the sole executive authority with the support of a Moderate Body which advised him on
his decisions. The Moderate Body was composed of the two metropolitans,
all of the bishops, the president of the Court of Cassation and 64 members
personally appointed by the prince. By electing the church hierarchy to the
Moderate Body, Cuza suggested that his regime would continue as previously, although it was now authoritarian.

Support of the church hierarchy was also evident in the administrative
organisation of the plebiscite. The church composed special prayers for the
event in which Cuza was portrayed as a defender of Romania and his political actions pleased God:

> God, you always looked upon the Romanian nation sending from time to
> time defenders and leaders and you freed her from the chains of those
> who wanted to destroy her and erase her from the book of Christian and
> civilised people [. . .] and we, like our ancestors, say: God is with us and
> nobody [will be] against us.[18]

In addition, the festivities held in Bucharest on the evening of 21 May 1864
(on the eve of the plebiscite) combined religious and political elements. At
the main reception an ode dedicated to Cuza, inscribed on a large display
board, had both a religious and political significance:

> To he who made unity; to he who took [our] wealth from the foreigners'
> hands; to he who gave land to the peasants; to he who made all
> Romanians aware of their civilian and political rights; to he who gave free
> and compulsory education; to he who killed death; to he who built the
> army; to the defender of nationality; to the protector of justice; to the
> father of the fatherland; to the ruler of Romanians; Alexandru Ioan I
> [Cuza]; eternal gratitude.[19]

Both in the prayers and in the ode, Cuza was presented as a special ruler
with supreme powers. The ode claimed he was the one 'who killed death',
making a subtle reference to Christ who, according to Christian doctrine,
rose from the dead so becoming the only human being to defeat death. Cuza
was thus the providential ruler who united the nation with the church, God
and history. His political acts not only had an impact *hic et nunc* but, moreover, were connected with the evolution of the history of Christianity.

However, Cuza's reign was short, and faced with the opposition of
Conservatives and Radical-Liberals, he was forced to abdicate on 11 February
1866. The interim rulers were Lascăr Catargiu, representative of the

Conservatives, General Nicolae Golescu, representative of the Liberals and Colonel Nicolae Haralambie, representative of the Army. The important element lacking from this composition was the church hierarchy, suggesting that even if the church was a representative force in Romania it had not acquired an independent political voice.

In order to avoid the dissolution of the state, which the Ottoman Empire had previously requested would take place after Cuza's regime, Romanian politicians chose Count Philip of Flanders, the brother of King Leopold II of Belgium, as the new ruler. However, afraid of France's opposition, Count Philip declined and, for this reason, the politicians turned to the Hohenzollern-Sigmaringen's family. Prince Carol of Hohenzollern-Sigmaringen accepted the throne and on 10 May 1866 was proclaimed Prince of Romania under the name of Prince Carol I. The Constitution of 1 July 1866 stated in its first article that 'The Romanian United Principalities are an indivisible state under the name of Romania', and in this way, the political work achieved by Cuza was continued by Prince Carol I who further strengthened Romanian unity.

Cuza is known in the history of Romania as the political ruler who founded the Romanian state. He used the financial resources of the church to consolidate his state both politically and economically and to further popular measures such as the state education and land reform.[20] By confiscating church land and controlling the Orthodox hierarchy, Cuza acquired a major role in Romanian national mythology especially after the First World War, when the country expanded its territory. He was perceived as an authoritarian ruler who had managed to remove church land from Greek control and to acquire international recognition of the Romanian state. Moreover, his reforms showed that the church could become an important ally in building the Romanian nation-state; power of the church hierarchy gave Cuza overall control of the church and, indirectly, of the Romanian faithful. As will be shown, this tenet increased during all subsequent political regimes starting with the monarchy and later in communism. It would become clear that the ruler who had control of the church would have the public support of the population.

Orthodoxy and politics during the reign of King Carol I

Political control of Orthodoxy, in its incipient stages during Cuza's regime, acquired more of a nationalist dimension during the reign of Prince Carol I. It would gradually increase further during the following decades. Romanian politicians started to use not only religious elements in propagating the unity of the country but also made increasing references to the main historical figures of the Romanian principalities, turning them into myths of national unity and identity. Political leaders made reference to popular mythologies which had a religious substratum. The main figures presented in this way in both political and religious discourses were the ruler of Moldavia, Stephen the Great (1457–1504) and the ruler of Wallachia, Michael the Brave (1593–1601). Stephen the Great was seen not only as a victorious ruler who defended his territory but also as a religious man who enjoyed one of the longest reigns in Romanian history. Building a church after every battle made him a popular and religious figure. Despite his complicated personal life, he was thought of as a saint for many years (and was officially sanctified by the Orthodox Church in 1992). For a few years Michael the Brave controlled Transylvania and Moldavia, uniting their territories with Wallachia. After 1821 he was considered a national genius who had managed for the first time to unite all provinces which had a Romanian majority. He supported the church and established a bishopric in Alba Iulia for the Romanian population in Transylvania. Both figures combined political with religious elements and, they were mythologised as examples of the glorious past of Romanians.

From the beginning of Prince Carol I's regime, political leaders employed elements of Romanian national mythology and religion. The telegram sent by Ioan Brătianu, representative of the Romanian government and later prime minister, to Prince Anton of Hohenzollern on 3 April 1866, indicating the decision to appoint Prince Anton's son as the ruler of Romania, suggested that Romanians were waiting for their ruler as they would a saviour:

> Five million Romanians are acclaiming as their Suzerain, Prince Carol, the son of your Royal Highness. All temples are open and the voice of the clergy is rising with that of the whole people towards the Eternal Sky, asking for the Chosen One to be blessed and to be made worthy of his ancestors and of the trust the whole nation that relies upon him.[21]

In his speech on 11 May 1866, after arriving in Bucharest, Prince Carol I used a nationalist and religious discourse, referring to Stephen the Great and Michael the Brave, selecting these two mythologised figures, one from each principality:

> In human destinies there is no more noble duty than to be called to maintain the rights of a nation and to consolidate its liberties. Such an important mission made me immediately leave an independent position, my family and country towards which I was bound with the most sacred laws, in order to follow your appeal. Receiving your election, which put on my head the crowns of Stephen the Great and Michael the Brave, gave me an important responsibility. I hope that with the help of God and with my whole devoutness I shall give my new country a happy existence worthy of its history. Romanians! I am yours with all my heart and soul. You can rely on me at any time as I rely on you.[22]

Prince Carol I was Catholic and, although he was not forced to convert, article 82 of the 1866 Constitution stated that his children and the future royal rulers of Romania would be brought up in the dominant religion of the state, Orthodoxy. Referring to the relationship between church and state in Romania, article 21, asserted that

> The freedom of conscience is absolute. The liberty of all religious confessions is guaranteed if they do not contravene public order and good custom. The religion of Eastern Orthodoxy is the dominant religion of the Romanian state. The Romanian Orthodox Church is and remains independent of any foreign Church, while maintaining unity with the Ecumenical Church of the East regarding its dogma. The spiritual, canonical and disciplinary affairs of the Romanian Orthodox Church will be regulated by a unique central synod authority, according to a special law. Metropolitans and bishops of the Romanian Orthodox Church will be elected according to a special law.[23]

The political regime continued to favour Orthodoxy as the dominant religion and Romanian politicians felt the need to explain to other European countries why Romania supported Orthodoxy and did not convert its population to a different Christian confession. In a letter to the French newspaper *L'Opinion Nationale*, Brătianu suggested that the attachment of the Romanian state to Orthodoxy and not Roman Catholicism was connected to two main elements. Firstly, all Orthodox states which had previously achieved political unity proclaimed their own national Orthodox Church independent from Constantinople. Brătianu emphasised the role of Orthodoxy in preserving the national identity of Romanians in the past and claimed that the new state has benefited from its activity. Secondly, by preserving Orthodoxy as the state religion, Romania could have a strong political influence in the Balkans among the other Orthodox countries.[24]

The ecclesiastical reform started by Prince Cuza was completed during Prince Carol I's regime. Because the church hierarchy was divided between those appointed before 1859 and those appointed during Cuza's regime, Prince Carol I decided that a law dealing with this matter should be issued with the approval of the Patriarchate of Constantinople, which he personally acquired on his visit to the Patriarch in 1866.[25] Prince Carol I wanted to reinforce the hierarchy's loyalty and his laws were directed towards ensuring that the church supported his political decisions and national unity. He confirmed that the bishops who had been elected by Prince Cuza would remain in their positions.

Complete control of the church hierarchy was achieved in the law concerning the election of the metropolitans and eparchial bishops promulgated on 19 December 1872 (*Legea pentru alegerea mitropoliților și episcopilor eparhioți, cum și a constituirii Sfântului Sinod al Sfintei Biserici Autocefale Ortodoxe Române*).[26] First, the system of electing the hierarchy changed.

The members of the hierarchy were elected by simple majority in a secret vote of an Electoral Collegium which included metropolitans, eparchial bishops, appointed vice-bishops and the Orthodox members of Parliament (art. 1). Second, the metropolitans were elected only from bishops; bishops only from appointed bishops by the Electoral Collegium and not by the prince (art. 2). The prince endorsed their election after official confirmation from the Ministry of Religious Confessions (art. 4). Third, the law stated that the highest ecclesiastical office would be the Synod which acquired the title of the 'Holy Synod'; it would be composed of 16 members including the metropolitans, six bishops and eight vice-bishops and would meet twice a year (art. 8). Although these measures might appear to have made the church hierarchy more democratically elected, in fact, they created a small circle of people who could be controlled more easily by the regime. The law also opened the possibility of political parties having their own representatives in the church hierarchy by electing their favourite candidates, as the number of members of parliament always surpassed that of the Holy Synod in the Electoral Collegium. The law led to an unstable ecclesiastical life in Romania and until its modification in 1909 the Orthodox Church was ruled by seven primate metropolitans and seven metropolitan deputies, an instability which would benefit the political regime.[27]

In addition to these provisions, the law had a nationalist dimension. Article 25 stated the establishment of 'appointed bishops' (*arhierei titulari*) which would be elected by 'the Holy Synod with the help of the government'. These new bishops would not have a bishopric see but would be named according to the most important cities in Romania.[28] Their main tasks were to participate in the meetings of the Holy Synod and in the election of the bishops. In this way, the government could guarantee having the majority in the Holy Synod. In addition, as the deposition of a bishop required the agreement of 12 bishops, the government would have complete control in maintaining its support; there would always be at least 12 bishops in the hierarchy who would endorse its position.[29] The government thus took advantage of the concept of *symphonia* and ensured that the church obeyed the political interests of the state.

Because Romania wanted complete independence from the Ottoman Empire, the government declared an independent state on 9 May 1877 and supported Russia in the war in the Balkans by invading the northern part of Bulgaria.[30] On 10 May 1877, Prince Carol I addressed the Romanian clergy enlisting support for the government's military efforts. In his speech he reinforced the idea that it was not only the Romanian government fighting against the Ottoman Empire but the whole nation, thereby connecting the church with the people:

> In the greatest days of Romania, the clergy was always with the nation, or, to put it better, was the nation itself. It could not be different today

when the Romanian nation proclaimed its decision to live its own life, to be the master of its destinies and to be self-governed.[31]

Winning the war, the Russian-Turkish Treaty of San Stefano declared official recognition of Romanian independence by the European powers on 19 February 1878.[32] In order to raise the country's position in the Balkans, Romania declared itself as the 'Kingdom of Romania' and Prince Carol I as its king. To strengthen its religious position, the Romanian Parliament proposed, on 9 March 1882, that the Romanian Orthodox Church should have the rank of Patriarchate, being equal with the other Patriarchates of Eastern Christianity. Furthermore, in a spiritually symbolic gesture, the Romanian hierarchy officiated a few days later, on 25 March 1882, the celebration of the Holy Chrism, which represented official jurisdictional separation from the Patriarchate of Constantinople. Chrism is an oily substance used immediately after the sacrament of baptism symbolising the gifts of the Holy Spirit upon the faithful.[33] According to the church's doctrine, only autocephalous churches were entitled to produce chrism and send it to churches of its jurisdiction. Before 1882, the Romanian Orthodox Church had received the chrism from Constantinople and so, by officiating independently this ceremony, it *de facto* announced its jurisdictional independence. Thus, after the Romanian military victory against Ottoman rule, the church proclaimed its own spiritual victory against Constantinople. The actions of the church had King Carol I's support and after intense governmental diplomatic concessions, the Romanian Orthodox Church was finally recognised as an Autocephalous Church but not yet as a Patriarchate by the official *Thomos* of the Ecumenical Patriarch Joachim IV on 25 April 1885.

At the opening of the meeting of the Holy Synod on 1 May 1885, King Carol I signalled the role of the government in gaining the autocephaly of the church and his personal vision of the role of Orthodoxy in Romania. In his opinion, the glory of Romanians was intrinsically linked to Orthodoxy and only by transforming the church into a national institution could the state and church coexist:

> I am happy to announce to your Holy Eminence that the Secular Autocephaly of the Romanian Orthodox Church received the blessing of His Sanctity Ecumenical Patriarch and that, thus, the position of the Romanian Church is equally defined with the other Autocephalous Orthodox Churches, its sisters of the same faith and ritual. This good result was obtained by the effort of my government and the patriotic support of your Holy Metropolitans and Bishops of the Romanian Church and due to the strong feeling of true Christian brotherhood of His High Sanctity Joachim IV and His Holy Synod [. . .] The Church, through the defence of which in past centuries Romanians made their glory, was always connected with the destiny of the country. Touched by this historical

truth, and knowing the strong faith of the people in their ancient religion, from the first day and during all the time of my reign I have had before my eyes a permanent aim: the strengthening of the Romanian Church, so that it could be that great state national institution, which the Romanian people could always rely on.[34]

The creation of Romanian Autocephaly meant that the Romanians now had international recognition of their national identity. A connection could be constructed to the glorious past in which mythologised figures such as Stephen the Great and Michael the Brave had fought for the Romanian nation. The process of reaching both religious independence and national unity would be achieved in 1925 when the establishment of Greater Romania would come together with the creation of the Romanian Orthodox Patriarchate.

The Orthodox Church considered itself at the core of Romanian national identity, with its main mission to absorb the ethnic minorities into the dominant religion. Thus, in a report from 1882 regarding the relationship between Orthodoxy and other religions in Romania, presented at the meeting of the Holy Synod, Bishop Melchisedec Ştefănescu stated that 'the Holy Synod has to do its best from a moral and legal point of view to ensure that the foreign nationalities who live in our country blend with the Romanian nationality, and that they become true and complete sons of the Romanian nation'.[35] The church saw a political opportunity to engage more actively in the life of the state by claiming that its main priority was to protect Romanian identity from the influence of other religions. However, as the 1899 census showed, the Romanian Orthodox Church comprised 91.5 per cent of the population, with the Jewish population at 4.5 per cent and Roman Catholics at only 2.5 per cent.[36] The church's position towards the Jewish population was supported by the Constitution which stated in article 7 that only people of Christian denomination could be landowners. Roman Catholics were seen as a potential threat, as the king was Catholic and their number increased from 45,152 in 1859 to 149,667 in 1899.

However, even if the political regime managed to control the church hierarchy with its 1872 law, there was a separation between the religious leaders and ordinary clergy. Most Orthodox priests were poor and uneducated; they memorised their daily religious ceremonials and lived in rural areas only with the material support of their congregations. Prince Cuza's laws, which had deprived the church of its vast land possessions, brought the clergy into financial dependence on the budgets of local city halls. Due to these changes, the number of priests dramatically declined. If the 1859 census indicated that there were 9702 priests in the principalities, in 1904 the Orthodox Church had only 4998 clergy while the number of monks dropped to 861 and the number of nuns to 2220.[37]

In a report from 1888, concerning the deplorable material and moral situation of the Romanian clergy, Bishop Melchisedec stated that

> the improvement of the priests' situation [. . .] is a vital element for a nation that wants to strengthen itself [. . .] As land would not exist without the sky which gives it light, warmth, air, rain, dew, etc., in the same way a civilised nation would not exist and progress without religion and morality, which are represented by the Church and its servants.[38]

Moreover, because the clergy did not receive financial support from the state but from local authorities, it was under permanent political pressure to support the party which promised more financial assistance. Bishop Melchisedec lamented those priests who supported the wrong candidates: 'Pity the priest who worked with the party that lost the elections'.[39] Moreover, Bishop Melchisedec argued that without financial support from the state, Orthodoxy faced a dangerous situation. Some priests started to organise themselves in associations with a hostile attitude towards the Holy Synod claiming that the hierarchy should be composed only of priests and not bishops. Because the bishops were elected from monks, while priests were married and had families, they argued that the church hierarchy no longer represent them.[40]

The political regime took advantage of the priests' financial crisis in order to use the church, with its strong influence on the masses, as a political tool. A financial solution was promulgated on 20 May 1893 in the law concerning the situation of the Orthodox clergy and theological seminaries (*Legea clerului mirean şi a seminariilor*) which would be subsequently reworked in 1906. By paying part of priests' salaries, the state would act as manager of the church, controlling the exact number of clergy and churches and limiting the number of those who wanted to be ordained. The law had strong political elements. First, the regime indicated that the number of parishes in Romania would be strictly maintained at 368 urban and 3,326 rural (art. 3), not more than 20 church buildings would be raised per year and these figures would not be changed except by law (art. 4). Second, Orthodox priests would gather once a year in pastoral conferences in which they would discuss ecclesiastical matters (art. 19). These conferences, which were later detailed in special regulations on 5–6 May 1910, were used by the state to present its position on political and religious issues.[41] Third, the law indicated that every parish should have a small library which would include 'moral-religious, economic and national-historical books'. In addition, the Ministry of Religious Confessions would print and send to these libraries sermons written by approved clergy (art. 87). Fourth, each priest had to keep in his parish library a copy of every sermon he preached (art 89), thus enabling the state to monitor and control his activity more easily.[42]

The law represented the complete and coordinated control of Orthodoxy at the parish level. For this reason, Primate Metropolitan Iosif opposed discussion of the law at the Holy Synod considering it against ecclesiastical traditions; but he was soon forced to resign. The law was then promulgated by King Carol I when the Orthodox Church did not have an officially installed primate metropolitan.

Furthermore, total financial management of church possessions by the state had been achieved by establishing the so-called 'House of the Church' in the Royal Decree No. 255 of 21 January 1902 (*Legea pentru înfiinţarea Casei Bisericii*). The House of the Church was under the direction of the Ministry of Religious Confessions and Education, and controlled the wealth of every church, monastery and ecclesiastical establishment in Romania. It was the department that offered financial support in paying priests' salaries and conserving church patrimony.

As the 1872 law had allowed the influence of political parties in the election of the church hierarchy, King Carol I decided to change the small circle of those eligible for hierarchical positions on 20 March 1909. The 1909 revised law introduced two main elements: it created a new institution, titled the Superior Clerical Consistory; and it allowed any Romanian to be chosen as metropolitan or bishop with the condition that they should be born in Romania and not naturalised (*Legea pentru modificarea Legii sinodale din 1872 şi pentru înfiinţarea Consistorului Superior Bisericesc*). This new institution was composed of all members of the Holy Synod, a professor from the Faculty of Theology in Bucharest, a professor elected from all professors in theological seminaries, two abbots of monasteries and 17 priests. The main aim of the Superior Clerical Consistory was to gather the lower clergy with the church hierarchy as a 'parliament of the clergy united with the Holy Synod'.[43] The creation of this forum provoked another crisis within the church led by Bishop Gherasim Saffirin who refused to recognise its authority and anathematised the Holy Synod and all those who were against his opinion. However, for his views on this matter, and following a now familiar pattern, he was soon deposed from the church hierarchy. During the communist period those who publicly opposed the regime would not only lose their positions but also faced the restriction of their individual liberties or even physical elimination. The pattern of removing members of the hierarchy who did not obey political leaders had already taken place before the installation of communism.

The 1909 law was widely seen as too permissive in allowing any Romanian to be a possible candidate for the church hierarchy. Elections proved to be very difficult and lengthy, so on 18 December 1911 the circle of those eligible was again restricted. Thus, metropolitans could be elected only from bishops, while bishops could be elected from any member of the Romanian clergy. The state continued to control church elections through the Electoral Collegium which was composed of members of the Holy

Synod, the Superior Clerical Consistory and all Orthodox deputies and senators in parliament.

The 1902 and 1909 laws are significant as they indicate a new phase in the development of the Romanian state. At the beginning of the twentieth century the Romanian Orthodox Church was transformed into a state institution which served the monarchy and the ruling political party. For this reason, political attention shifted not only to controlling the church but also to shaping it into a more viable national institution adaptable to new political situations. In this way, the election of the primate metropolitan was meant to reflect both the national unity of Romanians and the possibility that any Romanian citizen would be able to acquire the highest ecclesiastical position in the state. In addition, the establishment of the Superior Clerical Consistory was directed towards bringing the interests of the lower clergy closer to the decisions of the church hierarchy. However, the conflict started by Bishop Gherasim Saffirin, the most important clerical dispute since the Cuza period, indicated that there were people within the church unsatisfied with its role in society.

On the eve of the First World War, when Romania declared its neutrality, various voices attempted to give the church a new role. In a speech given to the opposition Conservative Party on 3 December 1916, Mariu Theodorian-Carada set out his ideas concerning the 'Religious Politics of Romania'. He mentioned the opinion of a high-ranking official from the Ministry of Religious Confessions and Education that the church was only an instrument of the state which held in control the masses and helped the state to bind together various nationalities.[44] Although he did not particularly agree with this official's opinion, his speech showed how widespread the view was regarding state control of the church. In Theodorian-Carada's view, the religious politics of Romania were 'very delicate'[45] because only half of Romanians lived within Romanian territory while the rest were spread in Transylvania, Bukovina, Bessarabia and in the Balkans.[46] For this reason, Theodorian-Carada asked the Conservative Party to promote a form of foreign politics which would lead to the inclusion of all Romanians in one national church. In this way, Romanians who were not living in the Romanian kingdom would be encouraged to retain their national identity.[47]

On 10 October 1914, King Carol I died and because he did not have an heir, the parliament elected his nephew Ferdinand as the next monarch. In 1916 Romania declared war on the Austro-Hungarian Empire with the political aim of liberating Romanians in Transylvania. However, they were unsuccessful and in subsequent battles lost the territory of Wallachia to the German army. During the occupation, the royal family and the Romanian government relocated to Iaşi. Primate Metropolitan Conon Arămescu-Donici remained in the capital and the Germans persuaded him to write a letter to the Romanian soldiers and faithful in Moldavia asking them not to oppose the German authorities. The letter had little effect on the population as the

Romanians remained loyal to the monarchy and did not support the occupying forces.

Orthodoxy and the 'Sacred Unity' of Romanians

The dissolution of the Russian Empire in October 1917, the end of the First World War and the appeal of Emperor Charles I of Habsburg to his 'loyal people' in October 1918 hastened the redefinition of the international system with the creation of the 'League of Nations' and the redrawing of European borders. The kingdom of Romania found itself in a difficult situation as most of its territory was occupied by German forces. In order to end the war in its country, the Romanian government signed a Peace Treaty with the Central Powers in Bucharest on 7 May 1918, transforming Romania into an economic satellite of Germany. The signing of the armistice at Compiègne on 11 November 1918 between the Allies and Germany, marking the end of military hostilities, brought Romania into a new position.[48]

On 9 April 1918, the Moldavian diet (*Sfatul Ţării*) at Chişinău had decided to unite the Democratic Republic of Bessarabia with Romania, and now, with German forces retreating, the Romanian government felt that there was a unique moment to achieve the national dream: the unification of all Romanians into one state. On 28 November 1918, the Romanian National Council of Bukovina declared union with Romania and on 1 December 1918, more than 100,000 people from all parts of Transylvania and Banat attended the Grand National Assembly of Alba Iulia also asking for unification.[49]

During the Paris Peace Conference that followed the end of the First World War, Hungary withdrew its troops from Transylvania; however, in July 1919, its army re-crossed the border and attempted to regain its former territory. Romania's counter-offensive resulted in the defeat of Hungary and the occupation of Budapest from August until November 1919. Romania promoted its political interests by combining extensive lobbying in the European capitals with military achievements, and, from this position, signed the Peace Treaties of Saint-Germain and Neuilly with Austria and Bulgaria on 9 December 1919. A few days later, on 29 December 1919, parliament passed laws ratifying the union of Bukovina, Bessarabia, Transylvania and Banat with Romania. Thus, at the end of 1919 the country had new geographical and population configurations with its territory more than doubled from 138,000 km² in 1915 to 295,049 km² in 1918. The 1899 census of the Old Kingdom showed Romania's population was composed of 5,956,690 people; with unification this figure reached around 14.7 million in 1919, and 18,057,028 in 1930.[50]

The Orthodox Church was a leading supporter of Romanian unity and many Romanian clergymen in neighbouring territories were members of local assemblies which asked for unification. The position of the church was evident in the composition of the delegation from Transylvania and Banat that went to Bucharest after the assembly of Alba Iulia on 1 December 1918 handing the official proclamation of unification to King Ferdinand and the Romanian government. Bishop Miron Cristea of Caransebeş was the leader of the four-member delegation. His thoughts on that moment were recorded in letter dated 5 December 1918 addressed to Metropolitan Pimen of Moldavia after meeting the king in Bucharest:

> We have bowed our motherland to his Majesty King Ferdinand, now the king of all Romanians. The king's eyes were filled with the holy tears of joy [. . .] I have lived here the most glorious days in the history of our nation.[51]

After territorial unification, on 23 April 1919 the church in Transylvania declared itself to be part of the Holy Synod in Bucharest. On 28 May 1919, at Câmpia Turzii, in a symbolic gesture signifying the unity of Romanians from these new territories with those of the Old Kingdom, the king and government went to the grave of Michael the Brave (1593–1601), the first ruler who, for a short time, had united all Romanian provinces. Bishop Miron Cristea led the religious service of commemoration, and in his speech drew a parallel between King Ferdinand and Michael the Brave:

> The earth of his body [Michael the Brave] is moving today, feeling how Your Majesty, as a good Romanian and Christian, came with your adored Queen and the enthusiasm of all Romanians [. . .] and the soul of the Great Voivode is happy there in the sky seeing that Your Majesty as a second Michael, did not stop your army at Turda [. . .] but continued victorious until [the river] Tisa, achieving his and our boldest dreams.[52]

Bishop Cristea suggested a direct link between the glorious past of Romanians and the new political situation after unification. King Ferdinand was the first leader chosen by God's providence since Michael the Brave to unite all Romanians while, by claiming that the 'soul of the Great Voivode is happy there in the sky', the bishop emphasised the idea that past rulers were acting as protectors of unified Romania's future. The nationalist discourse of Bishop Cristea was continued by King Ferdinand on 21 November 1919 who, in his address at the opening of the first parliament of Greater Romania, stated: 'Today more than ever, the sacred unity of all Romanians who love their country is necessary for the solution of internal and external problems.[53] Thus, the unification of the country achieved a 'sacred' character and from then on it was the Romanians' 'sacred' duty to preserve it.

The Orthodox Church was seen by the regime as an extremely important part of Romanian national identity in the construction of the state, and it needed to be properly represented in the new political context. Because of collaborating with the Germans during the First World War, Primate Metropolitan

Arămescu-Donici was forced to resign from his office on 1 December 1919. Only two days after the ratification of the union of Bukovina, Bessarabia, Transylvania and Banat with Romania by parliament on 29 December 1919, the Electoral Collegium, dominated by Romanian senators and deputies, chose Bishop Miron Cristea as primate metropolitan of Romania with 435 votes out of 447.[54] Romanians now had not only a unified country but also a spiritual leader who, coming from a previously occupied territory, represented all of those who before 1918 were under a different political authority. Primate metropolitan's position was endorsed by the king who claimed at his official enthronement on 1 January 1920 that 'you have the beautiful mission to realise the religious politics of Michael the Brave'.[55] As primate metropolitan, Miron Cristea continued the tradition of his predecessors in supporting the political regime.[56] On 11 March 1920, the church issued a statement in support of the government, asking the clergy to help the state financially by encouraging the faithful to buy bonds.[57] In this way, the clergy acted as agents of the state, carrying its message directly to the people.

With territorial enlargement, the Orthodox Church reformed its structure. In its reorganisation, the church incorporated elements of national identity indicating a special relationship between divinity and Romanian history. This type of discourse which asserted the prime position of national elements would later form the basis of the writings of Romanian intellectuals in the 1930s. In a booklet on the 'Fundamental Principles for the United Organisation of the Romanian Orthodox Church', Primate Metropolitan Cristea pointed out, firstly, that the church should continue to be autocephalous and independent from foreign jurisdiction.[58] Secondly, in his view, the church should be declared the 'National Church of the Romanian State' as the Orthodox faith was part of the Romanian 'soul'.[59] He suggested renaming some Romanian metropolitanates to make connections with symbolic elements in national history: the Metropolitanate of Bukovina acquired the title 'of Suceava' and the Bishopric of Cluj was renamed the 'Bishopric of Cluj, Feleac and Vad' because Stephen the Great (1457–1504) had his capital in Suceava and had established a bishopric in Vad.[60] Moreover, Primate Metropolitan Cristea emphasised that the church would always be a 'Romanian, national and patriotic Church' and '[its] interests would never be in conflict with the purest interests of the state and of the Romanian people'.[61] Referring to the position of the clergy, Primate Metropolitan Cristea stated that the clergy should have a stronger position in society being regarded as equal to civil servants although the structure of the church differed from that of state administration. In his opinion:

> The Church is not a democratic institution. It is based on a superior authority, on the corner stone, on Christ. The Christological principle is at the foundation of the Church [. . .] the clergy and bishops should not have fewer rights than their equals in lay, democratic administration.[62]

Setting out the role of the church in Romania, Primate Metropolitan Cristea claimed that

> the Church brings up the citizens in faith and morality, in fidelity for the throne and motherland, in patriotism and love for sacrifice for general good, in obedience towards authorities, in respect towards Constitution and laws, in love of order and fulfilment of all Christian and citizenship virtues; the Church exerts charitable acts in asylums, orphanages and other institutes, etc. Finally the Church prays for everything and everybody.[63]

In this way, he combined traditional religious values with patriotic duties, referring to 'faith and morality' alongside the 'motherland', 'patriotism' and 'obedience towards authorities'. He saw the church as having not only a religious but also a political role to play in society as part of the newly enlarged state.

The position of the church was endorsed by the king's presence at religious ceremonies, who was perceived as the head of state with divine qualities and as the elected ruler of Romanians. The combination of Orthodoxy and politics was particularly evident during Easter when the king blessed the masses,[64] and also during the most important religious festivals when students had to participate in compulsory religious ceremonies.[65] After the First World War, various societies were established with the support of the church. The most important was the National Orthodox Society of Romanian Women (*Societatea Ortodoxă Naţională a Femeilor Române*) whose main purpose was 'the development of culture and the education of Romanian children from a national and religious perspective according to patriotic interests'.[66] The society campaigned to build a symbolic church, 'the Church of the People', dedicated to those who fought in the war;[67] after 19 years of work, this was achieved at Mărăşeşti, the battlefield of the most dramatic Romanian military resistance.

The climax of Romanian unity was displayed in the combination of religious and political festivities on 15 October 1922 when King Ferdinand was enthroned at Alba Iulia as king of Greater Romania. He was crowned in Transylvania, rather than in the Old Kingdom, in a symbolic gesture of unity with the new territories. In his speech, the king declared that Romania was a nation blessed by God's grace and referred to the enlargement of the country in terms confirming its long temporal significance rather than simply as a result of recent politics:

> Through the grace of God and national will I have inherited the Crown of Romania after the glorious reign of the Founding King [Carol I]. Coming to the throne I asked the Sky to help the unceasing work which I decided to devote to my beloved country, as a good Romanian and king. The divine grace blessed us and through the strength of the people and

the victory of our soldiers gave us the possibility to enlarge the borders of the kingdom and to realise the longing of ancient times of our nation.[68]

The ceremonies represented an orchestrated combination of Orthodox and political messages, using religion to reinforce national politics. Making the connection between 'the divine grace blessed us' and 'the longing of ancient times of our nation' the king suggested that he had obtained the crown with divine help. The church hierarchy endorsed his speech by emphasising the divine participation at that moment:

> Glory to God in highest heaven, and on earth his peace for men on whom his favour rests. That God blessed his people, his people. Today is the most joyful day for Romanians.[69]

'The greatest day of joy' when 'God blessed his people' indicated a special relationship between the Orthodox Church and the political regime. Not only was the king sacrosanct but the whole nation was also seen as being divinely protected and the enthronement day was the culmination of the 'longing' of centuries. The ceremonies combined extensive references to the religious and national past of Romanians, during which the church hierarchs praised the king as being divinely protected and in their prayers made comparisons between Christ and Virgin Mary as being those who enthroned and protected the monarchy.[70] As the ceremonial suggested, the Romanian government and the church hierarchy worked closely together. With the active participation of the church, the government wanted to ensure that it would acquire the support of Romanians from all of the new territories. In this way, the ceremony showed that, despite being imported from a non-Orthodox country, the church 'nativized' the monarchy who acquired divine qualities. The Romanian monarchy was similar in grandeur to that of the Byzantine Empire and the nation was 'chosen' by God. Following the principle of *symphonia*, the church legitimised the monarchy which in turn legitimised the newly established nation-state.

The establishment of the Romanian Orthodox Patriarchate

Political leaders remained interested in church affairs throughout the interwar period. Political pressure on the church was visible when the Holy Synod agreed to incorporate elements which seemed to modernise it and to offer the country a civilised image in Europe. After attending a Pan-Orthodox Conference in Constantinople in 1923, the Holy Synod decided on 1 October 1924 to adopt the Julian calendar maintaining only the date of Easter with other Orthodox countries which had not yet converted.[71] The church hierarchy presented the decision of incorporating the new calendar not only as a religious necessity, but also due to the other Christian confessions which celebrated the same festivals but on different days. Consequently people were confused and refused to work on the days of both festivals. Transylvania had the largest number of religious confessions and through this measure the church aimed at placing Orthodoxy in a dominant position.[72] Thus, while the state gained economically as people were absent from work on only one day, the Orthodox Church increased its influence in the country by imposing its own dates over those of other confessions.

The political transformation in the Balkans after the war between Greece and Turkey brought Romania into a new position. In 1923, at the time of signing the Treaty of Lausanne between Greece and Turkey, the Romanian government took advantage of its historical connections with the Ecumenical Patriarchate in Constantinople. During the process of negotiations, Romania, which enjoyed good relations with both countries, asked Turkey to ensure that the transfer of population stipulated in the treaty would not lead to the abolishment of the Ecumenical Patriarchate.[73] Romania's stance on this cause would soon be rewarded as other Orthodox churches would regard the country as a main protector of its faith. Furthermore, the increasing Romanian influence in the Orthodox world was recognised by the visit of Patriarch Damian of Jerusalem to Bucharest in 1924. The patriarch urged the government and the church hierarchy to help financially his churches and monasteries at Holy Places.

Romania was regarded as one of the most important actors in the Orthodox commonwealth because it had the largest number of faithful which professed their religion freely. The largest Orthodox Church was the Russian Orthodox Church but, after the October 1917 Revolution, it suffered major persecution, while since 1878 the Bulgarian Orthodox Church had been in prolonged conflict with the Ecumenical Patriarchate over jurisdictional independence. The Romanian government and the church hierarchy took advantage of these international circumstances. A particular opportunity was offered by the fact that in 1925 the Patriarchate of Jerusalem celebrated 1600 years from the First Ecumenical Council. There were various voices in the Romanian Orthodox Church and in parliament claiming that the church should attend the celebrations but in a higher position than that of a simple autocephalous church. In addition, in 1917 the communist authorities in Russia re-established its Patriarchate and in 1920 the Serbians, with a population of only seven million, acted similarly. In order to raise internationally the status of Romanian Orthodoxy, the Holy Synod officially proposed on 4 February 1925 that the Romanian Orthodox Church should acquire the status of Patriarchate. By obtaining this title, it would be considered equal to other Patriarchate Sees and the country's position in the region increased. The proposal was voted by parliament (*Legea pentru ridicarea scaunului archiepiscopal şi metropolitan al Ungrovlahiei, ca primat al României, la rangul de Scaun patriarhal*) which sent a letter to all other Orthodox churches informing them of its decision.

Rather promptly, and recognising the Romanian government's support of the Ecumenical Patriarchate's place in Constantinople, on 30 July 1925 the Ecumenical Patriarch Basil II gave his *Thomos* of recognition. The religious ceremonies of Patriarch Miron Cristea's enthronement took place on 1 November 1925 and were attended by representatives of other religious confessions in Romania and the Patriarchates of Constantinople, Jerusalem, Serbia, Greece, Bulgaria, Poland and of the Russian diaspora. Patriarch's enthronement mirrored that of the king three years previously. After King Ferdinand gave the pastoral staff as the symbol of his new authority, Miron Cristea stated that the new mission of the church was to make 'our country a model for the state and a blossoming field of all Christian virtues: a Christian, happy Romania'.[74] The king showed his close support for the Patriarchate and presented his version of national history, referring to the role of the church in the past:

> From the time of Basarabs and Musats, founders of the country, who established the everlasting Metropolitanates of Wallachia and Moldavia, there was no such glorious page in the history of the Romanian Orthodox Church as that of this year by raising the Primate Metropolitan of Romania to Patriarch. National history proved that for us, Romanians, the nation and religion were always connected. The Church was founded slowly in the shadow of the forests, with the formation of language, of nationality and of the State. The State grew together with the Church.[75]

The establishment of the Romanian Patriarchate was the most important religious event after obtaining Romanian Autocephaly in 1885. The king indicated that the church was closely connected with the Romanian language and nationality, and was part of the state's development. According to his view, because of the intrinsic relationship between church, state, nationality and language, the Romanians were a nation. His discourse was an example of the concept of *symphonia* between church and state which would lead to a more active combination of religion and politics in the public sphere.

Relations between monarch and patriarch also become closer. On 4 January 1925, parliament sanctioned the decision of King Ferdinand's son, Prince Carol, to renounce the throne and recognised his grandson, Prince Michael (Prince Carol's son), although still a minor, as heir apparent. Prince Michael's regency was composed of three members: Patriarch Miron Cristea, Gheorghe Buzdugan, President of the Supreme Court of Cassation, and Prince Nicolae of Hohenzollern (Prince Carol's brother). The presence of Patriarch Cristea in the regency showed his influence in the political field. When King Ferdinand died on 20 July 1927, Michael, still only six year old, was proclaimed king and the regency officially started its prerogatives. The church mourned its ruler, comparing its loss to that of the Moldavian ruler, Stephen the Great, thereby locating King Ferdinand within the national pantheon of the most significant leaders:

> On the morning of 20 July, the day of Saint Prophet Elijah, the news that the King left for the eternal territories passed as a thunder over the whole country [. . .] But this is the month in which the Romanian people had another loss, when the great ruler Stephen the Great passed to the Lord [. . .] So the Romanian people would now have two mourning dates in July.[76]

Church and society

The newly established Patriarchate sought not only a close collaboration with the regime but also a stronger presence on the international stage participating more actively in inter-Orthodox and ecumenical dialogues. The church's actions followed the foreign policy of the state which was seeking its own interests from ecclesiastical meetings. From 19 August to 30 August 1925, Metropolitan Nicolae Bălan of Transylvania attended the meeting of 'Life and Work' in Stockholm and a Romanian delegation attended the meeting of 'Faith and Order' from 3 August to 21 August 1927 in Lausanne. In 1938 'Life and Work' and 'Faith and Order' united to form the World Council of Churches. The third ecumenical organisation, the 'World Alliance for International Friendship through Churches' held some of its regional conferences in Romania, including Bucharest from 14 May to 19 May 1933 and in Râmnicu Vâlcea from 1 July to 15 July 1936.

International Church activity was supported by the state which saw an advantage in using Orthodoxy to achieve its political interests in the Orthodox world. After Patriarch Damian came to Romania in 1924, Patriarch Miron Cristea visited Jerusalem, Constantinople and Alexandria in May 1927 and Poland in May 1938. The most important non-Orthodox contacts were with the Church of England which was independent in its ecclesiastical jurisdictions from the Catholic Church. The Romanian Orthodox Church found many similarities with the Church of England especially regarding the ecclesiastical organisation and the sacraments. Relations between these churches were helped by the fact that Queen Mary, the wife of King Ferdinand, was born in England and was the niece of Queen Victoria. Close contacts between their churches led to the organisation of a Romanian-Anglican Conference in Bucharest from 1 June to 8 June 1935 which resulted in religious recognition by the Holy Synod on 19 March 1936 of the validity of the ordination of priests in both churches.[77] Moreover, Patriarch Cristea deepened relations and visited London in June and July 1936. Contacts between these churches became stronger with the establishment of a branch of the Fellowship of Saint Alban and Saint Sergius in Romania on 17 September 1935.[78] However, despite good relations at hierarchical level, many priests and, in particular, many members of the monastic clergy regarded with suspicion any contact with the non-Orthodox.[79]

In an agrarian dominated society, the church was the key to winning general elections. The introduction of the universal vote in 1918 led the church to become more active in political debates as peasants could be easily influenced by the clergy's authority.[80] As the church's position strengthened in society, parliament wanted to ensure that it maintained control of the process which appointed the church hierarchy. On 6 May 1925, parliament voted for new church legislation (*Legea şi statutul de organizare a Bisericii Ortodoxe Române*) which set out its administrative composition. The law brought in more lay members who were elected from local parties, thereby increasing political domination of the church. Five metropolitanates were divided into 18 eparchial bishoprics.[81] In addition, since 1921 a Bishopric of the Army was already functioning in Alba Iulia. The law stated that the legislative body of the church was the Clerical National Congress (*Congresul Naţional Bisericesc*), which would meet once every three years, composed of six representative of each bishopric (two priests and four lay members) who were elected for six years.[82] The executive body was the Ecclesiastical Central Council (*Consiliul Central Bisericesc*) composed of 15 members, with three from each metropolitanate (one priest and two lay members). The patrimony of the church was supervised by the Guardians of the Church (*Eforia Bisericii*), composed of three members (one priest and two lay members). The same formula of 1/3 clerical participation and 2/3 lay members was then applied to the entire organisation of the church, from bishoprics to local parishes.[83]

The law would remain valid until 1948 and constituted the legislative method used by the Romanian authorities to impose its own people in the church hierarchy. Being elected by the majority of parliament, the church hierarchs were, as previously, the political products of the party winning the elections. In addition, the lay positions were mostly occupied by those elected according to the political interests of the government.

The state allowed the church to retain an important place in the country's health system and education. The church ran some hospitals, which did not promote public health and social welfare but merely ensured the exclusion of the ill from local communities. State regulations of the health system in the 1930s would lead to an increase in specialised staff and a reduction of the position of the church in this domain.[84] Regarding education, according to article 61 of the Law of Education, the clergy taught religion in primary schools. The subject was also an elective class in secondary education. The church had seminaries in Moldavia and Wallachia near major bishopric centres,[85] five theological academies in Transylvania, on the level of secondary education, and three faculties of theology.[86]

According to the 1925 legislation, the state covered part of priests' salaries. However, they faced major financial difficulties as they were mainly supported by their local congregations which were often poor, particularly in the countryside. For this reason, becoming a priest remained an unattractive profession and many were from the uneducated part of the peasant class.[87] The legislation led to dissatisfaction among the clergy, with the main place of dispute Transylvania. The clergy there complained that the state was not doing enough to support them while other confessions received more help. Comparing the situation of various confessions, Father Ioan Crăciun complained that Orthodox priests had a maximum of 12 land units (*iugăr*) and received their salaries from the state after five or six months, while a Lutheran pastor had 32 land units, forest and more financial support from his community and the state. His lament reflected the general opinion of the Orthodox clergy: 'After six years of Romanian rule [in Transylvania] what bad things did [the Orthodox clergy] do in the past and what sins do they have now to be [treated like this and] completely abandoned?'[88]

A general report in 1935 of the Romanian Patriarchate revealed that 12,375,850 people (3,432,541 families) declared themselves to be Orthodox. The church was structured at local level into 8474 parishes; there were 820 urban and 7654 rural parishes in charge of 10,740 church buildings. The Orthodox clergy numbered 8542 priests, 7868 of which received salaries from the state, 380 salaries from private funding, 188 were retired while 106 priests were without any state salaries. The composition of the clergy showed that most of them had only elementary theological education. Only 76 priests had doctoral degrees in theology, 1446 had university degrees, while 3287 had completed seven or eight years in a seminary and 537 only four years. In addition, the priests were helped by 10,452 cantors; 9166 received state support and 1268 private funding.[89]

This data has to take into account the massive territorial changes and population movements after the establishment of Greater Romania in 1918. If in 1899 Romania numbered 5,956,690 people, in 1930 this figure tripled reaching 18,057,028. A previous report in 1904 indicated that the Orthodox Church had 4998 priests and 6678 church buildings[90] while the increased numbers of the interwar period showed that large sections of the population remained illiterate living in rural areas (the 1930 census revealed that 6,029,136 people were illiterate and 14,405,989 of the population was rural). The large number of rural clergy with a relatively low level of education was directly connected with the political and economic trajectory of Romania, as due to the traditional character of Orthodoxy, peasants tended to support the political position of their local clergy.

The diminishing number of faithful attending religious services was particularly visible in urban areas where people attended church only twice a year for the main religious festivals, Easter and Christmas.[91] The increased secularisation of society remained the dominant trend throughout the interwar period, from 1922, when Primate Metropolitan Cristea complained that most students who completed a university degree in theology were not ordained,[92] to 1940 when only 10 per cent of Romanians attended church once a week.[93]

Orthodoxy and other religious confessions

The establishment of Greater Romania represented an important change in the religious composition of the country. The Orthodox Church remained dominant but its faithful dropped from 91.5 per cent in 1899 to 72.6 in 1930. The second most important confession in the country was the Greek Catholic Church with 7.9 per cent followed by the Roman Catholic Church which grew from 2.5 in 1899 to 6.8 per cent in 1930. Other major religious confessions were the Jewish Community with 4.2 per cent, the Reformed Calvinist Church with 3.9 per cent and the Lutheran Church with 2.2 per cent.[94]

Transylvania was the most ethnically diverse of all Romanian provinces in which the Orthodox Church and the Greek Catholic Church each represented around 30 per cent of the population. The Greek Catholic Church, also known as the Uniate Church, which was established in 1699, recognised the primacy of religious jurisdiction of Rome over that of Constantinople. It retained the Byzantine liturgical rite and incorporated new doctrinal elements from the Roman Catholic Church. Before the unification of Transylvania with the Old Kingdom, it supported the national identity of Romanians and was a key political player in defending the rights of Romanians in the Habsburg Empire. The 1923 Constitution declared the Romanian Orthodox Church as the 'dominant' church of the state but also offered the Greek Catholic Church an honorific first place among other religions (art. 22).

After 1918, relations between the Orthodox and the Greek Catholics became tense and conflict arose based on the financial and political positions of their churches. The Orthodox hierarchs saw the Greek Catholics as an attempt to attract their faithful to Catholicism. In a book published in Arad in 1923 which presented Hungarian data from 1910, the author, signed as 'a man of the Church', argued that the Roman Catholic Church in Hungary numbered 9,010,305 people, which represented 49.5 per cent of all religions. The Catholics were in possession of 88 per cent of the entire land of all religious confessions representing 1 land unit for 6 people. The Greek Catholic Church numbered 1,900,000 people and 7 people had 1 land unit. Other religious confessions were dispersed as follows: the Unitarian Church numbered 74,245 people and 78 people had 1 land unit; the Reformed Church, 2,603,381 people and 89 people had 1 land unit; the Lutheran Church, 1,306,384 people and 325 people had 1 land unit. The Orthodox Church numbered 2,339,979 people, representing 13.1 per cent of all religious confessions in Hungary and 115 people had 1 land unit. However, the Orthodox faithful in Hungary was divided between the Serbian Orthodox Church and the Romanian Orthodox Church. The Serbs numbered 454,431 people and 24 people had 1 land unit, but the Romanians accounted for 1,798,669 people and 1777 people had 1 land unit. The author concluded that the huge discrepancy in land possession between religious confessions was a direct measure against the Romanian people aimed at attracting them to Catholicism or to the Greek Catholic Church. Although the author of this book remained anonymous, his position reflected the concerns of the Orthodox Church as his findings were published in the church's main journal, *Biserica Ortodoxă Română* (Romanian Orthodox Church).[95]

Political debates in the Romanian Parliament on the status of religions in Romania led to dissatisfaction among Orthodox hierarchs. The signing of the Concordat with the Vatican in Rome on 10 May 1927 and the new general law of religious confessions in 1928, which offered better status to other churches and religions, were viewed by Orthodox hierarchs as threats to their spiritual authority. One Orthodox prelate even accused the politicians of being unaware that other religions were in fact 'Trojan horses' of other states which had the political intention of destabilising the country and obliterating Romanian unity.[96]

The most representative religious dispute was between the Greek Catholic Church and the Orthodox Church entering parliamentary debates, with each church claiming that they were the real protectors of the Romanian faith.[97] The Greek Catholic Bishop Iuliu Hossu stated in Parliament that 'We brought [to the people in Transylvania] national awakening by preserving the Latin soul of Romanians [. . .] [and] unity with Rome was made by faith'. The Orthodox reaction was expressed by Bishop Lucian Triteanul who responded rhetorically, 'Which faith? Did not the Orthodox Church have the ecumenical faith that was above that of Rome?'[98]

Orthodoxy and ethnocracy

The notion of Orthodoxy as a main factor in moulding the identity of Romanians arose with the controversy among scholars on the concept of Romanianness. As the church had acquired a prime place in the establishment of the nation, its nationalist discourse was incorporated by intellectuals in constructing the myth of Romanian exceptionalism. Nae Ionescu, professor of metaphysics at the University of Bucharest, promoted the idea of an inextricable relationship between Romanianness and Orthodoxy to the extent that professing, or being born to a different religion represented exclusion of Romanian nationality. He presented his ideas mainly in the highly influential journal *Gândirea* (Thought). Orthodoxy was not only a religious confession but led to a transfigured existence, a different way of living in the world. According to Ionescu,

> one cannot be Romanian, for instance, if one does not achieve, let's say, in the concrete, in the individual, the organisationally spiritual structure

whose depositary is, essentially, our nation. Nations are, nevertheless, historical realities. They are born in space and in time and are, therefore, conditioned by whatever gets into the category of individuation. A people's attitude to God, the way it experiences not only its relation with the divinity but even divinity itself is part of the intimate structure of the nation.[99]

Ionescu's ideas gained the support of some Orthodox theologians. Professor Dumitru Stăniloae from the Faculty of Theology in Bucharest suggested that Orthodoxy was the 'essential and major function in the life of Romanians. The permanent national ideal of our nation could not be conceived without its connection'.[100] Romanians were the only Latin people of the Orthodox faith and this fact 'enhances the value of our race'.[101] In his opinion 'Any ethnic group could raise itself from a people to a nation only through belief in God'.[102] Professor Gheorghe Ispir argued that only by belonging to the church and the nation could the individual reach its human fulfilment: 'Nations are social entities created by God through which man reaches his ideal. The Christian Church works within nations in complete harmony'.[103]

Ionescu's vision was developed by one of his closest admirers, Nichifor Crainic, also a professor of theology in Bucharest. He claimed the need to establish an 'ethnocratic state' which had at its foundation the Romanian nation and Orthodoxy.[104] According to Crainic, 'Ethnocracy is the political will of the indigenous race to make the state its property and its mission in the world'.[105] An ethnocratic form of government would combine Orthodoxy with the Romanian nation giving priority to the dominant religion, thereby excluding ethnic minorities, particularly the Jews. In his view, the main principles of the ethnocratic state were as follows:

> The law of Christ is the law of the state. The state is the supreme householder of the nation. Nationalisation of the national resources. Individual property for each Romanian worker. Not classes but professions are the organic functions of the nation. Numerical proportionality in all professions. Removal of the agent between producer and consumer. The destruction of Jewish parasitism.[106]

Crainic argued for an intimate connection between Orthodoxy and the Romanian nation in which the faithful had to completely obey both religious and political rulers which were seen as emanating from the same divine source.

Clerical dispute between religions, increasing secularisation and financial difficulties of the clergy would lead to a new employment of Orthodoxy, as, with the rise of fascism across in Europe, Romania created its own version. On 24 June 1927, a young radical, Corneliu Zelea Codreanu established a fascist organisation named the 'Legion of Archangel Michael' (*Legiunea Arhanghelului Mihail*), which would later become the Iron Guard and based its teachings on a combination of Orthodox and national elements.[107] The Legion claimed the superiority of Orthodox people in Europe and an intrinsic relationship between the Orthodox faithful and the Romanian nation; Orthodoxy was the major condition of being Romanian, while citizens belonging to religious minorities were considered potential enemies of the country.[108] The rise of fascism in Romania increased during the world economic depression from 1929 to 1933. On 8 June 1930, parliament decided to install Prince Carol as King of Romania with the title of Carol II while his son, King Michael, was made Prince Successor with the title of 'High Prince of Alba Iulia'.[109]

The Iron Guard used the scholars' debates and employed the church for its own political benefit. In November 1936 Codreanu sent a Guardist group to fight on Franco's side in the Spanish civil war. When Ion Moța and Vasile Marin, two prominent Guardist figures, were killed on 13 December 1936, their bodies were taken to Romania. They were declared national heroes who fought for the defence of Christ and their coffins were transported by train to major towns in Romania. At each station they were welcomed by Guardists while local clergy held religious prayers; even Nicolae Bălan, Metropolitan of Transylvania, celebrated a special religious service in their honour in Sibiu.[110] The political scene reached a sensitive point when the Iron Guard obtained the highest number of votes in elections since its establishment, from 0.4 in 1927 to 15.6 in 1937.[111]

The Iron Guard attracted numerous members of the lower clergy reaching almost 2000 priests[112] and four priests were elected senators representing it in parliament after the 1937 elections.[113] However, while some hierarchs sympathised with the movement, they remained officially opposed to the involvement of the church in its activities,[114] as it should remain loyal to the Crown. In a meeting of the Holy Synod on 8 March 1937, Patriarch Cristea criticised political parties for engaging priests in electoral disputes. The Holy Synod agreed that priests should not perform any religious ceremonies in which they would bless the insignia of political parties; the clergy should not take any political oath; in their sermons the clergy should refrain from political debates. However, these measures should not be seen as contravening the right of the clergy to have a political activity in society.[115] Discrepancies between the high number of clergy supporting the Iron Guard and the official discourse of the church were largely due to the low standard of living in rural areas and the attractive combination of Orthodoxy and nationalism promoted by the Iron Guard.

At the same time as this internal ecclesiastical crisis, in contrast with its previous position, the regime attempted to attract the lower clergy on its side by issuing on 10 March 1937 a new law concerning their salaries.

Orthodox priests would now receive their whole salaries from the state budget, at a level similar to that of a public clerk.[116] The funds for the Orthodox Church increased considerably in the following years. If in 1932 the church received 413 million lei and a reduced amount in the following years (353 million lei in 1933, 382 million lei in 1935),[117] in 1939 it received 571 million lei.[118] Moreover, the budget of the Ministry of Religious Confessions and Arts, which administered the funding of religious confessions, was almost equivalent to that of the Ministry of Health and Social Assistance, and surpassed that of the Ministries of Agriculture, National Economy and Foreign Affairs.[119] The increasing interest of the regime in the church was evident in December 1937 when the Octavian Goga government came into power with the slogan 'God, King and Nation'.[120] All these measures were directed to alleviate the dissatisfaction among the clergy and find the means of winning the support of the church for the regime.

In order to gain control of the Iron Guard, on 9 February 1938 King Carol II established a royal dictatorship which incorporated the church hierarchy at its core. The new government was formed by Patriarch Cristea who declared a national state of emergency on 17 February 1938. A new Constitution was promulgated on 20 February 1938 which assigned supremacy to the king and turned parliament into an auxiliary body. Article 4 of the Constitution stated that 'All Romanians without distinction of their ethnic origin have the following duties: to consider the fatherland as the highest meaning of their life, to sacrifice for its integrity, independence and dignity, to contribute through their work to its moral raising and economic development'. In addition, article 8 proclaimed that 'priests from all religious confessions are forbidden to use their spiritual authority for political propaganda [. . .] Any political association which has a religious basis or pretext is forbidden'. In addition, the church collaborated extensively with the Army in order to create a 'spiritual atmosphere in the Army units'.[121] The Constitution made explicit reference to those priests who were sympathetic to the Iron Guard or who were politically involved. The dictatorship deepened on 31 March 1938 when all political parties and associations were dissolved by a decree-law with the king, the only ruling authority in the country. On 16 April 1938, Codreanu and the other leaders of the Iron Guard were arrested, judged and sentenced to imprisonment. Most of them, including Codreanu, were shot on the night of 29–30 November 1938.[122]

Political instability increased with the pressure of communism on the Romanian border. Across the Dniester River on the border with the Soviet Union, people witnessed the removal of crosses from churches' belfries, while church bells were used in the early morning on collectivised farms to get people up and start work.[123] The presence of communism near Romania had a deep impact on the life of the clergy and rising tensions were visible.[124] Ordinary people perceived communism as a sign of the Anti-Christ on earth and Romania witnessed an increase of religious mysticism. Petrache

Lupu, a shepherd in Maglavit village, Dolj county, in 1935 had visions of the Virgin Mary and called on all Romanians to repent; otherwise they would face punishment from the East.[125]

Patriarch Nicodim Munteanu and the Second World War

Political and religious turmoil deepened just before the Second World War. On 26 February 1939, Patriarch Cristea left the country and went to Cannes in order to benefit from the climate; he died there on 6 March 1939 and was buried in the Patriarchal Cathedral in Bucharest. The Electoral Collegium elected Nicodim Munteanu, Metropolitan of Moldavia, as Patriarch on 20 June 1939.

The reasons for the election of Munteanu remain unclear. He studied in Kiev and after entering the monastic clergy, was elected Bishop of Huşi in 1912. During the First World War he held the position of Archbishop of Romanians in Bessarabia and represented the Romanian Orthodox Church at the Pan-Orthodox Synod in Moscow in 1917. Various accusations were made against him which affected his position, and in 1924 he withdrew from the church hierarchy being appointed the abbot of Neamţ monastery. In 1935 he was suddenly elected Metropolitan of Moldavia, a position from which he held the *locum tenens* in case of the patriarch's death. The ambiguity of his hierarchical trajectory was supported by the fact that on the day of his election as patriarch, two of his main opponents, Metropolitan Bălan of Transylvania and Metropolitan Visarion of Bukovina, publicly declared that they would not stand, leaving him as the only candidate. Munteanu received 458 votes while his opponents, despite their previous pleas, received 12 votes each; in addition, Bishop Tit of Hotin collected 8 votes, Bishop Nifon of Huşi 1 vote and 19 votes were declared null.[126] In an article dedicated to Patriarch Munteanu written in 1945 in the official church journal, Iuliu Scriban suggested that his knowledge of Russian Orthodoxy and his 'insights in diplomatic affairs' were the main reasons for his election.[127] In the interwar period relations between Romania and the Soviet Union were tense, and most probably the regime sought the use of international ecclesiastical channels in order to improve contact between their countries.

On 5 July 1939, Patriarch Munteanu was received by the king in a public ceremony. After Carol II offered him the pastoral staff as a symbol of his religious authority, the Patriarch stated his support for the regime emphasising close connection between religion and the national identity of Romanians. Making reference to the book of Genesis in the Old Testament, Patriarch Nicodim claimed that

> When God wanted to create our people, he kneaded Roman and Dacian clays and he made us. He blew the wind of the Gospel of his son and thus

the Romanian people became Christian from birth. All people around us have a date of their Christianization. The Romanian people do not have such a date, because they are Christians from birth. The political and religious lives of the Romanian people are closely bound.[128]

This type of discourse showed how closer the nationalism promoted by the church became to that of the state. In his reply, the king followed the same discourse and pointed out that 'to us, Romanians, the Church and the nation are the same [. . .] When the political battle diminishes, the Church takes up its role, that of being the protector and guide of the faithful'.[129]

The outbreak of war in Europe led to military preparations in Romania. The Patriarch encouraged the people to help the army and in January 1940 he personally donated it his monthly salary and asked each member of the clergy in his metropolitanate to contribute one month's salary.[130] On 26 June 1940, Romania received an ultimatum from the Soviet government to withdraw its troops from Bessarabia and Northern Bukovina which would now be considered part of the Soviet Union. Furthermore, on 30 August 1940 the German-Italian 'arbitration', known as the 'Vienna diktat', decided that the northern part of Transylvania would be taken from Romania and given to Hungary. The church mourned the territorial loss and the Patriarch sent a Pastoral Letter in which he asked the faithful to help the Romanian refugees from the former territories.[131]

Losing its territorial unity, a solution to the political crisis was reached on 5 September when General Ion Antonescu received dictatorial powers from King Carol II. On the following day, the king, now isolated on the political scene, was forced to abdicate, bestowing his royal prerogatives upon his son, Prince Michael, who again become king. On 14 September, Romania was declared a 'National Legionary State' and General Ion Antonescu was titled 'the Leader' (*Conducător*), ruling the new government with a fascist majority.

Antonescu associated the patriarch with the political regime of Carol II. Tensions between the government and the patriarch were visible on 30 September 1940 and 30 January 1941 when the patriarch twice sent his resignation; however, it was not accepted.[132] Seeking to gain control of the fascist group on 14 February 1941, Antonescu dissolved the National Legionary State. In the aftermath of his decision more than 9000 people were arrested including around 422 priests and 19 cantors who were members of the Iron Guard.[133] Despite these clashes between the church and the regime, the hierarchy continued to support the state policy and the Patriarch blessed the Army in fighting 'the holy war' against the communist 'nation without God'.[134]

On 22 June 1941, Antonescu ordered the Romanian troops to cross the Soviet border and in less than two months Bessarabia and Bukovina were reunited with the country.[135] The church organised a religious mission (*Misiunea Ortodoxă Română*) to the territory between the Dniester and

Dnieper Rivers, Transnistria, sending around 250 priests who helped the revival of the Orthodox faith after atheist domination of this region.[136] Antonescu sought the support of the Orthodox Church in a similar manner to that of previous regimes and in July 1943 issued a law declaring illegal the so-called neo-Protestant churches which were now perceived as a potential threat to national security.[137] Thus, the previous discourse of the Orthodox Church on 'Trojan horses' became national policy.

Romania continued to push its military forward, advancing with the Axis forces until Stalingrad.[138] However, victory of Soviet forces and their counter-offensive towards Romania in 1944 led to the establishment on 20 June 1944 of a National Democratic Bloc, composed of the former democratic parties, which started secret negotiations with the Allies. With the Soviet Army reaching the Iaşi-Chişinău line, King Michael led a coup d'état on 23 August 1944, arresting Antonescu and his government. Romania had an unstable political life until 6 March 1945 when the Soviet Union imposed the first government with a communist presence.[139]

Conclusion

If in 1859 the first Romanian Prince of the united principalities, Alexandru Ioan Cuza, had to convince the European powers that the populations of Wallachia and Moldavia were one nation, the subsequent monarchical regimes ensured that this united Romanian identity was built on a common language, on unified political institutions and on the substratum of the dominant religion. Politicians saw that only by controlling the church hierarchy would they have the support of the church and of the Romanian faithful. Even if the Romanian Orthodox Church suffered a drastic reduction of its economic and political influence in society, by following the concept of *symphonia* it adapted to its new role. This close relationship between Orthodoxy and politics had an impact on the evolution of Romanian politics in the interwar period.

After 1918, the Orthodox Church in Greater Romania was employed by the state as a nationalist tool, but it did not regain the economic status and political influence which it had enjoyed before the creation of the Romanian state in 1859. Internal political instability after the creation of Greater Romania was not only the result of developments in European politics, but was also the product of the Romanian elite. Perceiving its place in society to be in competition with other religions, the Romanian Orthodox Church promoted a combination of religious discourse and political nationalism, suggesting that the Romanian nation was intrinsically bound with Orthodoxy. The political regime benefited from including Orthodoxy as its main political ally and by transforming the church into a state institution. The regime used Orthodoxy to reinforce the connection between the Romanian past and contemporary politics.

In the following years, with the support of the Soviet Union, the communists would dominate political life, leading to the abolition of the monarchy and the installation of a regime of state terror. During the communist period the Romanian Orthodox Church changed its nationalist discourse and adapted to the new regime. The communists were not interested in using religion at the core of their public discourse. The employment of the church as a political instrument of the regime followed a tradition that went as far back as the first years of the Romanian state.

3
Orthodoxy and the Installation of Communism, 1944–7

23 August 1944 and the Romanian Orthodoxy

In the interwar period the Romanian communist party was a small faction without any political influence,[1] numbering only a few thousand members in 1944.[2] However, from 1944 to 1948 it managed to consolidate political power, supported by the Soviet Union.

A gradual increase of communist presence in Romanian political life began on 23 August 1944 when the leaders of the democratic opposition, acting with the support of King Michael, mounted a coup d'état against Marshall Antonescu and his dictatorial pro-German regime. From this date until 22 November 1944 Romania was ruled by two provisional governments, both led by General Constantin Sănătescu, a well known antifascist, and took part in the war against Germany with Romanian forces advancing through Hungary and Czechoslovakia until the official end of the war in 1945.[3] In the negotiations of 12 September 1944, Romania lost the territories of Bessarabia and Northern Bukovina to the Soviet Union. By 25 October 1944 all pre-war Romanian territory in Transylvania had been recovered by Romanian and Soviet armies. Under the control of the Red Army, the Soviet Union installed its own local officials and refused to return the territory to Romanian administration. Moreover, at the Moscow meeting in October 1944, Churchill and Stalin agreed to 90 per cent Soviet political interest in Romania.[4]

23 August 1944 was not only a historical moment for the Romanian nation state but was also considered an important day for the Romanian Orthodox Church. Among messages addressed to the throne and other political officials, on 24 August 1944, Patriarch Nicodim sent a telegram to King Michael reinforcing the church's support for his decisions, stating that 'the Church will be near the Romanian people and its king as always'.[5]

A few months later, on 9 October 1944, during the General Sănătescu government, the patriarch issued an official Pastoral Letter. This letter was the first official document of the church hierarchy publicly expressing support for the political regime of the Soviet Union whose military forces were present in the country. The Pastoral Letter is one of the most interesting official declarations of the Romanian Orthodox Church in showing how religion combined with politics in praising the new regime.

The Pastoral Letter begins by asserting that Romania suffered under the former dictatorial regime, which 'suffocated the free thoughts and the feelings of our people'.[6] The letter stated that 'the dictatorship is something strange for the soul of Romanian people'.[7] Although the former Orthodox patriarch, Miron Cristea, was prime minister of three governments of the dictatorship of King Carol II in 1938–9, the Pastoral Letter noted that 'the dictatorship is forbidden by the Church's teachings'.[8] The letter continued:

> The day of 23 August 1944 will remain written as one of the most important dates in the history of the Romanian people. [. . .] The historic act of 23 August 1944 has removed the dictatorial regime, which was inappropriate to the tradition of our country. [This day] gave us democratic freedoms. [. . .] These freedoms are precious to us because most of them are the product of our Saviour's teachings: 'Render Caesar what is due to Caesar, and render God what is due to God' (Matthew 22: 21). These words are the basis and the source of individual, political and religious liberties.[9]

Referring to the creation of the Romanian state, the Pastoral Letter added that Romania was a nation state with the help and support of great powers. The armistice between Romania and the Soviet Union demonstrated that Romania was a united country only with the military and political help of the Soviet Union. The USSR's 'generosity' in not conquering Romania 'imposes us to bind, in a spirit of absolute trust of Soviet Russia and our elder sister in the right faith'.[10] The letter emphasised the main reasons for following the Soviet Union's policy, briefly presenting the relationship between the hierarchies of the Orthodox churches in these countries. Finally, the letter asserted that 'the Church has been the mother of the Romanian people, an ethnic mother, because she united colonists with Dacians which led to the birth of the people and the creation of the Romanian states'.[11] From this special position the church had the authority to indicate to the king and the nation the political path to follow, suggesting the need to be closer to the Soviet Union.

This new attitude of the church towards the Soviet Union was due to two main factors. On the one hand, in the war against Germany, Soviet troops occupied Romanian territory, and a declaration such as this letter was aimed at pleasing the Soviet authorities. On the other hand, the Pastoral Letter conformed to the directives of the Soviet High Command of the Southeast European Front. Through the armistice, Romania was under the Allied Control Commission and the religious framework in the Balkans had been set by the so-called 'Vyshinsky Plan' which imposed the 'liquidation of undesirable clergy and replacing them with Soviet-trained or sympathetic clergy'; it also urged 'forming an alliance with Orthodox Churches under the leadership of the Moscow Patriarchate'.[12]

By sending this letter to the Romanian faithful, Patriarch Nicodim suggested *de facto* political obedience towards the new political regime after the events of 23 August 1944. The Pastoral Letter was not merely a declaration of the church hierarchy but described the role of the church in fostering better relations with the Soviet Union. Thus, the letter used religious symbols with political messages aimed at changing the faithful's perception of the Soviet Union.

The first reinterpretation of religious symbols is perceivable in the description of 'the individual, political and religious liberties'.[13] The Pastoral Letter indicated that these liberties were achieved both through the events of 23 August 1944 and Christ's words from the New Testament. With the change of the political system, the church hierarchy transformed the meaning of an important day of the Romanian nation state offering it a mythical dimension.

According to the letter, 23 August had acquired a symbolic dimension for the faith of people as the unique moment in history when Romanians were 'liberated from tyranny'. The freedom brought by this day was comparable with the divine revelation which came from Christ. The use of this paragraph from the Gospel of Saint Matthew emphasised the separation between the church and the state and also had a special significance in the history of previous church-state relations. In 1922, for the religious ceremonies enthroning King Ferdinand and Queen Mary as rulers of the kingdom of Great Romania, the Orthodox hierarchy created a special 'Doxology', a religious service for that particular occasion.[14] The most important part of this Doxology was the moment when the metropolitan read the Gospel. According to church symbolism, the metropolitan symbolised Christ himself, present there, teaching the masses and blessing the royal couple. The paragraphs from the New Testament used in 1922 were Matthew 22: 15–22, thus including the words that would be used 22 years later. By employing the same religious words in different political contexts, the church hierarchy asserted that church-state relations at the enthronement of King Ferdinand in Greater Romania were the same as in the new political regime dominated by the Soviet Union and implied its acceptance of the new authority. If in 1922, the metropolitan symbolised Christ who blessed the king and his nation, in this Pastoral Letter the church hierarchy suggested that the coup d'état of 23 August 1944 was akin to a divine revelation and had the same authority as Christ's words.

The second reinterpretation of religious symbolism is offered by the position of the church in public life. As the Pastoral Letter indicated, the Orthodox Church should be seen as the 'mother' of the Romanian people. The church claimed that its structure incorporated all generations of the

Romanian nation, including the dead, the living and the unborn. Moreover, the Pastoral Letter stated that even when the Romanian nation was not yet formed, the church had always existed. For this reason, according to the church's view, Romanians would not be a nation among other nations without this intrinsic link to the church.

The Groza government and the church

Because General Nicolae Rădescu had opposed the communist takeover, especially of key positions in the Romanian Army, by February 1945 he was accused by the Soviets of being a 'fascist'. This accusation led to an ultimatum to the king from Andrei Vyshinsky, the Soviet Deputy Foreign Minister, on 5 March 1945, who asked for a new government led by Petru Groza, a founder of the Ploughmen's Front and Rădescu's former deputy. Otherwise, Vyshinsky insisted, he 'could not be responsible for the continuance of Romania as an independent state'.[15] The Groza government was installed the following day and the communists acquired top positions in the ministries of the Interior, Justice, War and National Economy and undersecretaries in all other ministries.

Thus, in less than one year from King Michael's coup d'état against Marshall Antonescu, the communists, who formed an insignificant proportion of the Romanian electorate, were now an active political presence in all ministries. Groza was accepted as prime minister by the king and the other political parties mainly because he was considered the best solution at that moment, the right compromise between the traditional parties and the communists.

As prime minister, Groza began a campaign aimed at obtaining public support from the Romanian masses for the communist cause. He realised that only by acquiring legitimacy would his regime survive and attract electoral support. From his first days in power, Groza made reference to both Romanian nationalism and religion. His ability to manipulate the church was probably helped by the fact that he was the son of a priest, and had formerly been a layman in the Orthodox Church.[16]

The first act of the Groza government was the official transference of the Northern Transylvanian territories from Soviet occupation to Romanian administration on 8–9 March 1945. This gesture aimed to obtain public support for the Soviet Union and to promote the communist faction in the coalition government. Secondly, on 21 March 1945, Groza founded the Workers' University of the Romanian Communist Party, which was later transformed into the 'Ştefan Gheorghiu' Academy, named after a prominent communist of the time, which served as the party's school for communist activists. Thirdly, on 8 May 1945, four days before the official end of its participation in the 'anti-fascist' war, the government established the Superior Directorate of Culture, Education and Propaganda of the Romanian Army.

These institutions would later become the official means through which communist propaganda would be present in state educational structures.

In addition to these institutions, the Soviet leaders themselves supported the communist oriented leadership in Romania. Thus, on 6 July 1945, King Michael was decorated with the order 'Victory', the highest medal of the Soviet Army, indicating good relations between the Romanian monarchy and the Soviet Union. By accepting this decoration, King Michael recognised the authority of a Soviet symbol, indirectly supporting the growing communist influence in the country's political life.

On 17 and 19 August 1945, the representatives in Bucharest of the United States, Roy Melbourne, and of Great Britain, John Le Rougetel, sent diplomatic notes to the king informing him that their countries refused to recognise the Groza government. Facing international isolation, the king asked for the resignation of the government on 21 August 1945. Groza, after consultation with the Soviet representative in Bucharest, rejected this demand. For this reason, King Michael started a 'royal strike', in which he abstained from having contact with government officials and from countersigning their decrees and acts.

The Groza government and the communists then initiated new political actions suggesting the future course of Romanian politics. First, on 23 August 1945, the government issued a bill for agrarian reform expropriating the property of those who had collaborated with the Germans, who had more than ten hectares and who had not worked their lands in the previous seven years.[17] Second, on 24 August, Romania and the Soviet Union raised their diplomatic missions to the rank of embassies. Third, a Romanian governmental delegation led by the prime minister visited Moscow between 4 and 13 September 1945. As all of these changes took place during anniversary celebrations, one year after 23 August 1944, they can be seen as political activities which anticipated the future course of Romanian politics. On the one hand, connecting this major date in Romanian history with the communist regime made a direct relation between the Groza regime and the success of the coup d'état. On the other hand, raising the diplomatic missions on 24 August, the day after the official celebrations, suggested that the Soviet Union had become the closest ally of the Romanian government.

In order to influence the church's attitude towards the government and the communist party, the communist leaders installed figures inside the church hierarchy who were supportive of their politics. The most significant person was Justinian Marina, a priest from Râmnicu-Valcea, who was promoted to bishop in the Metropolitanate of Moldavia and Suceava.[18] Some scholars have suggested that he was helped by Ana Pauker, the secretary in charge of recruitment for the Romanian Communist Party[19] or by Gheorghe Gheorghiu-Dej, a leading communist at that time. Justinian Marina, having leftist inclinations,[20] had probably hidden Gheorghiu-Dej when he escaped from prison during Marshal Antonescu's regime. Justinian's unusual progress is proved by the fact that in normal circumstances, in order to be appointed bishop, a person has to live a long period of his life as a monk. However, Justinian was made a monk on 11 August 1945 and a bishop on the following day. Another person supported by the communists was Father Constantin Burducea, who was appointed Minister of Religious Confessions.[21] Before joining the communist party, Burducea was a member of the extremist nationalist organisation, the Iron Guard;[22] because of this compromising past, he was easy to manipulate into allegiance with the regime.[23]

The first public action of the Groza government concerning religion was the organisation of a General Congress on 16–17 October 1945 in the Chamber of Deputies in Bucharest, bringing together representatives of all religious confessions in Romania and participants from the Bulgarian, Serbian and Russian Orthodox churches. This Congress was important not only because it united different religious leaders, but also due to its suggestive employment of religion and politics.

Father Burducea, as Minister of Religious Confessions and the President of the Democratic Union of Priests, declared in his inaugural speech that the congress represented 'a unique act in the religious history of Europe, which, after two thousands years, has not had this kind of Synod, bringing together the Old Testament, the New Testament, the Koran and the Protestant Reform'.[24] Referring to the Orthodox church, Father Burducea emphasised that 'the Church has an important role in uniting the people from the Soviet Union, Romania and the Balkan states'.[25] For this reason, he saw

> the Church as precious collaborator of the state, because the Romanian Church is older than the state and because it conducts the souls of the Romanian people [. . .] Today, this collaboration between church and state is something normal, because the government belongs to the people, the Church is of the people, and all these three belong to God.[26]

By bringing all religious leaders together in a congress, the authorities had two aims: on the one hand, at a time of political tension with the king, it presented an image of the communist government as different from previous governments; on the other hand, it offered a new dimension to the relationship between church and state. Thus, according to Burducea's words, the congress acquired a mythical role: being a 'unique Synod', an exceptional event that had never happened anywhere in the last 2000 years in Europe. Offering a symbolic dimension to this synod, the communist leaders tried to gain electoral support for the Groza regime influencing the Romanian Orthodox faithful. Moreover, the General Congress of the Romanian Communist Party was held from 16 to 21 October 1945, also in Bucharest, at a time when a new Central Committee and a new general secretary, Gheorghe Gheorghiu-Dej, were elected. Taking place during the same days, in the same city, these two congresses had a direct message for the Romanian people: one clearly representing the country's political future and the other representing its spiritual path.

The end of 1945 brought new international pressure on the government. At the request of the United States and Great Britain, on 16–26 December 1945, the foreign ministers of these countries met in Moscow with Vyacheslav M. Molotov and agreed that the Groza government would only be recognised upon inclusion of representatives of the opposition followed by free elections. The agreement in Moscow on 5 February 1946 between the ministers of foreign affairs stated that two representatives of the opposition would be included in the Groza government; this development led to its international recognition and the end of the royal strike.

Because elections would take place later in the year, the government began seeking stronger support from the church. Thus, 1946 would represent a new stage in the competition between church and government for the electoral soul of the Romanians. For the communist leaders, the church hierarchy would acquire a stronger, mythical, transcendental position as representatives of pan-Orthodoxy in general and of the Romanian nation. If other political enemies could be eliminated by force, as the next years would demonstrate, domination of the Romanian soul could only be achieved through influence over the church. The best image of how the political authorities perceived the Orthodox hierarchy during those years is offered by reports gathered by the *Securitate*. Written by officers of the Groza government, the reports set out what was said in conversations with church hierarchy and their analyses of what the government should do.

A report from 6 February 1946 reveals that during the meeting between Patriarch Nicodim and the Apostolic Nuncio Andrea Cassulo, the latter indicated that the pope would have been interested in appointing Nicodim as 'cardinal of the East'.[27] If he accepted the title, the patriarch would represent the whole of Eastern Orthodox Christianity thereby achieving unity between the Orthodox and the Roman Catholic churches. The report stated that the patriarch refused the pope's proposal fearing the Soviet reaction.

There are no other sources which could verify this assertion but the information is suggestive of communist perception of the Orthodox leaders. The message indicated that the patriarch represented the most important figure of the Orthodox hierarchy, and implied that having his support the communist regime would have access to the spiritual realm of the Romanian nation. On the one hand, the patriarch acquired a special mission, being perceived as the person who could potentially achieve unity between eastern and western churches at this historical moment. On the other hand, this report expressed the communists' fears regarding a more active religious and political intervention of the Roman Catholic Church in Romania, thus diminishing the role of the communist party. During this uncertain period, there were suggestions that the end of the war could lead to the union of the Orthodox and the Roman Catholic churches. From this perspective, if this project was achieved, the communists would face great obstacles in acquiring the main position in government and political control of state structures.

Another report from February 1946 suggested concrete ideas of how the communists interfered in the election of church hierarchy. The report indicated that 12 new bishops could be appointed from 26 members of the Holy Synod of the Orthodox Church to fill the places of bishops who were over 70 years old and other vacant positions in the church hierarchy. The report urged that if those 12 were devoted to the regime 'the supreme leadership of the Church would be at government's disposal'.[28] In this way, the communists would be able to consolidate their positions in the state. Firstly, having their own people in the church hierarchy would diminish the political opposition's accusations of them being against religion. Secondly, before the general elections, which would be held that year, the government would have the possibility of attracting the sympathies of the Romanian electorate with the help of the Orthodox hierarchy. Thirdly, having its own people in the Holy Synod, the government could have access to the most important 'fortress of electoral battle and to the spiritual forum of the Romanian people'.[29]

The church hierarchs who sympathised with the government started to use a combination of religion and politics in their discourses. Bishop Justinian Marina, in his speech on the commemoration of Victory Day of the Romanian War of Independence on 9 May 1946, mixed religious teachings with an interpretation of contemporary politics, supporting the regime. Justinian made a parallel between Christians who were sad but hopeful for the Resurrection Day after the crucifixion of Christ and the Romanian people who after being sad in the last 'two thousand years' could now be confident that the new regime would bring them 'the Ascension Day, the ascension of the country and of the people'.[30]

As the Groza government wanted to achieve stronger control of the church hierarchy, in the summer of 1946 the Ministry of the Interior issued a law imposing censorship on religious publications. In order to avoid general opposition, the state continued to subsidise the priests upon the condition of publicly supporting the regime. As one of the *Securitate* reports suggests, the Groza government also wanted to pass a new law obliging bishops who were over 70 to retire. The law would eliminate the patriarch's potential successors and allow younger bishops who could be more easily influenced to obtain higher positions in the church. Furthermore, in order to strengthen the relationship between the Romanian and the Russian Orthodox churches, in the fall of 1946, on a government initiative, a mission of the Romanian Orthodox Church visited Moscow. The delegation was composed of Patriarch Nicodim; the youngest of the Orthodox hierarchs, Nicolae Popovici, bishop of Oradea; Justinian, vicar bishop of Moldavia and Suceava and Father Ioan Vasca, president of the Union of Democratic Priests.[31]

Figure 3.1 Prime Minister Petru Groza welcoming Patriarch Nicodim Munteanu back to Romania after his visit to Moscow, 1946 (courtesy of the Romanian National News Agency ROMPRES)

Returning from Moscow, Justinian had a meeting at Dalles Hall (*Sala Dalles*) in Bucharest with priests from the capital and from Ilfov County and with Gheorghe Gheorghiu-Dej, the secretary of the Romanian Communist Party. Holding a meeting at Dalles Hall signified its importance. As Justinian recalled in a speech in 1948, Gheorghiu-Dej was invited to present 'the relationship between the communist party, whose materialist conception was clear and the spiritual institution of the Church'.[32] In a diplomatic manner, Gheorghiu-Dej indicated that the government would continue to help the church in order to achieve its mission within the state. He suggested that

> It is wrong to think that the Church should only have cultic and spiritual concerns. If the Church and clergy would only have empirical interests, the relationship between Church and people would be affected. The village priest should not only be the confessor and the soother of souls, but in every moment should also unite the teaching from the pulpit with the leadership of an enlightened and true father and guider.[33]

The Groza government needed the church's support and, for this reason, political leaders and the church hierarchy continued to be on good terms publicly with government representatives participating in important religious festivities.[34] For example on 6 October 1946, Patriarch Nicodim consecrated the new cathedral in Timişoara in the presence of King Michael and other governmental officials.

In order to fully obtain political control of the government, the communists falsified the results of the general election held on 19 November 1946.

The official figures would indicate that the coalition of leftist parties representing the government bloc obtained a majority of 84 per cent with 348 deputies in parliament while the opposition had only 66 deputies.[35]

Communist control of the church

Romanian-Soviet relations intensified after the elections. On 10 January 1947, a Romanian delegation led by Gheorghe Gheorghiu-Dej, now the Minister of the National Economy, visited Moscow and signed an economic agreement between the two countries, and on 10 February 1947, the Groza government signed the Peace Treaty in Paris with the Allies. Article 3 of the treaty stipulated that in being offered international recognition, the Romanian government should 'take all necessary steps to ensure to all its citizens, without distinction, the full enjoyment of human rights and fundamental freedoms, including freedom of opinion, of the press and of information, of worship and of public meetings'.[36]

An increased role of the communist party in political life came about due to the implementation on 14 June of 'The Proposals of the Romanian Communist Party in order to raise the economic and financial situation of the country' and the refusal of the Marshall Plan on 9 July. The Groza government deepened not only economic and political relations with the Soviet Union but also strengthened ties, which had symbolic significance. Thus, on 10 March the government opened a House of Friendship between Romania and the Soviet Union and on 26 March 1947 a Romanian-Soviet Museum. On 18 March Vyacheslav Molotov, the Soviet Foreign Minister, was granted honorary citizenship of Cluj, one of the most important cities in Transylvania, and on 21 March 1947, Stalin was declared an honorary citizen of Romania. Targeting the political resistance of the opposition, in May 1947 the government started a state terror campaign and arrested many of its opponents.[37] On 17 July 1947 Iuliu Maniu, head of the opposition, was arrested and in August his party, the National Peasant Party, was outlawed. Maniu was labelled a 'fascist', an 'agent of American and British imperialism' and received a life sentence, remaining in prison until he died in 1953.

Having the legitimacy of fraudulent elections and using terror against their political enemies, the communists started to act more directly towards the Orthodox Church. A *Securitate* report from June 1947 indicated that some priests believed that the Groza government would soon have to resign, as American forces would liberate Romania.[38] In attempting to manipulate further the church hierarchy, Patriarch Alexius of Moscow was invited by the Groza government to visit Romania between 29 May and 12 June 1947. The main aim of his visit was the intensification of ecumenical relations between the Romanian and Russian Orthodox churches. Patriarch Alexius visited the most important religious centres in the country, and had meetings with Orthodox hierarchs and governmental officials in Bucharest. His visit was meant to demonstrate that during the communist regime religious communities enjoyed complete freedom. In addition to the spiritual dimension of his trip, the patriarch's visit also had a direct political purpose. Thus on 3 June, in the presence of church and government officials, he inaugurated the Institute of Romanian-Soviet Studies in Bucharest. As some *Securitate* reports indicate, the visit of Patriarch Alexius was perceived as an important opportunity for communist propaganda. The main concerns of the communists were twofold: what political message would Patriarch Alexius bring back to Moscow after his visit, and how would the Romanian Orthodox hierarchy perceive his visit in the context of new Romanian-Soviet relations?

A report dated 5 June stated that during the liturgy celebrated by Patriarch Alexius and Patriarch Nicodim, the former invited his Romanian counterpart to join him in Moscow with other Orthodox patriarchs to discuss ecumenical issues. Patriarch Nicodim declined the invitation asking 'why in Moscow?', but did not receive an answer. The report suggested that this question 'was interpreted as a gesture of great courage' by the clergy and the faithful.[39] From the communists' point of view, by daring to ask this question, the Romanian patriarch was disobeying Soviet authority and putting at risk the relationship between these two countries. A favourable response from Patriarch Nicodim would have meant total obedience towards both the Soviet regime and the Groza government. By challenging Patriarch Alexius's request, Nicodim was jeopardising the communists' effort to have good relations with the Soviet Union and the communists' position in Romanian society.

Another *Securitate* report of 5 June asserted that the Romanian clergy perceived the visit of Patriarch Alexius as if 'he had come to his own parish' thereby diminishing their independent status.[40] The report also stated that he asked that all official gifts from Romanian hierarchs should only be in gold, thus further upsetting the clergy who felt that he was only interested in economic gain rather than ecumenical and spiritual relations. The communiqué of the official Romanian communist Uni-Press stated at the end of the patriarch's visit to Romania that the Groza government aimed to foster relations with the Soviet Union. In Groza's words, the Catholics had started a campaign to create a 'world front' directed against Eastern European states. For this reason, the Uni-Press communiqué suggested that 'Orthodoxy should also start an action, because we are informed that there is a plot which would like to use the faith for other reasons and for imperialist tendencies'.[41]

If in previous years, the government had attributed a prestigious role to the religious hierarchy, after winning the elections in November 1946 it became more directly interested in manipulating the church. The visit of Patriarch Alexius changed the attitude of the communists. Patriarch Alexius was now the person who became the 'Prince de l'?glise'[42] while Patriarch Nicodim lost his former privileged place, and was now considered undesirable. In addition, the government's official position regarding the visit was a reason both for strengthening relations with the Soviet Union and for asking the Orthodox hierarchs to take a more active position in supporting the new regime. According to government policy, the Romanian Orthodox Church should have been more involved in a 'common Orthodox front' against those who opposed the regime. This attitude was immediately visible in the speech of some church hierarchs. On the occasion of his enthronement as Metropolitan in Ia?i on 28 December 1947, Justinian asked the faithful again to obey the government and called the people to 'an unprecedented work for the establishment of the social man, the man of peace, love, honour and work'.[43]

By issuing a new law in November 1947, the government obtained stronger control of the church hierarchy giving itself a more important position in its electoral process. On 19 November 1947, Bishop Justinian was appointed Metropolitan of Moldavia and Suceava by the National Ecclesiastical Congress in Bucharest. The prime minister and other governmental officials attended the Congress and Groza specified that the government wanted 'to delineate the spiritual role of the Church under the sign of the new times'.[44] In his speech at the official ceremony of his investiture, following the regime's line, Justinian used a strong combination of communist terms with Orthodox teachings, offering a communist interpretation of Christianity: 'Christianity has been and is, first of all, the comradeship of hearts and the voluntary socialisation of souls produced by the law of love and mutual respect'.[45]

After some of the opposition members of the Groza government resigned, on 7 November 1947 the communists filled their positions. A few days later, on 12 November, King Michael and his mother, Queen Elena, left Romania for the wedding of Princess Elizabeth in London. The communists believed that the king would not come back; however, to general surprise, he returned on 21 December 1947. This precipitated the course of political events as the communists wanted to eliminate the final political obstacle to obtaining total control of the political scene as soon as possible. On 28–29 December, communist leaders finalised the details of a coup d'état in order to abolish the monarchy. King Michael was invited to Bucharest by Groza to discuss 'a family matter' but on 30 December in an audience with Groza and Gheorghiu-Dej he was forced to abdicate and fled the country into exile. On the same day, the Deputy Assembly declared the country as the Romanian People's Republic. The abdication of the king and the establishment of the People's Republic represented the official installation of communism in Romania which survived until 22 December 1989. From 30 December 1947 the communists continued to eliminate any form of opposition both in politics and in the church. With its people in the church hierarchy, the communists had absolute political power. The establishment of the Romanian People's Republic occurred just before the suspicious death of Patriarch Nicodim on 27 February 1948. In his last Pastoral Letter, Nicodim considered the Romanian monarchy as a burden to the Romanian people and urged the faithful to obey the communist regime.[46]

Conclusion

The installation of communism in Romania was mainly achieved with the military and political influence of the Soviet Union and also with the participation of interior forces. In an Orthodox environment, in which the majority of Romanians were religious and practicing, the communists tried to control the church by gradually promoting its own people into the hierarchy, thus engendering a favourable and supportive attitude from it. They also exploited the religious brotherhood between Romania and the Soviet Union, and distanced the country from the Catholic and Protestant West.

As speeches of political and religious leaders and *Securitate* reports suggest, the regime regarded the Orthodox Church as 'the spiritual forum of the Romanian people', through which it could gain access to the 'fortress of the electoral battle'. Before 1947 the Orthodox hierarchy and especially the Romanian patriarch had a 'mission' over all Eastern Christianity, however this role clashed with the political interests of the Romanian communists. The Orthodox Church collaborated with the regime within the Byzantine tradition of *symphonia* between church and state. The hierarchy's combination of communist vocabulary with Orthodox teachings represented a step towards accommodation with the regime. The ambivalent position of the church in the public sphere would continue and its nationalist position would form one of its main elements of resistance during the process of Sovietization in the first years of the Romanian People's Republic.[47]

4

'The Light Rises from the East': Orthodoxy, Propaganda and Communist Terror, 1947–52

Romanian Orthodoxy and the communists

The installation of the communist regime on 30 December 1947, through Law 363 which abolished the monarchy and proclaimed the Romanian People's Republic, represented the start of a new era. The country was ruled by a Presidium[1] while on 4 January 1948 King Michael was forced to leave the country. The abolition of the monarchy affected the Orthodox Church which now came under the power of the new rulers. The transformation of the church and the rupture with the past were visible from the first days of 1948. Every year on 6 January, on the religious celebration of Christ's baptism, the patriarch had gone to the royal palace to bless the royal couple with holy water, invoking the divine grace over the monarch for the new year. But, of course, the communists were not interested in any religious ceremonials which might associate them with the 'bourgeois' values of the past. On the contrary, the spirit of the new regime was to find new materialistic forms and to praise the force of the working class.[2]

The communists, attempting to assess the new structure of Romania and the complete reform of the economy, began a national census on 25 January 1948[3] which was supported by some parts of the church. Metropolitan Justinian of Moldavia asked the priests under his jurisdiction to help the authorities in this matter and gave a speech in the Faculty of Law, Iaşi, on the relationship between the Romanian and the Russian Orthodox churches.[4]

Political and economic consolidation between the Soviet Union and Romania become more visible on 4 February 1948 when a Romanian-Soviet treaty of friendship, cooperation and mutual assistance for 20 years was signed in Moscow. The Romanian delegation, composed of Ana Pauker, Minister of Foreign Affairs, Gheorghe Gheorghiu-Dej, Minister of Industry and Commerce, Vasile Luca, Minister of Finance and Prime Minister Petru Groza met Stalin and made Romania the first People's Republic in signing this kind of treaty with the Soviet Union, a pattern which would shortly be followed by the other Eastern European states. The communists sought to become politically stronger and unified the Romanian Communist Party (RCP) with the Social Democratic Party (SDP), leading to the creation of the Romanian Workers' Party (RWP) in a congress from 21 to 23 February in Bucharest. Gheorghiu-Dej was elected general secretary of the Central Committee of the RWP which had 41 members, 31 from the RCP and 10 from the SDP. Furthermore, with a general election in March, on 27 February the RWP entered an electoral alliance with smaller parties which were later absorbed into the RWP, establishing the Popular Democratic Front (*Frontul Democratic Popular* – PDF).[5] Prime Minister Groza was elected president of the National Council of the PDF.

When the communist electoral alliance was created, Patriarch Nicodim was ill and, in circumstances still not fully elucidated, died on 27 February 1948.[6] Nicodim did not publicly oppose the installation of the regime in December 1947. Since 1946, when the communists first held ministerial positions, the patriarch was reticent about the rise of communism and the regime saw him as an obstacle to its complete control of the church hierarchy. After his death, according to Orthodox canonical tradition, Metropolitan Justinian was appointed deputy patriarch until the Electoral Collegium of the church elected the new patriarch.

The declaration of King Michael in London on 4 March, in which he stated that the results of elections in recent years had been forged by communists and that he was forced to abdicate, came too late and had little effect on Romanian politics. The elections on 28 March 1948 marked the triumphal political progress of the communists in the Romanian legislative and executive bodies. The PDF won 405 seats out of 414, while two small parties were attributed the remaining nine seats.[7] The 1948 elections would initiate a tradition in which the communists would always win elections with close to 100 per cent of the vote.[8] Complete political subservience to communist domination was achieved on 13 April 1948 when the Grand National Assembly (GNA) (*Marea Adunare Naţională*) endorsed a new constitution which copied the Stalinist Soviet Constitution of 1936. Romania was now ruled by a new Presidium which was led by Constantin I. Parhon as its president with Mihail Sadoveanu, Ion Niculi and Petre Constantinescu-Iaşi as vice presidents, while the Council of Ministers continued to be headed by Petru Groza.

The new regime's official attitude towards religion followed the Soviet propagandist model. Rather than prohibiting religion, as might be expected from an atheist regime, article 27 of the constitution stated that

> Freedom of conscience and freedom of religious worship shall be guaranteed by the State. Religious denominations shall be free to organise themselves and may freely function, provided that their ritual and practice are not contrary to the Constitution, public security and morals. No religious denomination, congregation, or community may open or maintain institutions for general education, but may only operate special theological

schools for training ministers necessary to their religious service under State control. The Romanian Orthodox Church is autocephalous and unitary in its organisation. The method of organisation and the functioning of the religious denominations will be established by law.[9]

The new constitution represented a break with the previous constitutional legacy regarding the position of religion. In theory, religious denominations were considered 'free to organise themselves' and to 'function', however, in practice the state found numerous motives to suppress those who threatened 'public security'. The constitution continued to declare the Romanian Orthodox Church as 'autocephalous and unitary' but did not make any reference at all to the other major church in Romania, the Greek Catholic Uniate Church. After the unification of all Romanian provinces into one state in 1918, all constitutions mentioned both the Orthodox and Uniate churches due to their important roles in the construction of the Romanian state. Not including the Uniate Church, which was under the Vatican's jurisdiction, implied that the communists saw it as an unimportant church in Romania, connected with foreign interests in the country.

The election of Patriarch Justinian Marina

In these new political circumstances the Orthodox Church was in a delicate position. As documents from *Securitate* archives show, in the first months of the People's Republic the church hierarchs continued to be in close contact with the communist regime but also praised the previous monarchy. On New Year's Eve, Metropolitan Nicolae Bălan of Transylvania declared that 'we should be grateful for the past' and that priests should not think about new forms of government but rather continue their pastoral mission according to Christ's gospel which does not have an earthly ruler. In February 1948, Metropolitan Bălan had an audience with Prime Minister Groza and, as *Securitate* files suggest, Groza asked that all priests under his jurisdiction enter his party, the Ploughmen's Party, offering Bălan the position of patriarch. Neither happened; however, after meeting Gheorghiu-Dej and having 'a good impression' of the communists, Metropolitan Bălan changed some of his diocesan priests, a gesture which was interpreted as favouring communist policy. Moreover, in March, during the congress of Transylvanian priests in Sibiu at which Stanciu Stoian, Minister of Religious Confessions, was present, Metropolitan Bălan continued to gradually change his discourse, now speaking 'in the democratic spirit, assuring the government of the clergy's support'.[10]

A *Securitate* report on Vasile Lăzărescu, who was elected Metropolitan of Banat on 18 November 1947, showed that he was not against the government but more passive to communist policy. The report claimed that he personally declared he was a good friend of Prime Minister Groza and the communists thought that he might like to become patriarch or at least a future metropolitan of Transylvania.[11]

Analysis of these reports reveals that at the start of the communist regime the Orthodox hierarchy was generally on good terms with the party leadership. The hierarchy took neither a confrontational attitude towards the communists nor completely supported them. While Metropolitan Bălan was cautious with his remarks, Metropolitan Justinian continued with the discourse that had helped him to be promoted in the church hierarchy. According to some reports, Justinian was seen as 'established and activating intensively on democratic lines, being considered the only element from the Orthodox Church hierarchy who enjoys the trust of the present regime'.[12] Together with Firmilian Marin, archbishop of Craiova, Justinian wrote a Pastoral Letter on how the Romanian faithful should respond to the new Romanian Constitution.[13] In addition, Justinian held a conference with all bishops from the Metropolitanate of Wallachia instructing them on how they should inform the people regarding the constitution.[14]

Two months after the Romanian communist delegation signed the treaty of friendship with the Soviet Union, in April 1948, the Romanian Orthodox Church made its own trip to the Russian Patriarchate. The Romanian delegation was led by Stoian, Minister of Religious Confessions, who, together with Bishop Nicolae Popovici and Father Petre Vintilescu had various meetings with Soviet officials and met Patriarch Alexius of Moscow. A *Securitate* report suggests that Patriarch Alexius was interested in who was most likely to be appointed the next Romanian patriarch; Stoian indicated 'certainly Justinian'.[15] However, the proposal that Metropolitan Justinian should be named patriarch was not backed by the other hierarchs, with the sole exception of Firmilian.[16] The position of Justinian as the 'best' candidate become more visible on 8 April 1948 at the burial of Irineu Mihălcescu, the former Metropolitan of Moldavia, when Justinian gave his panegyric. Until then Metropolitan Irineu had been considered the most important hierarch of the Orthodox Church and a possible next patriarch but he was forced to resign due to communist pressure in 1947 and the circumstances of his death are still unclear.[17]

In competing to achieve the highest ecclesiastical office, hierarchs also took the opportunity to improve their own personal positions in the eyes of the communists. One of the most effective ways of doing so was at the expense of other religious confessions in Romania. Thus on 15 May 1948 at the centennial celebrations of the National Assembly at Blaj, Metropolitan Bălan attacked the Greek Catholic Uniate Church and declared, in the presence of the prime minister, that the Habsburgs had encouraged its establishment in 1698 in order to diminish the religious unity of Romanians. The same type of discourse appeared in the mass media, accusing the Uniate Church of promoting the interests of the Western European powers because of its connection with the Vatican.

On 21 May Justinian published the first volume of a collection of speeches in which he officially presented the new doctrine of the church, the 'Social Apostolate', forging collaboration between the church and the communist government. According to this theory, the church was subservient to the state as the 'servant Church of the people', while the state assured religious liberty. The social apostolate meant a mutually beneficial relationship: on the one hand, the church enjoyed the preservation of its religious dominance and its influence in society, while on the other hand, the state benefited from continuing to control the church and using its spiritual authority as a political tool.[18]

The application of this doctrine become visible on 24 May when the Electoral Collegium, dominated by the new communist parliamentarians, elected Metropolitan Justinian Marina as patriarch; he was enthroned on 6 June 1948. Justinian's election reinforced communist control of the church. Some reports of the time suggest that Justinian was appointed patriarch due to pressure from Moscow which rejected any other nomination.[19] As a further sign of close church–state relations, the patriarch received his pastoral staff, the religious symbol of ecclesiastical power, not from the king as his predecessors had, but from Constantin I. Parhon, the President of the GNA, and was congratulated by top-communist leaders. In his speech, Parhon pointed out that 'religious life in our country is inextricably bounded to the Romanian democratic state and [. . .] the state and the church should help each other'.[20] The patriarch thanked the communists for their support and spoke against the Vatican drawing particular attention to a paragraph in the New Testament that 'There are other sheep of mine, not belonging to this fold, whom I must bring in; and they will too listen to my voice. There will then be one flock, one shepherd.' In this way, Justinian indirectly predicted the course of the Uniate Church in Romania in the next months when it would be abolished and forcibly unified with the Orthodox Church. Justinian's enthronement was also a moment of reaffirmation of connections with the other Orthodox countries which were under Soviet influence. The ceremony was held in the presence of members of the Romanian government as well as Metropolitan Nikolai of Krutik, the Soviet emissary who was previously delegated to bring the Russian diaspora under Soviet influence, and delegates from the Bulgarian Orthodox Church, led by Metropolitan Kiril of Plovdiv and Bishop Parteni Levkiischi.[21]

The date of the election of Justinian as patriarch was staged in contrast with the abolition of the Romanian monarchy. A few days after his election, on 27 May 1948, the government seized all royal property and, on 28 May, King Michael and other members of the monarchy were deprived of their Romanian citizenship. Thus, Romania had a new religious leader at the very time when it made a complete break with its former political rulers. The communists managed to affect the very core of the previous relationship between church and state, that between the patriarch and the monarch.

Figure 4.1 Gheorghe Gheorghiu-Dej congratulating Patriarch Justinian on his enthronement with Metropolitan Nicolae Bălan of Transylvania behind him, 1948 (courtesy of the Romanian National News Agency ROMPRES).

Under these new political circumstances, Romania had a new patriarch while the monarch lost even his citizenship and all of his property.

In addition, on 9 June, the Romanian government reorganised the Romanian Academy and removed 98 of its members. Some of them were important religious figures such as Patriarch Nicodim, Metropolitan Nicolae Bălan, Uniate Bishop Iuliu Hossu, and political leaders such as the imprisoned leader of the Liberal Party, Iuliu Maniu, or representatives of Romanian culture such as philosopher Lucian Blaga. Moreover, more than 8000 book titles were considered inappropriate for communist ideology and were not permitted to be republished or read.

The government pressed ahead with its domination of political, economic and social life. On 11 June 1948, the GNA introduced a law on nationalisation of the main industrial, banking, insurance, mining and transport enterprises under which 1060 factories became state property, representing 90 per cent of Romania's economic production. Furthermore, a State Planning Commission was set up as the governmental body for planning the national economy with Gheorghiu-Dej as its president. The Romanian economy, which was mainly agricultural, went through a major transformation on 6 July 1948 when the commission set quotas on cereal production. These quotas were reinforced by the state police who imprisoned those who refused to hand their cereal over to the state. Most peasants were unable to reach their quotas and were forced to buy cereal to give to the state. As a reward for political obedience, on 29 July the Romanian communists obtained from the Soviet Union a reduction by 73.2 million dollars of war debts as part of war reparations. This reduction was followed by a restructuring of the Romanian economy and on 28 December 1948, the GNA adopted the first General Economic Plan for 1949.[22]

The abolition of the concordat with the Vatican

The government's intention to control all aspects of social life led to conflict with the country's other religious confessions. If the Orthodox leadership was penetrated by people from the regime, the Catholic Church, with its hierarchy imposed by the Vatican, was seen as a major threat to the construction of communism in Romania. This perception was clearly acknowledged by Gheorghiu-Dej when he publicly attacked the Vatican during a meeting of the GNA stating:

> The Pope will undoubtedly find occasion to assail our Constitution because it does not tally with the Vatican's tendencies, which are to interfere in the internal concerns of various countries under the pretext of evangelizing the Catholic faithful [. . .] Who knows whether the Vatican will not consider anathematising us on the pretext that our Constitution does not provide for the submission of our fellow countrymen of Catholic persuasion to the political interests of the Vatican or because we do not allow ourselves to be tempted by America's golden calf, to the feet of which the Vatican would bring its faithful.[23]

Following this line, on 17 July 1948 the government decreed the abolition of the Romanian Concordat with the Vatican, which had lasted since 1927, thereby affecting the organisation of the Catholic Church in Romania. Pursuing this policy, only a few months later, a further decree on 18 September reduced the Roman Catholic sees from six to two and all its schools were taken by the state.[24]

Discussions among party leadership on the abolition of the concordat show how the communists regarded the role of religion in society and political control of the church. The minutes of the meeting of the Council of Ministers on this matter are extremely suggestive. In the meeting which decided the abolition of the concordat, Ana Pauker, Minister of Foreign Affairs stated that 'the so-called Vatican State is an unfortunately strong agency [. . .] of world imperialists and American monopolists'.[25] Stanciu Stoian, Minister of Religious Confessions, argued that while the Orthodox Church had 18 bishops and 12 million faithful, the Vatican had 11 bishops in Romania (5 Roman Catholics, 5 Greek Catholics and one Apostolic

Administrator for Armenia) for only 2 million faithful, and its hierarchy received their salaries from the Romanian state. He suggested that in this way, the Roman Catholic Church was in fact 'a State within a State'.[26] Emil Bodnăraş, Minister of National Defence, proposed how the government should act on this matter. In his opinion, the government should take a different attitude towards the Catholic hierarchy and its faithful. It should not attempt to convince the hierarchy to obey it but rather should speak directly to the masses. If the faithful saw that the government maintained contact with the Catholic hierarchy, then they would still believe in their religious leaders.[27] Prime Minister Groza indicated that he had recently met Bishop Hossu, one of the leading hierarchs of the Uniate Church, to convince him that the government did not want any martyrs. In conclusion, Groza suggested controlling the hierarchy financially by refusing to pay their salaries, and in the following months the government would decide what course it should take for every individual case.[28]

At the same time as the denunciation of the concordat with the Vatican, a delegation of the Romanian Orthodox Church participated in the Pan-Orthodox Congress celebrating 500 years of the autocephaly of the Russian Orthodox Church in Moscow. The Romanian delegation was composed of Patriarch Justinian, Archbishop Firmilian of Craiova, Bishops Antim of Buzău and Nicolae Colan of Cluj, Father Petre Vintilescu, the Dean of the Theological Institute in Bucharest, Father Simion Neaga as translator and Ovidiu Marina,[29] son of the patriarch, as secretary.[30] The congress was attended by the most important Orthodox leaders of those countries under communism: Patriarch Alexius of Moscow, Patriarch Kalistrate of Georgia, Patriarch Justinian of Romania and Patriarch Gavrilo of Yugoslavia. In discussions the Romanian delegation was active and presented reports on the most current issues in Orthodoxy and politics.[31] At the end of the conference, on 17 July 1948, on the same day when Romania denounced the concordat with the Vatican, Patriarch Justinian signed with the other members of the congress an 'Appeal towards all Christians from all over the world' urging them to be united in 'good understanding, unity, peace and brotherhood'. The appeal asked the faithful to be prepared for a new war, dividing the world in two parts, between the Catholic and Protestant West and the Orthodox East, the aggressors and the peace makers.[32]

The Law of Religious Confessions

The gradual control of religion in Romania took a dramatic turn in August 1948. On 2 August all foreign schools were banned[33] and on the following day the government introduced a law on educational reform closing all religious schools and imposing seven rather than four years of compulsory free education.[34] On 4 August, the government issued a law 'regarding the general regime of religious confessions', which remained at the core of church–state relations throughout the communist period.[35]

The law provided the means to control every aspect of religious life and the most pertinent points indicated the place of religion in Romanian communist society. The law stated that 'any person could profess any religion or religious faith if this does not contravene the Constitution, security, public order or morals' (art. 1). The dialogue between governmental and religious bodies officially stipulated that 'all recognised religious confessions should have a central organisation which represents the faithful, regardless of its number' (art. 12). The state exercised its authority by imposing that 'the confessions should be recognised through a decree of the GNA after the proposal of the government and preceded by the recommendation of the Minister of Religious Confessions. In justified cases this recognition could be revoked' (art. 13). The regime ensured that 'the organisation of political parties according to a confessional basis is forbidden' (art. 16), while the names of all leaders of local religious communities should be made available in a special register in the local city hall (art. 17). Article 21 stated that the leaders of religious confessions had to be recognised by presidential decree and before beginning their activities had to take the following solemn oath:

> As a servant of God, as man and citizen, I swear to be faithful to the People and to defend the Romanian People's Republic against any enemy from outside and within. I swear that I shall respect and that I shall ensure that my subordinates respect the laws of the Romanian People's Republic. I swear that I shall not allow my subordinates to start or take part and I shall not start or take part in any action which would affect public order and the integrity of the Romanian People's Republic. So help me God.

The law stated that for the 'creation and functioning of bishoprics' a religion should have at least 750,000 faithful (art. 22). This requirement affected the organisation of the Orthodox Church which on 18 September 1948 had to abolish the Metropolitanate of Suceava and the Bishopric of Maramureş establishing a new eparchy for the wider region, the Archbishopric of Suceava and Maramureş. This example was followed in 1949 when the Bishoprics of Argeş, Caransebeş, Huşi and Tomis-Constanţa were incorporated into other regions. The law abolished the Orthodox Military Bishopric with its seat at Alba Iulia (art 58), and its Bishop Partenie Ciopron was at the disposal of the Holy Synod (art. 60).

The law stated that the Orthodox Church would have only three theological institutes while the Catholic Church and the Protestant churches were allowed only one seminary each (art. 49). The Orthodox Church was forced to close its theological seminaries of secondary education and to transform its theological academies into three theological institutes in

Bucharest, Sibiu and Cluj[36] which were directly subordinated to the Ministry of Religious Confessions rather than the Ministry of Education. In addition, the Orthodox Church was allowed to have a seminary for monks at Neamţ monastery[37] and two seminaries for nuns at Plumbuita and Agapia monasteries.[38] The law led to extensive control of the church, carried out by inspectors from the Ministry of Religious Confessions, including the reduction and the censure of any published religious material.

The Orthodox Church did not largely respond publicly to these changes. However, as *Securitate* reports indicate, they came as a shock to the general public. Some believed that the regime would probably be changed by American and British forces after another war[39] while a few hierarchs believed that the government's policy would fail.[40] The most active voice in the hierarchy was Nicolae Popovici, Bishop of Oradea, who preached against the communist authorities and declared that 'the Church could not be humiliated by any political regime, being superior to any temporary government'.[41] For his public stance against the communists, Bishop Popovici would soon face exclusion from the hierarchy.

The gradual control of every aspect of Romanian society intensified with the official establishment of the General Direction of the People's Security within the Ministry of Internal Affairs on 30 August 1948. Following the model of the Soviet Police, the *Securitate* would become firmly rooted in public life and special offices would deal with the activities of religious confessions.[42] The gradual control of Romanian society would become more visible in the following years when the *Securitate* grew from 4000 employees in 1948 to almost 14,000 agents and 700,000 informers in 1989.[43]

'Unification' of the Greek Catholic Uniate Church

Having abolished the concordat with the Vatican and secured political subservience of the Orthodox Church, the government turned its focus to the second largest religious confession in Romania, the Greek Catholic Uniate Church. On 3 September 1948, the Uniate Bishop Ioan Suciu, Apostolic Administrator of the Archdiocese of Alba Iulia and Făgăraş, was removed from his position. On 18 September, three further bishops had the same fate: Valeriu Traian Frenţiu of Oradea, Alexandru Rusu of Maramureş and Ioad Bălan of Lugoj. The only Uniate hierarchs who were allowed to remain in their sees were Bishop Iuliu Hossu of Cluj-Gherla and Vasile Aftenie, vicar bishop of the Archdiocese of Alba Iulia and Făgăraş with residence in Bucharest.[44]

The abolition of the Uniate Church was intended to represent the climax of 'religious freedom' in communist Romania and the historical moment when Romanians were fully united. According to communist propaganda, unification with the Romanian Orthodox Church was due to a popular people's movement which started from within the Uniate Church. It was an action which represented the freedom of the people to unite with their 'mother Church', something that could only happen in the new Romania. In addition, the union was designed to be a symbolic act of the people who were liberated from the suffering of the past and who would achieve unity again. On 1 October 1948, under *Securitate* pressure, 38 Uniate priests gathered in the sports building of a high school in Cluj, 37 of whom signed in the name of 430 priests the union of the Uniate Church with the Orthodox Church, appealing to their faithful to follow their decision.[45] The action of these priests was not recognised by the Uniate leadership and on the same day they were excommunicated by Bishop Hossu. Furthermore, on 7 October Gerald Patrick O'Hara, the Apostolic Nuncio in Romania, sent a letter of protest to the Presidium of the GNA,[46] while other Uniate leaders sent another letter of protest to Prime Minister Groza.

The Orthodox Church supported the government and on 3 October 1948 Patriarch Justinian received the delegation of 36 priests in Bucharest welcoming the Uniate Church within the Orthodox Church. The minutes of the Central Committee of the communist party on 8 October showed that Bishop Hossu started to organise conferences in six Transylvanian counties, excommunicating those who were in favour of unification with the Orthodox Church. In addition, even within the Orthodox Church some priests were against the leadership's decision and continued to support Uniate freedom.[47] In a symbolic gesture, on 17 October, Patriarch Justinian replied to Uniate anathematisation of those who accepted unification by issuing his own blessing.

The Uniates' protests were disregarded by the government and on 21 October in the Cathedral in Alba Iulia, on the 250th anniversary of the establishment of the Uniate Church, Patriarch Justinian read the proclamation of the union of the Greek Catholic Church with the Orthodox Church in which he stated that

> It is a miracle of history that our Romanian nation was born and appeared in history as a Christian Orthodox nation [. . .] 21 October 1948 should become the day on which our good God gave our faithful people justice which our ancestors were yearning for the last 250 years.[48]

On 27 and 29 October, one week after these festivities, for publicly opposing unification, six Uniate bishops were arrested and were given permanent domicile at a resting house of the Romanian Patriarchate, in Dragoslavele in Argeş County, while 25 Uniate priests were arrested and sent to Neamţ monastery. Fearing a mass reaction as occurred after the arrests of religious leaders in Bulgaria and Hungary, the government refused to organise an immediate public trial of these clergy.[49] Suppression of the Uniate hierarchy continued when these clergy were moved to Căldăruşani monastery on 27 February 1949 and then to the Sighet prison on 24 May 1950. Bishop Vasile

Aftenie was sent to Văcărești prison where he died on 10 May 1950. In September 1950 Bishop Ioan Suciu was sent to Sighet prison where he died on 27 June 1953. The Uniate Church was completely abolished and in the following years 11 Uniate bishops died in prison.

Despite the Orthodox Church's public support for the abolition of the Uniate Church and communist propaganda, large parts of the Greek Catholic faithful continued to preserve their faith. A *Securitate* report from 6 November 1948 indicates that some Uniate priests clandestinely celebrated religious ceremonials in private houses in Oradea, Brașov and Ciuc while Uniate nuns dressed in peasant clothes preached a Pastoral Letter of Bishop Suciu regarding an alleged divine vision in Blaj. According to this revelation, the Virgin Mary appeared to a Uniate nun asking all to remain faithful to the Pope and to their bishops. Moreover, in other cities, such as in Beiuș, the Uniate faithful refused to attend Orthodox ceremonials participating instead in Roman Catholic masses.[50] The unification also failed in monastic life as out of 146 Uniate monks only 10 joined the Orthodox Church.[51] However, officially, the government completed its domination of the Uniate Church by confiscating all of its properties on 1 December 1948 and offering most of its churches and other buildings to the Orthodox Church.[52]

The statute of the Orthodox Church

Almost one year after the installation of communism, the government's anti-religious policy was wide-scale. The minutes of the Central Committee of the communist party of 25 November 1948 reveal how the communists behaved in this matter. In September 1948 the Ministry of Education ordered the removal of religious objects from schools replacing them with photographs of the communist leaders. For this reason, in many villages parents refused to send their children to school at the beginning of the new academic year, visibly acting against the communists, while some teachers asked their students to pray and to resist even further. In addition, *Securitate* reports indicate that when peasants asked communist activists about their own religious affiliations, most of them replied that they were religious. The Holy Synod found a compromise, instructing the priests to write to the education authorities asking for the icons to be returned to them. In this way the icons would be given back to the church rather than be taken publicly out of schools. The communists were aware of the place of religion in society and their lack of progress in setting the Orthodox faithful off against the other religions. As a *Securitate* report points out

> most peasants, 76 per cent of country's population, 12 million of people are religious and we can not make all of them suddenly materialists [. . .] instead of using the Orthodox Church against the Catholics, both of

them [these Churches] have a common platform, the priests are going from house to house and [. . .] amplify the religious mysticism.[53]

The contrasting responses between the Orthodox Church hierarchy who collaborated with the communists and the ordinary people showed that there was a deep separation between the church leadership and the masses. While most peasants were suffering from poor economic conditions, restricted quotas of agricultural products and religious persecution, the leaders of the church continued to have good relations with the leading communists. A *Securitate* report from 18 November 1948 even states that the children of Gheorghiu-Dej often went to the Patriarchal Palace and were treated with 'sumptuous' meals.[54]

However, despite these official good relations some *Securitate* reports claim that the patriarch started to use the social apostolate as a means of independence in selecting the people around him and began to employ, in the patriarchal administration, people who were well known for their political activities during the previous regime. Thus, a report of 8 December remarks that the patriarch started 'a personal dictatorship without taking into account the past of his protégées'. The most important changes in this respect were the appointments of Abbot Efrem Enăchescu, former metropolitan of Chişinău during Marshall Antonescu's regime (Chişinău was then within the Soviet Union) who was appointed deputy-bishop (*locotenent*) of Râmnicu Vâlcea; Archimandrite Benedict Ghiuş, former assistant professor of Nichifor Crainic and Bishop of Bălţi during Marshall Antonescu's regime, was appointed priest of the Patriarchal Palace;[55] Protosigel Barancea, former chief cantor of the Orthodox Cathedral in Odessa and close friend of Metropolitan Visarion Puiu, was appointed the second priest in the Patriarchal Palace; Tit Simedrea, former metropolitan of Bukovina and a friend of Ion Mihalache, the leader of the Peasant Party during the interwar period, was named abbot at the Cernica monastery near Bucharest.

Justinian's decisions were seen as controversial by the communists. As all of these *Securitate* reports show, the communist agents around the patriarch feared that he had begun to act against the regime especially because he also started to discriminate against some priests in Bucharest who he considered were preaching the 'democratic line' of the government too strongly. Due to the constitution of the Orthodox Church and the central role of the patriarch in the life of the church, Justinian was inclined to eliminate any opposition within the hierarchy. Rather than following the official communist line, he seemed to have had his own strategy. By promoting people who had had troubled pasts, Justinian might have reassured himself that they could easily be manipulated and would not act against him. Further *Securitate* reports also suggest that the patriarch chose people who had previously been involved with the monarchical regime in order to ensure his position in case the communists fell.[56]

Even after the official unification of the Orthodox and Greek Catholic Church, the Orthodox hierarchy remained conscious of the religious implication and also of the people's reactions. For this reason, the patriarch continued to maintain relations with the Uniate hierarchs. Thus, Justinian and Archimandrite Teoctist Arăpaşu visited the arrested Uniate bishops at Dragoslavele on 4 December and a few days later the Holy Synod discussed the new statute of the Orthodox Church.[57]

On 1 January 1949, Patriarch Justinian surprised the communists once again by appointing other people with compromised pasts to the church hierarchy. A *Securitate* report shows that Bishop Valeriu Moglan, vicar of the Metropolitanate in Iaşi, was pensioned and replaced by Archimandrite Teoctist Arăpaşu who, according to the note, 'claimed to be a member of the RWP even if he had been a legionary'; Bishop Ilarion Băcăoanul from the Bishopric of Roman, Bishop Galaction Grodun, secretary of the Holy Synod and Bishop Veniamin Pocitan from the Metropolitanate of Wallachia were pensioned; Partenie Ciopron, former bishop of the Army, Eugeniu Laiu, abbot at the Căldăruşani monastery, Pavel Şerpe from the archbishopric of Bucharest and Atansie Dinca from the Antim monastery were appointed professors in the Theological Seminary at Neamţ monastery. Moreover, in public conversations Justinian stated that he was not concerned with priests' pasts but encouraged them to conform to the present regime in order to continue to receive their salaries from the state.[58]

These independent actions of the patriarch continued through 1948 and into 1949 as he took further steps to ensure that he had a say in those around him on all levels. Thus on 22 February 1949 he appointed new advisors, some of whom were considered 'relics of the former peasant and liberal parties',[59] and he was even personally involved in the appointment of priests and cantors in all eparchies of the Orthodox Church.[60] The actions of the patriarch led to conflict among various factions of those around him. On 3 March 1949, Father Grigore Cristescu of the Theological Institute in Bucharest was allegedly removed after a conflict with Nicolae Vintilescu regarding the position of the Russian Orthodox Church in the Soviet Union,[61] while on the same day the patriarch employed more controversial people such as Father Nae Popescu, former member of the liberal party and former confessor of King Michael, and the painter Gheorghe Rusu, nephew of the late Patriarch Miron Cristea.[62]

Despite these controversial appointments, the patriarch continued to follow the party line which created a new form of propaganda by instituting the so-called 'missionary courses'. According to Decision no. 1574 of the Ministry of Religious Confessions, these courses were held for Orthodox priests at the theological institutes for one or two months.[63] The courses were extremely important and ended in written and oral examinations. The average grade of these exams was calculated together with the grade given for the priest's pastoral administrative activity in his parish thus forming the final grade.[64] If a priest wanted to move from his parish he needed a grade as high as possible. In implementing this new law, in February and March 1949 in Bucharest 500 priests attended 84 lectures: 66 were on pastoral theology subjects and 18 had social themes. There were also various conferences within the Romanian Association for Friendly Relations with the Soviet Union and films on religion in the communist bloc.[65]

In order to show the importance of these missionary courses for the church, on 30 January 1949 the patriarch spoke at the inauguration of the first session at the Theological Institute in Bucharest. Also present were Metropolitan Efrem Enăchescu; Stanciu Stoian, Minister of Religious Confessions; Sebastian, archbishop of Suceava; bishops Antim of Buzău, Chesarie of Constanţa, Iosif of Argeş and 400 priests from metropolitanates of Wallachia and Moldavia.[66] In line with communist discourse, the representative of the priests gave a speech which comfortably combined Orthodoxy with communism:

> We came to these courses with the strong decision to learn the social principles which are at the basis of the development of our dear country, the Romanian People's Republic, the country in which everybody works with their minds and arms in towns and villages. [. . .] we, all student priests, are committed to do everything possible for this purpose, because we are convinced that – being aware of our mission in the Romanian People's Republic and returning to our parishes and our responsible positions – we will consolidate the foundations of the Romanian People's Republic for the goodness, happiness and prosperity of our loving people, in which we all belong.
> Long live our loving His Holiness Patriarch Justinian!
> Long live our guiding Government!
> Long live the Romanian People's Republic!
> Long live the Romanian Orthodox Church![67]

In his speech on the role of the missionary courses for the Orthodox clergy, Patriarch Justinian emphasised:

> My heart is full of joy today and the happiness which is pouring over my soul is infinite [. . .] because today God fulfils one of the oldest and most arduous desires of our Holy Romanian Church, restoring its rights to prepare itself the servants of altars according to its education, system and necessity.[68]

The patriarch referred to communist ideology and combined religious paragraphs from the New Testament with political themes and pointed out that

> although materialism, the ideological basis of the Romanian Workers' Party, is foreign to our doctrine, to the Christian doctrine, it is the leading

political force in the Romanian People's Republic, [and] we can not but recognise everything that is good, [everything] that arises from the spirit of social justice in the activity of our regime and [we can] not [but] support [it] because 'The well-doer is a child of God; the evil-doer has never seen God' (III John 1: 11). These ideas should be your guiding light in your new activities after these missionary classes.[69]

The patriarch continued the same discourse the following day at the meeting of priests in Sibiu on 31 January 1949 presenting 'the attitude of the Orthodox Church towards the achievements of the working class'.[70] Furthermore, implementing the regime's policy, on 1 February he opened the missionary courses at the Theological Institute in Sibiu and on 6 February at the Neamț monastery.

After issuing the decree of religious confessions and missionary courses, on 23 February the government approved the 'Statute for the Organisation and Functioning of the Romanian Orthodox Church', which remained at the basis of the Orthodox Church throughout the communist period. The statute had already been adopted by the Holy Synod in its discussions on 19 and 20 October 1948 but the evolution of events led to the postponement of its official implementation by the government until 1949.[71] The statute indicated that, according to the law of religious confessions which limited the number of faithful for a bishopric, the Orthodox Church had five metropolitanates composed of 14 eparchial sees (art. 5).[72] The most significant symbolic change was the renaming of the Bishopric of the Lower Danube as the Bishopric of Galați. The name of this bishopric had historical importance as in 1878, when south Bessarabian districts were occupied by Russia, the eparchial see moved from Ismail to Galați but maintained the same name. In the new political situation the church adapted and renamed its see in an attempt to please the Soviet Union.[73]

According to the statute, the Orthodox Church had two legislative bodies: the Holy Synod (*Sfântul Sinod*)[74] and the National Clerical Assembly (*Adunarea Națională Bisericească*)[75]; and two executive bodies: the National Clerical Council (*Consiliul Național Bisericesc*)[76] and the Patriarchal Administration (*Administrația Patriarhală*) (art. 8). The Holy Synod continued to remain the highest ecclesiastical body being composed of all bishops and metropolitans (art. 11) who met at least once a year (art. 12) but only with the authorisation from the Ministry of Religious Confessions (art. 13). In order to function, the Holy Synod needed at least 12 members present and its decisions were taken by the majority (art. 14).

According to the 1949 statute, the church had the same structure as in 1923 with two-thirds lay members and one-third clergy. The lay members were mostly elected on the recommendation of the communist party which thus imposed its own people in the national clerical bodies and was able to dominate elections of the hierarchy. An innovative aspect of the 1949 statute

concerned the position of the monasteries which were assigned an economic function as craft centres (producing icons, leather goods, etc.) while
future monks and nuns were required to have at least seven years education
(art. 78).

Another novel measure of the statute was that all priests were installed in
their positions by bishops on the recommendation of the Eparchial
Councils or the Parochial Councils after the candidate passed a clerical
examination (*examen de capacitate preoțească*) (art. 120). The priests were provisional in their parishes for five years; they had to attend missionary
courses and pass a final examination at a theological institute (art. 123).
Only those who obtained acceptable grades were allowed to hold permanent
positions and they were compensated with better parishes according to their
grades. In addition to the missionary courses, article 138 stated that all
priests had to attend conferences each year in diocesan or eparchial centres.
The statute instructed that all cantors would also have to attend similar conferences once a year in diocesan centres and the themes for discussions
would include theological issues and 'the explanation of present problems
regarding the Church'. Although the missionary courses and conferences
were run by the church, they included political themes; thus in order to get
the best grades and be rewarded with a good parish, a priest was not only
judged on his religious knowledge and suitability but also on his wider pastoral awareness under the new regime.

Financial dependence on the state was a political measure which had been
applied in Romania since 1893 but the communists gave it an ideological
dimension. The clergy received some financial support from the state, but
because this was insufficient they were allowed to receive donations from
their faithful (art. 189–90). Some parishes were extremely poor, so priests
were particularly motivated to succeed in the missionary courses and be promoted to better parishes where the faithful were known to be generous in
their donations, or local traditions assured a better life.

According to the new statute, the election of higher clergy was similar to
the previous. Bishops were still elected only from monks or from celibate
priests whose wives had died, while the patriarch was elected from members
of the hierarchy (art. 129). The church hierarchy was elected in a secret vote
by the Electoral Collegium which was composed of members of the National
Clerical Assembly, the vacant Eparchial Assembly, the president of the
Council of Ministers, the minister of Religious Confessions, a delegate of the
GNA and the rectors of the theological institutes (art. 130). The Holy Synod
verified if the candidate had been elected canonically. Afterwards, he was
confirmed through a decree of the president of the Presidium of the GNA at
the proposal of the government according to the recommendation of the
Ministry of Religious Confessions. Although the statute seemed to grant the
church greater autonomy, in that not all members of parliament automatically became members of the National Clerical Assembly as it happened in

the interwar period, the church would ensure that the candidate had the complete support of the government before being elected. In this way there was no public conflict between church and government as the hierarchy maintained good relations with the communists while officially the government seemed to support the church by not publicly blocking the accession of unsuitable candidates.

In addition to all these measures, the church was allowed to publish two new national religious journals to which all parishes had to subscribe (art. 199). From 1949 the journal *Ortodoxia* (Orthodoxy) specialised in inter-confessional matters and the foreign affairs of the church, while *Studii Teologice* (Theological Studies) included the most important papers of professors of theological institutes. The official journal which continued to reflect the main decisions of the church was *Biserica Ortodoxă Română* (Romanian Orthodox Church), while most eparchial sees were allowed to publish religious journals which reported local activities.

Religious resistance and collectivisation

The government planned and implemented land reform according to the Soviet model of collectivisation. The church was an important land owner and the communists wanted to avoid direct confrontation with it. The minutes of the Central Committee of the RWP on 21 February 1949 showed that the hierarchy agreed to offer some land to the government, but with the following conditions: to give each priest and cantor possession of five hectares, to pay back the money invested by the church in agricultural activities in autumn 1948 and to include all priests in the category of peasants paying taxes for one to three hectares instead of five hectares.[77]

Although these demands were rejected, they suggest that the church was prepared to stand up for its own position. The minutes indicate that Vasile Luca advised that the communists should not agree to put the priests in possession of the proposed land while Ana Pauker argued that the communists should not hurry in making a decision on this matter. Pauker agreed with Luca in rejecting the idea of giving the clergy possession, but pointed out that the communists should be cautious and should not confiscate all of their land. If the clergy did not want to give up their land then they should be allowed to work on it. In her opinion, the communists should wait and make case-by-case decisions, knowing that the clergy would not cope with the high quotas imposed by the government. The priests should go to their local authorities, publicly state that they were unable to work on the entire land of the church, and because of the communist transformation of society they would like to offer the land to the people and to the state.[78] In this way the priests would act as agents of communist propaganda as the peasants would see and hopefully follow their examples. The government's official statement on the collectivised transformation of agriculture was declared at the Plenum of the Central Committee of the RWP from March 3 to March 5 1949[79] and peasants were forced to give their land to 'collective agricultural farms' (*gospodării agricole colective*).[80]

Despite the rejection of its proposed conditions and loss of land, the church hierarchy indirectly supported the communists in this policy. A few days after the party Plenum, the patriarch sent a Pastoral Letter on religious freedom. The particular timing of this letter might be seen as the church's response to collectivisation, suggesting that as the spiritual force of the mainly rural country, the church enjoyed cordial relations with the government. The letter was first read by the patriarch to 500 priests on completion of the first missionary courses on 27 February 1949, and then on 13 March on the symbolic first Sunday of the Easter fasting period at the religious celebration, 'Sunday of Orthodoxy'; in addition, the message was broadcast on radio.[81]

In this Pastoral Letter the patriarch stated that under 'our eyes' new transformations were taking place in Romania which would bring the country under 'new better rule in which a man would no longer be the servant of another man'. In this context the people were waiting for words from their spiritual rulers 'which would guide them in their search for happiness and salvation'.[82] The patriarch indicated the high level of religious freedom in the country, highlighting article 27 of the constitution and the fact that the church had 3 theological seminaries for monastic clergy, 12 schools for cantors and 3 theological institutes.[83] The letter emphasised that the government helped the Orthodox Church by denouncing the concordat, 'an act which removed a previous injustice which had placed the Orthodox Church in an inferior position'.[84] In the patriarch's words, 'in our regime, the Church enjoys all freedom, which is not only guaranteed but also defended' by the state.[85] Furthermore, a few days later on 25 March, Bishop Firmilian of Craiova was enthroned as metropolitan of Oltenia suggesting that the church was completely free.

With international intensification of the ideological competition between east and west, the use of Orthodox churches in the communist bloc would change, from the time of the first World Congresses of Peace which took place in Prague on 20 April and in Paris on 25 April 1949. Patriarch Justinian sent a peace message which was read in all churches in Romania, thereby initiating a new type of discourse. Already at the Pan-Orthodox Congress in Moscow in 1948 the Romanian Orthodox Church had issued a statement on peace, but from April 1949 the discourse and activity of the church would gradually focus on promoting peace. The participation of the church in the international and national Peace Movements would be central to the church during communism. Peace was a major communist theme and was used extensively in propaganda in attributing a negative image to the Western world. Thus, the communists presented only themselves as promoting peace, while the 'capitalist imperialist countries' were interested in conquering the east. The church became part of communist peace propaganda and participated in all major actions regarding the defence of peace.

At the same time, the Orthodox Church became more involved in the unification of the Uniate Church and in May 1949 the hierarchy sent a special mission of Orthodox priests to their parishes to convince the population to convert to Orthodoxy.[86] Priests trained in missionary courses were now well versed in communist policy. In order to increase religious pressure and to offer an example to those who still opposed unification, the Ministry of Religious Confessions decided on 29 May that financial support from the government for Bishops Aron Marton and Anton Durcovici, 3 canons, 132 priests and other administrative officials of the Roman Catholic Church was to be retrospectively confiscated from 1 February.[87]

The Orthodox Church emphasised its position regarding Uniate unification at the meeting of the National Clerical Assembly on 6 June 1949. The meeting now included for the first time former members of the Greek Catholic Uniate Church while the communists were represented by Prime Minister Groza; Stoian, Minister of Religious Confessions; V. Mârza, Minister of Health, with Anton Alexandrescu and Romulus Zaroni as delegates of the GNA.[88] One of the most important decisions of the assembly was the election on 8 June of Teofil Herineanu, a Uniate priest for 17 years who agreed to collaborate with the Orthodox Church and the communists, as bishop of Roman and Huşi. Teofil was elected by 58 votes while the other candidates received an insignificant number: Archimandrite Teoctist Arăpaşu received 4 votes, Archimandrite Benedict Ghiuş 3 votes and 6 votes were invalid.[89] In his speech Bishop Teofil thanked the Orthodox hierarchy for his election and asserted pro-communist discourse, stating that

> A political regime that takes care of its people and helps their material progress is worth collaborating with and not opposing. For this reason I value and respect the wisdom of those hierarchs of the Romanian Orthodox Church who did not reject the hand held out towards collaboration, who did not refuse the hand of the popular democratic regime at a time when other hierarchs, from abroad, improvident of the progress of the times, rejected this hand as an excommunicated one.[90]

Justinian reinforced Bishop Teofil's position in the Orthodox bishopric and on 28 August 1949 participated in the city of Roman at his enthronement, congratulating him in the name of the Holy Synod. At the end of the assembly, the Holy Synod sent official telegrams to communist officials assuring them of the continuous support of the church.

Religious pressure on those who continued to resist unification increased after the assembly when the Uniate Bishops Marton and Durcovici were arrested on 20 and 26 June 1949.[91] Furthermore, the other religious leaders in Romania were called to a meeting in Bucharest proposed by Justinian.[92]

On 24 June these religious leaders, headed by the patriarch, had an audience with Prime Minister Groza who assured them that obeying the regime was the condition of good relations. In his speech, Groza defined the broad role of religion in Romania and stated that

> Your meeting [. . .] proves that there is religious freedom in this country and that the representatives of these Churches and religions support the hard work of reconstructing the country, the hard work of building a new social order, socialism.[93]

In addition, Vasile Luca, vice-president of the Council of Ministers and Minister of Finance, suggested that their meeting had 'a historical significance' both for the 'development of the national economy' and for 'Romanian culture'. He pointed out that that there were significant differences between the religious leaders and the communists, but all of them were searching for the same thing: the economic and cultural development of the country.[94]

Despite the church's support for the regime and the meetings between religious leaders and the communists, the peasants began to revolt against collectivisation. In May and July 1949, peasants from villages in Botoşani county, Lunca sector in Năsăud county and from Arpăşel village in Bihor county clashed with the *Securitate*. The conflict took a dramatic turn when on July 29 in Girişu Negru village, Bihor county, 16 peasants were executed and others were arrested; while on 6 August in Călăfindeşti village, Suceava county, 4 peasants were shot and 13 were arrested. In its attempt to improve the organisation of the economy and following the Soviet model, on 6 September 1949 the communists adopted a law which reorganised Romania into 28 regions, 177 departments (*raioane*), 148 cities and 4052 communes. Imitation of Soviet policy also took place in international relations when on 1 October, on the recommendation of Moscow, the government renounced its treaty of friendship with Yugoslavia, a country seen too independent in the Eastern bloc.

On 21 October 1949, one year after unification with the Greek Catholic Church, Justinian gave a speech in Bucharest in which he stated that the Uniate bishops had been invited to stay in an Orthodox monastery and to celebrate their religious services there.[95] He gave the impression that unification had taken place at the people's request and that the former hierarchy had been offered good conditions. However, at the same time, religious persecution continued on 24 October when a faction of the Orthodox Church, which was seen as too liberal and close to Protestantism, the so-called 'the Lord's Army' (*Oastea Domnului*), was officially abolished and its missionary activity came under the control of local priests.

In addition, in order to strengthen communist propaganda, on 1 December 1949, the Holy Synod announced that from 1 January 1950 conferences similar to missionary courses would take place in all diocesan centres once a month at which the clergy would study articles from theological journals and other publications sent by the hierarchy. The conferences were mandatory and in the case of three absences the priests or deacons could be revoked from the clergy. Apart from the theological themes for discussion, the priests also had to study political themes such as: 'The Church Should Not Be Static' (consulting an article written by Groza); 'The Orthodox Church and the Russian Orthodox Church', 'The Vatican's Anti-Progressism', 'The Catholic Problem in the Romanian People's Republic' and 'Religious Liberty in the Romanian People's Republic'.[96]

A controversial gesture of the patriarch came on 5 December when he began visiting the eparchial sees in Transylvania, Crişana, Banat and Oltenia, all areas in the central and western parts of the country which had large religious minorities. As *Securitate* reports indicate, there were various opinions on the real reasons for his sudden decision to visit these places. Tit Simedrea, former metropolitan of Bukovina, believed that the trip was due to a government request which wanted to start an anti-Catholic movement,[97] while Professor Nae Popescu believed that the patriarch wanted to ensure a favourable position in the eyes of the communists. Professor Liviu Stan from the Theological Institute in Bucharest and a close friend of Prime Minister Groza thought that the patriarch wanted to strengthen his position among the other bishops. Professor Justin Moisescu from the Theological Institute in Bucharest was critical of the patriarch accusing him of losing any 'decency' by forcing the bishoprics to spend money on the trip and taking the time of the faithful during the ploughing season.[98]

Justinian visited the most important monasteries in the region, such as Sâmbăta de Sus and Păltiniş and the eparchial sees in Sibiu, Cluj, Arad, Timişoara and Craiova. In the evenings he showed Soviet films on the union of the Greek Catholic with the Orthodox Church in Ukraine and the celebration of 500 years of autocephaly of the Russian Orthodox Church. The patriarch was irritated when the local communist authorities, even if they attended his speeches and meetings, did not come in the evenings to watch the film he had brought.[99]

A *Securitate* report suggests that through the trip Justinian wanted, in fact, to gain the friendship of Metropolitan Bălan who was still in contact with anti-communist resistance abroad.[100] Referring to the Uniate priests who refused to become Orthodox, the patriarch pointed out that they would suffer the judgement of God and would be responsible for the souls of those who did not come to the Orthodox Church.[101] Justinian stated that the church's activities were regarded with great sympathy and that even the Soviet Ambassador, seeing the successful transformation of the Orthodox Church in Romania, asked for more details on the missionary courses as he wanted to suggest similar activities in his country. However, the report indicates that the general perception of the patriarch's trip was that he had became more diplomatic, not preaching too strongly against Roman Catholics and not praising the government excessively.[102]

The patriarch's ambivalent position was reflected by the actions of other church hierarchs. A report mentions how some hierarchs refused to completely accept censorship of their Pastoral Letters by the Ministry of Religious Confessions. Thus, Bishop Nicolae Colan of Cluj wrote a Pastoral Letter without taking into account the recommendation of the Ministry of Religious Confessions, while Bishop Nicolae Popovici of Oradea declared that he did not want to receive any advice on how he should write his letter. Metropolitan Bălan refused to insert economic or political issues into his letter while Bishop Andrei Mageru of Arad stated that he did not receive any advice except from the patriarch.[103]

One of the main aims of the regime was to transform Romania into a communist atheist society in which religion was considered the ideology of the bourgeoisie. Thus, in 1949, the Society for the Popularisation of Science and Culture was established. The main objective of this anti-religious society was 'to propagate among the labouring masses political and scientific knowledge to fight obscurantism, superstition, mysticism, and all other influences of bourgeois ideologies'. By 1951, it included leading communist intellectuals, reaching 28 regional organisations and branches in 37 districts with around 3000 propagandists organising 42,000 lectures.[104] In the following years, the progressive Sovietization of the country would affect the church more strongly. If in 1948 it seemed to benefit from its collaboration with the regime by incorporating the Uniate Church, in the long term it would suffer as the regime's anti-religious campaign aimed to discredit the church and to reduce the influence of religion in society.

Saints and communists

The regime's domination of social life intensified as state terror became its main method of imposing authority. On 14 January 1950, a decree was issued which stated the imprisonment of any person who 'through their activities, directly or indirectly threatened or attempted to render more difficult the building of socialism in the Romanian People's Republic and of those who, in the same way, libel state power or its organs'. In this way, the *Securitate*, the People's Militia or the communist activists acquired new privileges, being able to arrest any person on the simple basis of denunciation rather than after full examination of evidence.

The Orthodox Church hierarchy conducted activities which publicly seemed to confirm its closeness to the communist authorities as well as those which strengthened its autonomy by referring to its historical background. Thus, on 19 February 1950, the patriarch attended festivities organised by the Romanian Association for Friendship Relations with the Soviet Union at the Theological Institute in Bucharest emphasising the relationship between the Romanian and the Russian Orthodox churches while on

25 February the Romanian Orthodox Church celebrated 25 years of the establishment of its Patriarchate.[105]

This anniversary provided an occasion on which to promote elements of Romanian history. It began with the religious commemoration of previous patriarchs, Miron Cristea and Nicodim Munteanu, in the presence of the Holy Synod, Stanciu Stoian, Minister of Religious Confessions, and three representatives of the Orthodox churches of Armenia, Bulgaria and Russia. The former religious leaders, despite their tenure during the 'bourgeois regime', were considered as integral figures in the church's past. In his speech Justinian claimed that the establishment of the Romanian Patriarchate in 1925 was only a symbol for the political rulers of that time, while 25 years later, in the new regime, 'when the people themselves [. . .] took the helm of their own destinies in their hands', the church achieved its complete mission.

The patriarch named some religious figures from church history which were considered saints in popular tradition and stated the church's intention to canonise them.[106] In the patriarch's words, 'we consider it the duty of the Holy Synod of the Romanian Orthodox Church [. . .] to sanction what the people already decided a long time ago, considering [these figures] among the righteous'.[107] The suggestion to canonise some saints only a few years after the installation of communism in Romania emphasised the position of the Romanian Orthodox Church and particularities of Romanian communism in the Eastern bloc.

The communists did not initially oppose the patriarch's naming of these figures because the actions of the hierarchy fitted in with the new politics of cohabitation with the West. Supporting the patriarch's speech, Stoian referred to the recent developments in international relations and stated that

> The government of the Soviet Union believes that despite the differences between their ideological systems and ideologists, coexistence between these systems and peaceful relations between the United States and the Soviet Union are not only possible but are also absolutely necessary for the general interest of peace.[108]

The naming of Romanian saints was a means of asserting a particular voice within the communist bloc. The use of the church by the regime become more visible the following day, on 26 February, when Justinian sent a Pastoral Letter to the Orthodox faithful on the theme of the defence of peace in which he praised the government and the Soviet Union and criticised capitalist countries. Justinian's discourse changed from criticising the Roman Catholic Church to a new theme, that of blaming the enemies of the Soviet Union and of communism in general:

> During our times we see that the enemies of peace are gathering in the camp of those who machinate against the lives of those who love peace

and of workers all over the world. At the forefront of this camp are the money bags, the United States and England. The friends and defenders of peace are also rallying closer together before the danger threatened by the enemies of peace. Their camp is the 'camp of peace and democracy' having at its leadership the strong Soviet Union, the great neighbour and friend of the Romanian people.[109]

The church blessed peace and those who made peace, calling all Christians to be around their leaders who were 'steadily defending the flag of peace and the happiness of humanity'.[110] Justinian's Pastoral Letter suggested concrete actions to be taken by the people. Firstly, the faithful should participate in all activities organised by the national committee for the defence of peace. Secondly, the clergy should perform special ceremonies and prayers dedicated to maintaining peace in the world and should be active members in the committees for defending peace. Thirdly, the clergy should inform their faithful about the actions of the government in the defence of peace and this letter should be read on the third Sunday of the Easter fasting period and any other time which was considered appropriate.[111]

Communist control led to the restructuring of the administrative organisation of the church according to the new law of religious confessions of 1949. Thus, the National Clerical Assembly decided that the Archbishopric of Suceava would unite with the Archbishopric of Iaşi composing the Metropolitanate of Moldavia and Suceava and Sebastian Ruşan, archbishop of Suceava would become its metropolitan. The Archbishopric of Iaşi was vacant and according to church tradition its leader should become the next patriarch of Romania. The election of Sebastian Ruşan, who was also the godson and spiritual relative of Prime Minister Petru Groza, confirmed that the candidate for the position of patriarch was a person known to the regime.

In addition, the Bishopric of Constanţa was joined with the Bishopric of Galaţi and renamed as the Bishopric of Lower Danube, having its eparchial see in Galaţi with Chesarie Păunescu elected as its bishop.[112] The decision to restore the previous name to the bishopric, only two years after the Holy Synod symbolically changed its name in an attempt to please the Soviets, could be seen as a step towards attempting to promote a new type of nationalist discourse in the Orthodox Church.

The church hierarchy made sure that the Holy Synod did not have any members who were problematic for the communists and pensioned off a number of previously leading figures, making them abbot or a simple monk in a monastery. On 27 February Bishop Nicolae Popovici informed the meeting of the Holy Synod of the decision to pension some hierarchs and the new positions in the church.[113]

Continuing the theme presented only a few days earlier, on 28 February 1950 the Holy Synod publicly announced for the first time in the history of the Romanian Orthodox Church the canonisation of some saints.

The Romanian Orthodox Church was the only church in the Orthodox commonwealth without any saints of Romanian origin. Romania had the relics of some saints on its territory (Saint Parascheva in Iaşi, Saint Ioan the New in Suceava, Saint Filoftea in Argeş, Saint Dimitrie the New in Bucharest, Pious Nicodim Decapolitul at Bistriţa monastery and Pious Nicodim at Tismana monastery) while a saint of Romanian origin was even celebrated in Greece (Saint Ioan Valahul) but not at that time in Romania. The Holy Synod decided that the veneration of saints whose relics were in Romania should take place throughout the country rather than only in the local region, that the saint celebrated in Greece should be introduced into the Romanian Church calendar and that the church should sanctify its own people who should be venerated either at local or at national level (Metropolitan Sava Brancovici, Hieromonk Visarion Sarai of Ardeal, Metropolitan Iorest, peasant Nicolae Oprea of Transylvania, Calinic of Cernica and Râmnic, Ioan of Râşca in Moldavia and Iosif the New of Partoş in Banat). This list of proposed saints is particularly striking. It included a large number of people who had been active in Transylvania in previous centuries against the Roman Catholics and the Habsburg Empire in attempting to preserve the identity of Romanians in the region. It also included a peasant, suggestive of communist propaganda which promoted the lower classes in society.

The communists disagreed with the Holy Synod's proposals until 1955, by which time the Romanian communist faction led by Gheorghe Gheorghiu-Dej had attained complete power and used the canonisation in promoting a Romanian road to communism in the Eastern bloc. The church hierarchy's intention to canonise some saints as early as 1950 was a way of asserting its position in Romania.[114] It used nationalist references as a means of protecting it from Soviet influence and of imposing its domination over other religions in the country. By asserting the decision to canonise only saints of ethnic Romanian origin at a time when the regime was pursuing a strong Sovietization policy, the church hierarchy wanted to indirectly promote autonomy within the Orthodox commonwealth. After the installation of communism in Romania in 1947, the Romanian Orthodox diaspora accused the communist-appointed hierarchy of intending to transform the Romanian Orthodox Church into a branch of the Russian Orthodox Church.

In addition to canonisation, in what appears to be an attempt to dispel suspicions about being surrounded only by people from the former regime, on 28 February 1950 the patriarch removed Benedict Ghiuş, vicar bishop of the Patriarchate, from his position and sent him to become a professor in Neamţ monastery. In his place the Holy Synod elected Teoctist Arăpaşu, vicar of the archbishop of Iaşi, who was enthroned on 5 March in Bucharest.

Despite collaboration between the church and the regime, *Securitate* reports show that there was a general tension among the 'bishops, clergy and lay members' and in all their actions there was 'the impression of fear in expressing their thoughts and desires'.[115] The hierarchs too were very careful in everything they said and did as they could lose their positions on the slightest grounds. Even the tiniest remark was carefully recorded by the *Securitate* as exemplified in an amusing incident described in one of their reports. During a speech given to a meeting of the Holy Synod, Stoian, Minister of Religious Confessions, began to explain how priests should not obstruct collectivisation. Just at that moment Metropolitan Bălan wanted to go out to the toilet but he was stopped at the door by the minister who wanted him to hear his words. Bălan was surprised to be stopped and remarked that 'the Church and its clergy are interested only in the salvation of the souls of the faithful. If kolkhozes could have a role in the salvation of the soul, this remains to be proved'.[116] His unexpected public response at an awkward moment revealed his reluctance towards collectivisation and collaboration with the communists. Furthermore, the strain among the clergy had risen also because there were rumours that there were demonstrations in Leningrad and Ukraine against Stalin and that the communist regime may fall in three months.[117]

Orthodoxy in the battle for peace

After the enthronement of Sebastian Ruşan as metropolitan of Moldavia and Suceava on 25 March, and of Chesarie Păunescu as Bishop of Lower Danube on the following day, the Romanian Orthodox Church received confirmation from the Moscow Patriarchate that the Russian Orthodox Church was offering as a gift the component parts to produce the Holy Chrism. Celebration of the Holy Chrism was an act of canonical independence of an Orthodox Church while in communion with the other Orthodox churches. The Romanian Orthodox Church celebrated the Holy Chrism for the first time in 1882 when it declared its autocephaly and previously during the tenures of Patriarch Cristea in 1934 and of Patriarch Nicodim in 1942.

In order to bring the substances which composed the Holy Chrism (weighing almost 100 kilograms), Justinian sent vicar bishop Teoctist and Ioan Negrescu, professor of Russian at the Theological Institute in Bucharest, to Moscow from 30 March until 3 April 1950. Teoctist took with him a telegram from the patriarch and copies of all church journals published in Romania in 1949 to invite Russian theologians to publish more articles in future numbers of the journals. The religious ceremony of the Holy Chrism was led by Justinian on 6 April in Bucharest. By accepting the components of the Holy Chrism from Moscow and not from Constantinople as in the past, the Romanian Orthodox Church underlined its connections with the Soviet Union, confirming it as the leading spiritual centre at that time.

The Soviet Union not only controlled the Orthodox churches under communism but used ideological methods which would favour communism. The so-called 'World Congress of Peace Artisans' held in Stockholm from

15 to 19 March 1950 issued an Appeal for Peace which was signed by many people from East and West, an act which would be particularly problematic for Americans during the McCarthy era. The appeal opposed the use of the atomic bomb and stated that the first government to use the bomb was criminal.

The Holy Synod followed the communist directive by engaging in a mass campaign in which all Romanian faithful would also sign this document. Patriarch Justinian reinforced the message of this appeal and attended the conference of the Committee for the Defence of Peace in the RPR on 15 April. The propagandistic message of the appeal can be seen as aimed at diverting the attention of the population from the country's internal economic and political problems towards the less tangible idealistic image of peace. The church supported the signing of the appeal with special articles in its journals which used both religious and political messages, combining paragraphs from the New Testament with politics, declaring that 'those who fight for peace are the sons of light'.[118] For his actions and loyalty to the regime, the patriarch was decorated by the government with the order 'The Defence of the Fatherland' (*Apărarea Patriei*), second class.

Securitate reports show that the patriarch was concerned not only with his image in Romania but also with how he was perceived in Western Europe. As one of the reports mentions, in a sleeping compartment on a train journey to Galaţi with other hierarchs, at 11 p.m. the patriarch and the Minister of Religious Confessions 'listened to Radio London'.[119] In the broadcast Justinian was labelled as 'the Red Patriarch' and the Romanian Patriarchate compared with a Sovrom.[120] The report mentions that in discussions that night, Metropolitan Ruşan suggested that one day people from Western Europe might accuse them of 'Bolshevising' the church in Romania.[121] Church hierarchs were afraid that the fall of the regime would also mean the loss of their own church positions.

Communist control of religion reached a new level when some parts of the Roman Catholic Church also started to conform to propaganda by reorganising their church in Romania, thus creating a distance from the Vatican.[122] Andreas Agotha, a Catholic priest who was later excommunicated by the pope, accepted the leadership of a 'national Roman Catholic Church' at a congress in Târgu-Mureş on 27 April 1950, urging the faithful to support the regime by signing the Stockholm Peace Appeal.[123] His gesture was immediately condemned by the Roman Catholic hierarchy in Romania. On 18 May, Monsignor Glaser, assistant to the Roman Catholic Bishop of Iaşi, asked Catholic priests to read a declaration denouncing the Târgu-Mureş decision and forbidding people to sign the Stockholm Peace Appeal.[124] For his opposition to the regime, a few days later, on 25 May, Monsignor Glaser was arrested and shortly after 'died of heart failure'.[125]

The regime's absolute control of the political sphere continued to deepen when between 5 and 6 May all former ministers and deputy secretaries of state between 1920 and 1947, regardless of their politics, were arrested. Moreover, on 26 May agricultural quotas were fixed for the whole country at such a high level that they led to peasant uprisings in some areas.

Due to increasing political pressure, on 25 May 1950 Justinian went for the third time to Moscow to give an account of recent activities of his church. Justinian was accompanied by Metropolitan Ruşan, Bishop Teofil Herineanu, Diocesan Father Traian Belaşcu (vicar of the Metropolitanate of Transylvania and former president of the Greek Catholic Committee of Uniate priests which had declared unification with the Orthodox Church) and Father Mihail Mlădan as their translator.[126] The composition of the delegation looks to have been carefully made to clearly indicate that communist policy on religion in Romania was in accord with the Soviet perspective. Metropolitan Ruşan wanted to have better contact with the Soviets as he held the most important position in the hierarchy after the patriarch. The other members who represented the achievements of the Orthodox Church in the matter of unification were Bishop Teofil, a former Uniate priest who became an Orthodox bishop, and Father Belaşcu, president of the Uniate group that signed 'unification' in 1948. In strengthening the relationship between their churches, Patriarch Justinian was made an honorary member of the Theological Academy in Moscow on 28 May and two days later he signed with Patriarch Alexius of Moscow a document regarding the common position of their churches in defending world peace.[127] Justinian and Alexius visited some churches in Moscow, attended a concert commemorating 200 years after the death of Bach and met G. G. Karpov, chairman of the Council for the Russian Orthodox Church affairs under the Council of Ministers of the USSR. While the visit represented a significant moment for the Romanian Orthodox Church, showing its dependence on the Soviet Union, the meeting between the Romanian and Russian patriarchs was reported in only one Soviet newspaper, *Izvestyia*, on 3 June.[128]

Despite his public collaboration conforming to the regime's policy, Patriarch Justinian was not completely trusted by all communists. As a *Securitate* report claims, the patriarch 'intends to make the clergy into a mass of obedient people who listen only to him because in this way he can negotiate on equal terms with the party'.[129] The communists had believed that the special missionary courses would have stronger political messages; however, the church had preserved a significant theological content in them. Recently the patriarch stated at the opening of some courses that 'the clergy should not put the accent on reason, but should return to apostolic times and teach the people to pray'.[130] Dissatisfaction was especially visible after the patriarch returned from Moscow, when on 13 June 1950 he was asked by the communists to force priests to participate in collecting the harvest. The patriarch responded that he had hardly managed to convince the clergy to be engaged in the campaign to sign the Appeal for Peace and he did not think that he would manage to persuade them this time. According to the patriarch, if the communists really wanted him to do this, they should give an order from the Ministry of Internal Affairs.[131]

The government officially ceased its relationship with the Vatican on 7 July. Just a couple of days later, on 9 July, the patriarch blessed a new church in the hometown of the previous Patriarch Nicodim, in Pipirig-Neamţ. By making reference to a religious leader of the interwar period, the patriarch re-emphasised the Romanian religious past, a gesture which can be seen as sending a nationalist message due to close relations between church and state at that time.

The development of Romanian communism took a new shape as the internal battle for power among the communist leadership intensified. On 20 July 1950, the Central Committee of the RWP decided that all party members should be checked regarding their previous activities. Gheorghiu-Dej, in aiming to achieve complete control of the party, assigned responsibility for carrying out this task to Ana Pauker, Minister of Foreign Affairs, making her unpopular among the other Romanian communists.[132] In the following months, the party would eliminate 190,000 people from its structures representing 20 per cent of its total number and Gheorghiu-Dej's policy would bring good results for him as one year later he would become the sole leader of the party.[133] In addition, on 12 August 1950 the state strengthened its control of any opponents and introduced the death penalty for 'treason to the nation, [. . .] service to the enemy, or causing prejudice to the power of the state' or transmitting secrets to a foreign country.[134]

The Orthodox Church was affected by internal conflicts in the party. At the request of the communists, on 20 August the patriarch sent a Pastoral Letter titled 'For the Fatherland'.[135] Although directed against Josip Broz Tito's regime in Yugoslavia, the underlying message of the letter emphasised the need for the country to be united, thereby indirectly supporting the changes and controls within the party. The letter produced dissatisfaction among the clergy; as a *Securitate* report reveals, Professor Justin Moisescu declared that it was clear proof of the patriarch's collaboration with the communists, while Ovidiu Marina, the son of the patriarch, was disappointed, stating that his father 'did everything the people from the Ministry of Religious Confessions told him to do without thinking about his future'.[136]

Political clashes were also reflected in the internal life of the church. Thus the subject of the final examination of missionary courses in Bucharest on 5 August was 'Christianity and the problem of war. The just or unjust positions of Christian Churches in the last two [world] wars and in the future war.' A *Securitate* report shows that the clergy was unsure what future war meant, while the communists would prefer the subject to clearly state 'the danger of a possible war' or 'the threat of a new war'. The uncertain course of politics was indicated by the fact that some handwritten papers were spread in Bucharest which suggested that the people should be prepared for a new war which would start soon and that the Americans would liberate Romania.[137]

Although, the church hierarchy was largely obedient to the regime, one member lost his position for publicly opposing the communists and preaching against the government.[138] On 2 October, Justinian was called to Prime Minister Groza who asked for Bishop Nicolae Popovici of Oradea to be pensioned, as otherwise he would be arrested. Groza was irritated by Popovici's sermons and stated that he had not arrested him until then because he did not want to create a bad image for the church. The patriarch also met Gheorghiu-Dej in attempting to save Popovici's position but without success and even sent other influential members of the church hierarchy, Metropolitan Bălan and Metropolitan Ruşan, to speak to Groza. At their meeting Bălan suggested that Bishop Popovici should take a six month holiday but the prime minister rejected his proposal. Following the communist pressure, on 4 October the Holy Synod, at which Bishop Popovici was present, voted for his retirement. Two days later Bishop Popovici was allowed to travel in the patriarch's car to visit Cheia monastery, his future residence suggested by the Holy Synod, but he complained that it was too isolated. He requested another monastery in Moldavia and to be allowed to stay until the end of October in his eparchial see in Oradea. However, in the end he was forced to live at Cheia monastery until 1960 when his family took him home for a few of months before he died on 20 October 1960.[139]

The government's decision concerning one member of the church hierarchy affected the perception of the wider clergy towards collaboration with the regime. Bishop Nicolae Colan of Cluj declared that 'The trial of Popovici was the trial of each of us, the condemnation of Popovici contains the condemnation of each of us. If we were what we were supposed to be, we should have abandoned our pastoral staffs, rather than behead Popovici.'[140] In other words, Bishop Colan suggested that the hierarchy should resign rather than concede to communist influence. Most hierarchs were disappointed by the fate of Popovici, but other priests were confident that this was in fact the start of a new stage of the church. According to *Securitate* reports, Father Ioan Gheorghe, member of the former Peasant Party, declared that Bishop Popovici was dismissed because of Justinian, and after the fall of the regime Popovici would be the next patriarch: 'We, the Orthodox have a victim of the communist terror.'[141]

Control of the Romanian Orthodox diaspora

Political control of the church was not only a matter restricted to internal affairs but also an issue with international significance. Orthodox communities abroad represented the most significant centres of resistance as they gathered religious and political elites in exile. They not only criticised the communist regime but also lobbied Western governments regarding their positions towards the communist authorities. Thus, control of the religious leadership of the Orthodox diaspora was a crucial matter for the communist regime. Religious resistance to Romanian hierarchy has been mainly gathered around two sources, namely Metropolitan Visarion Puiu and the Romanian Orthodox Episcopate in the US.

Visarion Puiu was the former Metropolitan of Bukovina and Exarch of Transnistria during the Second World War. At the time of the coup d'etat in August 1944 he was in Croatia and, fearing that he would be arrested, did not return to Romania. In February 1946 he was condemned to death by a People's Court in Bucharest. He lived in Austria, Italy and Switzerland, before settling at the Romanian Chapel in Paris. He acted as head of the Romanian diaspora in Western Europe and former political leaders in exile supported him;[142] however, his jurisdiction was not recognised by all of the clergy outside Romania. His limited influence was exemplified by a saying at that time that someone who wanted to become a priest 'could always get Visarion do it'.[143] After 1955, facing health and financial difficulties, he asked the Romanian hierarchy to be allowed to return to the country. He received letters of encouragement from top communist leaders, including Mihail Sadoveanu, vice-president of the GNA, and from other hierarchs promising him that he would be reinstated and that his sentence would be abolished. However, he never returned and died in Paris in 1964.[144]

The most important challenge to the Romanian Orthodox Church was the Romanian Orthodox Episcopate in the US. The decision to establish an Episcopate in the US was taken by the local Romanian community in April 1929. In October 1932 in Detroit, Michigan, a constitution written by the community stated the rights of Romanians to elect their own bishop and recognised the jurisdiction of the Romanian Orthodox Church. Patriarch Miron Cristea approved the organisation in June 1935 and sent Bishop Policarp Moruşcă who was enthroned in Saint George Cathedral in Detroit. Moruşcă remained bishop until August 1939 when he travelled to Romania on a private visit. Due to the outbreak of the Second World War he was unable to return; the new communist government deposed him in 1948 and he was kept under surveillance by the *Securitate*.[145] During his absence, the Episcopate in the US was administered by a council. In February 1947 the Romanian Legation in Washington informed the council that according to the Holy Synod in Bucharest, they would soon receive a new bishop, Antim Nica; however, as he was seen to be imposed by the communist regime, the council refused to accept his appointment and voted for administrative autonomy from Bucharest.[146]

The community lacked a spiritual leader and in 1950, eight parishioners from Detroit elected Andrei Moldovan, a parish priest from Akron, Ohio, to be its new bishop. Moldovan was not recognised by the council and the split within the Episcopate provided fitting conditions for communist infiltration of the Orthodox diaspora. The *Securitate* dossier on Andrei Moldovan provides valuable material on this episode.

Moldovan was a Romanian-born American citizen who left the country in 1923 and last visited in 1935. He had a son who died during the Second World War and a daughter, Valeria, who married an engineer and was living in Romania in Stalin city (Braşov); Moldovan maintained regular contact with her and was eager to see her again. The *Securitate* opened his letters to Valeria; it noted that he seemed to be 'without political knowledge' and could 'easily be manipulated'.[147] The fact that his daughter resided in Romania meant that, should he fail to obey the regime's orders, she would face repercussions. Moldovan was invited to come to Bucharest to be officially enthroned in his new position. Before going he sent a few letters to Justinian expressing concerns regarding his possible arrest in Romania.

Moldovan's nomination by the Romanian authorities came very suddenly. Although Patriarch Justinian, in a meeting with Stanciu Stoian, Minister of Religious Confessions, decided not to answer Moldovan's concerns regarding his journey to Romania, on 25 August 1950, without having official approval from the communist authorities, he sent a telegram asking Moldovan to travel urgently to Bucharest. Justinian sent two more letters assuring him that he would be received with great honours and that he would be ordained by Metropolitan Nicolae Bălan, his former professor in Sibiu, Bishops Nicolae Colan and Veniamin Nistor, and also by the former leader of his community, Policarp Moruşcă. In his last letter, Justinian made sure to include a reference which suggests the real reason for his nomination: 'bring with you a list with the addresses of Orthodox and Greek-Orthodox parishes and the names of the priests [in the US]'.[148]

The *Securitate* placed a large number of agents around Moldovan who reported his comments and activities in detail. The officer in charge of his case suggested the following measures during his visit: Justinian should be briefed on how he should speak to him; the other members of the hierarchy should be instructed by Justinian on this matter; the other *Securitate* agents should ensure that they present a favourable image of Romania which would gain Moldovan's sympathy for the regime and future collaboration; that he should be extensively monitored as he might be an American spy; as means of propaganda, the Romanian Patriarchate should send more journals abroad which presented religious freedom in the country; and that a film with the ceremony of his ordination should be taped and sent to the Romanian community in the US. In addition to these actions, in order to dispel any doubts regarding the intentions of the church hierarchy and the regime, Justinian suggested that he should be accompanied only by the clergy and not by any civilians from the Ministry of Religious Confessions.[149]

Moldovan arrived in Bucharest on 3 November 1950 and was welcomed at the airport by his daughter, his son-in-law and a delegation from the Patriarchate which included Vicar Bishop Teoctist, Professor Liviu Stan and Stanley George Evans, vice-president of the British-Soviet Association. He travelled immediately to Neamţ monastery where he was made a monk on

5 November. Moldovan's activities were severely limited and he was not allowed to speak freely in public. He was given a speech to read at the end of the ceremony; a *Securitate* note states that he questioned why he should read it, but he did so. The atmosphere was tense as evident in the evening after the dinner. Another *Securitate* report states that Liviu Stan got drunk and asked everyone to be quiet as he had to hear everything they said and immediately write a report for the *Securitate*.

Moldovan made a good impression. He became friendly with the other hierarchs and when political issues were raised in discussions he did not comment, saying only 'yes' and 'no' to various questions. Justinian assured the *Securitate* that Moldovan seemed to be genuine and repeated his words from one of their private conversations: 'Your Eminence, if all priests and monks I have seen are not true guardians of the altar, then I am not. If these are false priests – as they say in America – then I am a false priest. If in Romania there is no religious freedom, then I don't know where there is'.[150]

On 12 November 1950, Moldovan was ordained as a bishop in Sibiu and over 1400 people attended the ceremony. The clergy and the regime were extremely surprised to see so many people there, with some priests even stating that this was the highest number of the last ten years. After the ceremony, Moldovan was asked to read another pre-written speech to which he did not object. Metropolitan Bălan performed the ceremony together with the other bishops promised by Justinian in his letter, with the exception of Bishop Moruşcă, who was said to be upset and refused to attend; in reality he was under house arrest.[151]

Moldovan's real intentions remain unclear; most probably he was an opportunist without interest in political affairs. A report from the British Legation in Bucharest stated that he hoped to secure the emigration of relatives of his parishioners in Detroit, however, he was unsuccessful.[152] There are no references in the *Securitate* files regarding this issue. Some suggestions as to his personal thoughts are in a file written after he visited Văratec monatery, accompanied by his daughter Valeria. She had an intimate conversation with a nun there who was a friend of hers, and who later disclosed their converstation to the *Securitate*. Valeria claimed that her father told her in private that another war was inevitable and that all factories in the US were producing bombs. She assured the nun that 'public opinion there is against communism and that very soon we will all get rid of the communist regime'.[153]

After Moldovan returned to the US on 21 November 1950, he did everything in his position to impose his jurisdiction as he was still not recognised by the majority of the Romanian community. He appealed to the American civil courts but lost all cases.[154] In this way, the Romanian community in the US was divided between those who accepted the independence of the council and those who supported Moldovan as their religious leader. In the following year the council elected a new bishop, Valerian Trifa, while Moldovan remained the leader of a small group of Romanians until his death in 1964. Moldovan's case was the first to show the systematic infiltration and control of the Romanian Orthodox diaspora. It revealed the importance of religious communities abroad and the ways in which the regime sought to manipulate its leaders.

The climax of Sovietisation

If outside Romania the regime started to extend its control over the diaspora, within the country new measures were implemented. On 15 December 1950, the GNA adopted two major laws: the 1951–5 Five Year Plan for the Development of the National Economy, the first five year plan of social-economic development, and the Law for the Defence of Peace. On 19 December, at the patriarch's proposal, the leaders of all religious confessions in the country met at the Patriarchal Palace to declare their common vision regarding the involvement of all religions in Romania in the battle for peace. The patriarch asked all priests to help their local committees in the defence of peace and to read a speech to their parishioners written by Metropolitan Ruşan, who had attended a peace congress in Warsaw.[155] The fact that all clergy were not completely obedient was evident in that around 60 priests were arrested in Bucharest for refusing to carry out these directives indicating strong grass roots opposition.[156]

Traces of an increasingly anti-religious policy of the regime were evident in patriarch's Pastoral Letter at Christmas when for the first time he referred not to a religious day but that the faithful should enjoy 'the winter celebrations'. At the same time, as a demonstration of religious freedom in the country, the government approved the statutes of Baptist, Adventist, Pentecostal and Gospel Christian communities who agreed to refrain from having contact with similar churches abroad.[157]

The economic burden of the peasants and the negative image of collectivisation were ongoing points of turmoil in society. For propagandistic reasons, the Central Committee of the RWP issued a statement on 30 January 1951 through which 'any economic or administrative pressure towards working peasants in order to force them to enter collective farms would be severely sanctioned'. This statement was followed on 1 March 1951 by the modification of fixed quotas and the obligation that every member of the collective agricultural farms should work 80 days annually for the state. This policy brought even more dissatisfaction in the country which was already suffering major economic problems. Moreover, the purges against any possible enemy of the country reached a mass scale on 18 June at Stalin's and Gheorghiu-Dej's orders, when 10,288 families (34,037 people) were deported, allowed to take with them only essential personal goods. Thus, all peasants who were suspected of being against the regime and a large number of the Serbian population from the Yugoslavian-Romanian border were sent to

Bărăgan and Moldavia where they were forced to live in mud huts. The fate of these deportees would be sealed until 1962 when Gheorghiu-Dej decided that those who had survived should be freed. The people who were considered most dangerous for the regime were sent on work at the Danube-Black Sea Canal where most of them died.

The manipulation of religion reached a new stage when the government succeeded in its control over the Catholic Church. On 15 March 1951, under communist pressure, 225 representatives of the clerical and laic Roman Catholic faithful gathered in Cluj and set up an organisation independent of the Vatican's authority, the so-called Roman Catholic Status (*Statusul Romano-Catolic*). At their meeting a council of 27 members led by Archpriest Grigore Fodor was established which agreed to cooperate with the government. Clergy who accepted this new Catholic hierarchy were offered state salaries, while those who resisted were imprisoned.[158] In order to fully impose its policy of distancing the Roman Catholics in Romania from the Vatican, the government even replaced Stoian as Minister of Religious Confessions with Vasile Pogăceanu who was considered to take a tougher line on this matter and would bring the Catholics under the complete control of the state.[159]

While the Catholics were suppressed, the Orthodox churches in the Eastern bloc continued to assert the dominance of the Russian Patriarchate. On 8 July 1951, the Romanian Orthodox Church sent a delegation to the enthronement of Metropolitan Makary Oksaniuk of Poland. Delegates from the Soviet Union and Bulgaria were present and Romanians were represented by Metropolitan Firmilian of Oltenia, Antim Târgovişteanul, vicar bishop of the Patriarchate and Father Ilie Bacioiu as their translator. The Polish Orthodox Church was declared autocephalous by Moscow on 22 June 1948 and Makary was one of the bishops who was active in bringing the Uniates to the Polish Orthodox Church after 1946. The Polish Church was small, having only three bishoprics with around 400,000 faithful, 202 churches, 2 monasteries and 160 priests, but Makary's enthronement was politically significant for the position of religion in the Eastern bloc. The granting of autocephaly to the Polish Orthodox Church was another means of controlling religion in the region and of using the churches in the propagandistic message of peace.[160]

Collaboration between the Russian and Romanian churches extended through personal contacts between their leaders. The Romanian Orthodox Church responded to Moscow's invitation to attend a pan-Orthodox meeting with the most important hierarchs under Soviet rule, and from 15 to 27 July 1951 Patriarch Justinian visited Patriarch Alexius for the fourth time. Orthodox leaders from Syria, Bulgaria, Georgia, Romania and Russia met to discuss the engagement of their Orthodox churches in the battle for peace. The Romanian delegation was composed of Justinian, Bishop Andrei of Arad, Vicar Bishop Teoctist and Father Simeon Neaga, the administrative advisor of the Archbishopric of Bucharest and translator of the delegation. On 16 July, Justinian met G. G. Karpov, president of the Committee of Clerical Affairs and presented the achievements of the Orthodox Church in Romania. According to Justinian's statement, monks were working in 60 monasteries which had been transformed into workshop centres; 80 per cent of the clergy was active in local committees for the defence of peace; the church hierarchy and all the other Romanian religious leaders met in December 1950 and decided to support the signing of the appeal for peace; furthermore, the church completely supported the political agenda of the Romanian government.[161] On the following day the Romanian delegation visited a special exhibition of the presents received by Stalin for his 70th birthday and on 18 July all of the patriarchs celebrated a liturgy for the religious day of Saint Sergii Radonezhsky, the protector of the Russian people. In addition, Justinian visited five churches inside the Kremlin, the residence of the Council of Ministers, the former palace of the Tsar and the mausoleum of Lenin. On 23 July the delegations of Patriarch Alexander III of Antioch, Patriarch Alexius of Moscow, Patriarch Justinian of Romania, Catholicos Patriarch Kalistrate of Georgia and Metropolitan Kiril of Plovdiv and president of the Holy Synod in Bulgaria signed a statement calling on Christians all over the world to promote peace and to oppose the war in Korea. On his way back to Bucharest, Patriarch Justinian visited Kiev on 25 and 26 July.

Collaboration between the Romanian Orthodox Church and the other Orthodox churches continued after the Moscow meeting when Patriarch Alexander III of Antioch visited Bucharest from 22 August to 3 September 1951. Together with Justinian, he celebrated the liturgy on 26 August and was offered the title of *Doctor Honoris Causa* of the Theological Institute in Bucharest.[162] Moreover, in December 1951 Justinian met Archbishop Paisi of the Orthodox Church in Albania who had participated in Prague at the declaration of the autocephaly of the Orthodox Church in Czechoslovakia, and, on his return to Albania, stopped in Bucharest to inform his Romanian counterpart about the events he had attended.[163]

At a time when the Orthodox Church closely followed Soviet directives, the last members of the Roman Catholic hierarchy who still opposed the communists went on trial in one of the strongest acts of religious persecution during communism in Romania. Despite forced unification of the Greek Catholic Uniate Church with the Orthodox Church, only 400 out of 1800 Uniate priests obeyed the government's decision. Even if it had officially ceased to exist, the communists felt that unification of the Uniate Church could not be complete without the removal of the Catholic hierarchy from their positions. Moreover, some communists even asked for the unification of the Roman Catholic Church with the Orthodox Church. Between 10 and 17 September 1951 the regime put on trial the most important leaders of the Uniate and Catholic churches. They were accused of treason, actions against the state and of working for Italian espionage. The most important leaders, Bishop Augustin Pacha of Timişoara,[164] Monsignors Iosif Schubert, Albert Borosh, Iosif Waltner, Father Pietro Ernesto Gatti and Ion Heber and other laymen were judged by a military court in Bucharest which sentenced them to imprisonment.[165]

The communists continued to condemn those religious leaders who refused to collaborate and spread a regime of terror against any religious manifestation. People were encouraged to cremate their dead relatives instead of traditional internment while the communists who were seen attending church services were excluded from the party.[166] Anti-religious propaganda was not only directed against adults but also affected children. Those, for example, who made the sign of the cross upon hearing the church bells near their school in Bucharest on Saint Nicolas Day, 6 December 1951, were expelled while their parents were criticised by the Romanian authorities.[167] As a Radio Free Europe report points out, ordinary people claimed that 'to believe in God is looked upon as a disgrace and to manifest faith is still worse'.[168]

During these sensitive times Patriarch Justinian largely kept out of public life, being present only at the opening of the theological courses and the celebration of 1500 years after the Fourth Ecumenical Council in 451 in Calcedon on 11 November.[169] Communist pressure on him intensified as the National Clerical Assembly elected Valerian Zaharia as Bishop of Oradea in place of the forcibly retired Bishop Nicolae Popovici on 14 November 1951. Zaharia was widely perceived as a communist and the patriarch disagreed with his appointment.[170]

Communist control of religion did not face any public opposition except from the Vatican, with all of its bishops in prison or exiled. On 27 March 1952, Pope Pius XII wrote an Apostolic Letter to the Catholic bishops and faithful in Romania, *Veritatem facientes*.[171] The pope encouraged Catholic Romanians to remain strong in the face of religious persecution and hoped that his letter would reach them. The report of the British Legation to the Holy See which monitored the religious situation in Eastern Europe emphasised that 'the tone of this letter, compared with that of the Pope's last letter to the Church in Poland or even the Church in Czechoslovakia, shows how much more successful the Romanian government has been than the Polish or Czech governments in suppressing the Catholic Church'.[172]

On 26 January 1952 the government implemented monetary reform by introducing new monetary bills. The exchange rate varied depending on the status of the owner, for example, a private person or a state cooperative enterprise.[173] This was the second monetary reform after 1947 and another attempt to confiscate all currency in Romania. The reform would have political implications in the clash among the communist leadership. The first signs of this conflict were at the plenum of the Central Committee of the RWP from 9 February to 1 March 1952 during which Vasile Luca, Minister of Finance and one of the leading communist party members, was criticised for mistakes in applying this monetary reform.[174]

The church did not publicly responded to the internal battle for power between communists; on 31 March the patriarch sent a Pastoral Letter regarding the harvest which did not mention the communist clashes.[175] The extension of Soviet control of every aspect of life in the country was evident again when Patriarch Alexius asked religious leaders to participate in a conference for peace in Moscow between 9 and 12 May 1952. The Romanian Orthodox Church sent Metropolitan Ruşan as its representative and on his return he gave a speech in the Theological Institute in Bucharest on 24 May, a sermon in Iaşi cathedral and presented a special report on 15 July 1952 to the National Clerical Assembly, all emphasising the relationship between the Russian and Romanian churches.

The communist method of using religion was demonstrated again on 3 April when the Romanian newspaper *Universul* published a letter of protest from Justinian and four metropolitans against the use of bacteriological warfare. Justinian put together some words from the New Testament in justifying government policy and criticising the West. His letter stated that the Americans were the servants of the 'golden calf' and cautioned the faithful that the prophecy in I Peter 3:12 would soon be fulfilled: 'for the face of the Lord is against those who do wrong, that they may perish from the face of the earth'.[176]

The apogee of the battle for power between communists took place at the Plenum of the Central Committee of the RWP on 26 and 27 May 1952 when Gheorghiu-Dej managed to remove the Ana Pauker-Vasile Luca group from party leadership. Luca was sent to face the control commission of the party while Pauker, due to her connections with the top Soviet communists, was only reproved. However, she was not re-elected to the Secretariat and the Political Bureau of the party, remaining only in the Organisational Bureau and as Minister of Foreign Affairs. Teohari Georgescu, Minister of Internal Affairs from 6 March 1945 to 28 May 1952, was removed from the Secretariat and from the Political and Organisational Bureaus of the party, and sent to do 'lower work'. In imposing Gheorghiu-Dej's dominance of the party, the whole structure of the communist leadership was modified.[177] On 2 June 1952 the GNA elected Petru Groza as president of the Presidium of the GNA, replacing Constantin I. Parhon, and Gheorghiu-Dej became chairman of the Council of Ministers, serving in this position until 3 October 1955.

The victory of Gheorghiu-Dej represented a short period of religious freedom. Thus on 6 June, Monsignor Vladimir Ghica, the Catholic archbishop of Bucharest, was released from his 'forced residence' in which he had been confined since May 1950. In a sermon he mentioned the assistance of the Minister of Religious Confessions Pogăceanu in achieving his release and addressed the topic of peace without preaching an anti-Western campaign.[178] However, he was soon reconfined to forced residence.[179]

On 14 June 1952 the Holy Synod sent a Pastoral Letter which supported the new course of Romanian politics advising the faithful of its attitude towards the battle for peace and of Gheorghiu-Dej as the leading figure of the party: 'Guided by the example of the Soviet Union and led by its best son, Gheorghe Gheorghiu-Dej, our people are building a new and happier life'.[180] After a special meeting, the Holy Synod sent telegrams congratulating Groza and Gheorghiu-Dej on their new political positions. It also sent a special message to Gheorghiu-Dej:

> The Holy Synod of the Romanian Orthodox Church assures you, the loving leader of our people, that we will engage in the future more actively to make every member of the Romanian Orthodox Church's faithful and clergy, a devoted son of the country, a loyal citizen and a firm fighter for the defence of the country and for national independence which are today threatened by the enemies of liberty and of the Romanian people, the American-British imperialists.[181]

Following these political changes the church was forced to restructure its own organisation, particularly regarding religious training and the Holy Synod adopted a new statute of religious education.[182] The number of schools for cantors was reduced from 12 to 6 (in Argeş, Neamţ monastery, Cluj, Craiova, Caransebeş and Buzău), the Theological Institute in Cluj was closed with only those in Bucharest and Sibiu functioning, while the monastic clergy would only be trained at Neamţ, Agapia and Hurezu monasteries.[183] In addition, in order to increase communist propaganda, a new law was introduced which reorganised the diocesan conferences of priests. Each month a diocesan priest had to organise a conference; these would alternately be an administrative conference, which he led, and one of 'social character', led by the bishop or his representative.[184] Each priest was obliged to prove at these conferences that he had read recent church journals, presenting abstracts of the articles published with communist indoctrination. Moreover, the new law adopted standardisation of religious music for all bishoprics in the country despite the particular traditions of each region.

Gheorghiu-Dej's dominance of the party was fully achieved when Pauker was removed from her position as Minister of Foreign Affairs on 11 July and as vice-president of the Council of Ministers on 16 September 1952. A draft version of a new onstitution was published on 18 July and officially the church agreed with its modifications.[185] The constitution represented the apogee of Sovietization of Romanian politics, from its first articles stating that

> The Romanian People's Republic is a state of working people in cities and villages. The basis of people's power in the Romanian People's Republic is the alliance between the working class and the working peasants, in

which the leading role belongs to the working class. The Romanian People's Republic was born and strengthened as a result of the liberation of the country by the armed forces of the Soviet Union from under the fascist yoke and imperialist domination, as a result of the removal of land owners' and capitalists' power by the popular masses in cities and villages together with the working class, under the leadership of the Romanian Workers' Party.[186]

By including a reference to the Soviet Union in the constitution, the communists stated *de facto* its role in establishing the Romanian People's Republic and the course of the country's politics.[187] From this moment, control of the party centred on Gheorghiu-Dej, and decisions were taken by a reduced number of communists.

For the first time since the creation of the Romanian state, there was no mention of the Orthodox Church in the constitution. The only point which referred to religion was article 84 which stated that

The freedom of conscience is guaranteed to all citizens of the Romanian People's Republic. Religious confessions are free to organise themselves and to function. The freedom of participation in religious confessions is guaranteed to all citizens of the Romanian People's Republic. The school is separate from the Church. No confession, congregation or religious community can open or support general educational institutions, but only special schools for the training of personnel of that confession. The organisation and functioning of religious confessions is regulated by law.

The Orthodox Church was considered by law as equal to the other religious confessions and any citizen could, in theory, profess his own religion. Officially there was no favoured religion in society even if the patriarch and church leaders collaborated with the communists, and other churches had been persecuted. The constitution referred explicitly to the separation between church and education thus implying the leading role of the communist party in this field.[188] The position of the Orthodox Church towards the constitution and communist propaganda was visible on 23 July 1952 when the patriarch himself received a group of Scottish workers, members of the Transport Union, and answered their questions regarding the course of religious freedom in Romania. In their discussions, the patriarch claimed that the new constitution offered the perfect model of church-state relations and that the Orthodox Church enjoyed complete freedom.[189]

Between 15 and 16 August 1952, the state carried out further arrests, focusing its attention on all intellectuals who had been connected with the royal palace during the interwar period. State terror increased on 25 August when Alexandru Drăghici established a commission within the Ministry of Internal Affairs in charge of internment in forced labour camps.[190] State control reached its climax on 20 September with the establishment of the Ministry of State Security, with Drăghici as its head, which deepened the regime's campaign of eliminating any potential threat.[191]

On 25 November, on the patriarch's initiative, all religious leaders in Romania including the first time representatives of the Roman Catholic Church, Prime Minister Groza and Pogăceanu, Minister of Religious Confessions, attended a large meeting at the Theological Institute in Bucharest. The discussions referred to the forthcoming elections of the GNA on 30 November 1952. The religious leaders signed a declaration in which they asked their faithful to show 'their love for the country' and actively participate in the elections. Furthermore, the declaration was read to all religious congregations on the day of the elections.[192] After the meeting the participants sent a telegram to Gheorghiu-Dej assuring the support of their confessions.[193]

The use of religion in the battle for peace was exemplified at the National Congress for the Defence of Peace when around 700 participants gathered in Bucharest from 5 to 7 December 1952 in a meeting which restated the complete authority of Gheorghiu-Dej in Romanian politics.[194] Symbolically, on the main wall of the meeting room, the portrait of Gheorghiu-Dej was hung between those of Lenin and Stalin with the inscription: 'Long live peace among the people'.[195] The meeting presented the achievements of communist peace propaganda in recent years, in that the state had created 22,165 local committees for the defence of peace and the appeal for peace had been signed by 11,060,141 people. Moreover 1500 Korean children had been brought to Romania and the government had sent aid to help Korea. The patriarch, a leading figure on the podium alongside the main communist participants, gave a speech in which he emphasised the support of the church for the defence of peace.[196] For his collaboration with the regime, a few days later Justinian was decorated with the medal celebrating 'The fifth anniversary of the Romanian People's Republic'.

The communists made sure that the Orthodox Church was not the only religion which praised the regime. In his Passover message, the chief rabbi of Bucharest, Dr Moses Rosen, compared the Prophet Elijah to a communist and a fighter for peace in a similar manner to that in which Patriarch Justinian compared Christianity to communism.[197] However, extreme propaganda was visible even in the most sacred religious materials. The Jewish communities received their unleavened bread for celebrating Passover wrapped in paper which had the inscription: 'We fight for peace against the Americans and the British imperialists and their Zionist lackeys'.[198]

Communist indoctrination against religion was most evident in schools. Teachers were instructed to teach that Christ was a communist and that Christianity was the forerunner of communism. In this way the Bible became a simple collection of myths and parables which were reinterpreted from their religious meanings. Negative figures from the New Testament, the

Pharisees, were now compared to the British and American imperialists who were exploiting people.[199] Teachers suggested that the regime allowed freedom of religion because all religious leaders were encouraged to participate in the building of socialism.[200] For this reason, schools organised activities for pioneers on Sundays while the Christmas holiday was renamed as 'the winter holiday'. In 1952 Christmas Day was a working day[201] and students had their usual holidays later from 30 December 1952 until 11 January 1953,[202] in a concerted attempt to discourage the public from attending religious celebrations. Priests were not allowed to preach against the aggressive anti-religious actions of the regime. In addition, those who published articles against communism in the interwar period[203] or who had children in exile were arrested.[204]

Inside the patriarchal palace

The communists were interested in using religion not only for ideological purposes in their own countries but also abroad. Radio Free Europe reported that in a meeting with Patriarch Alexius, the Romanian and Bulgarian counterparts complained that the archbishop of Athens was inviting students from Egypt, Syria, Lebanon, Palestine, Ethiopia and other countries to study in Greece, while their churches were not allowed to have foreign students. They argued that the students in Greece were returning home anti-communist and they should be permitted to invite students to counter the balance.[205]

Education was one of the most important spheres of propaganda of the regime and the Orthodox churches sought to influence the future religious hierarchs in other countries.[206] In Romania the regime allowed a very limited number of foreign students to attend courses at the Theological Institute in Bucharest.[207] These students came from countries with which the communists were interested in maintaining good relations or where the regime wished to encourage dissidents in their churches. From 1947 to 1957 only four foreign students studied religion in Romania; however, the regime's intentions were not accomplished as these students kept a low political profile after returning to their countries.

Two of them were from the Finnish Orthodox Church led by Archbishop German who promoted an independent position from Moscow. The first student, Väinö Kalev Ihatsu, studied in Romania for four years, obtaining a Bachelor's degree in Theology. In 1954 he returned to Finland and was ordained a priest in Nurmes.[208] He learnt Romanian with the help of Professor Alexandru Ciurea[209] and in summer 1955 led the Romanian delegation composed of Metropolitan Ruşan and Father Alexandru Ionescu at the Helsinki Peace Congress acting as translator. The second Finnish student, Iuhani Homanen, spent only four months in Romania.[210] He was the son of a priest and director of the Orthodox Cathedral choir in Helsinki. In Romania he studied religious music and violin at the Romanian Conservatory.

A third student, Archimandrite Anton Malek from Beirut, came to Romania in 1952. He arrived as a result of an agreement between Patriarch Alexander III of Antioch and Patriarch Justinian. He attended some religious courses and was offered a scholarship from the Romanian Patriarchate. The fourth student, Hieromonk Apostolos Panaiotis Dimelis, from the Theological Institute in Halki, was sent by the Ecumenical Patriarchate after Bishop James of Malta visited Romania in 1955.[211] He studied on the Masters programme and learnt Romanian with Professor Ciurea.[212]

The most controversial of these students was Archimandrite Anton Malek. After his return home he met a member of staff from Radio Free Europe and gave a unique confidential account of his stay in Romania from November 1952 to April 1953. The Radio Free Europe report mentions that an Arab priest, named only 'the Archimandrite', without giving his full name for security reasons, came to Romania on the invitation of Justinian and stayed at the Patriarchal Palace for a few months.[213] As Anton Malek was the only Archimandrite from Beirut who studied in Romania during that period it is highly likely that he was the source. In the report, the source describes aspects of his experiences in Romania and his own opinion on some members of the church hierarchy. His detailed account offers an intimate, although partial, eyewitness angle on life inside the Patriarchal Palace, presenting the internal situation of the church during those years and relations between the communists and the church hierarchy.

At the time of his arrival in Romania, Malek was 28 years old and had been selected by Patriarch Alexander III because he was known to be pro-communist. In Beirut he wrote numerous articles praising communism in Eastern Europe and was even under police surveillance. After the invitation from the Romanian Patriarchate he spent almost one year completing visa formalities and arrived in Romania in November 1952. However, soon after his arrival he became disappointed with the communism that he knew from propaganda, which contrasted sharply with Romanian reality. As he pointed out,

> I was overjoyed to be going to a Communist country [. . .] I even had the idea that I might remain for good in Romania and contribute something towards helping the workers achieve Socialism. Actually, after less than two weeks of life in Bucharest, I was longing to breathe the air of a free country again. Now I am cured of the Communist fever forever.

At their first meeting he gave Justinian an open introductory letter from Patriarch Alexander. Upon Justinian's sign, Bishop Teoctist, vicar of the Patriarchate, took the letter and sealed it in front of him thereby making sure that nobody else could read the content. The Archimandrite was shocked by the incident but found out later that any letter for the patriarch from abroad was supposed to be read first by officials from the Ministry of Religious Confessions before he would have the 'privilege' of knowing its content.

The report mentioned that inside the church hierarchy 'there prevails a mutual hatred and rivalry – carefully fostered by the communist authorities – which poisons the atmosphere of the Patriarchate', especially between Bishop Valerian Zaharia, Bishop Teoctist and Justinian. The report described how the 'Archimandrite' witnessed two major incidents which showed how the patriarch struggled for his own credibility among hierarchs who were known communist sympathisers:

> Once when the Archimandrite was in Justinian's office, the Bishop Teoctist entered and handed an envelope to the Patriarch; the latter drew out of the envelope some papers which he read while his face reddened with anger. With a glare at Teoctist, the Patriarch finally tore the papers in two pieces and threw them on the floor. Teoctist gave the Patriarch an insolent smile and remarked 'I have understood Your Holiness'. Then he picked up the torn papers and walked out of the office.
>
> On another occasion the Archimandrite and a Romanian priest named Ciurea happened to be in the Patriarch's office when the Bishop Valerian Zaharia sauntered in, drew up a chair without asking permission and began to speak to the Patriarch in mocking tones. 'Did you listen to the broadcast of Free Romania last night at 10 o'clock?' asked Zaharia. The aged prelate's hand began to tremble and he replied with agitation: 'no, no, no! I never listen to foreign broadcasts!' Unperturbed, Zaharia answered 'I know that you listened to that broadcast, as I did myself'. In the discussion that followed, Zaharia spoke loudly and aggressively, while the Patriarch measured his words despite his own nervousness. At the height of the discussion the priest Ciurea asked permission to leave the room; he told the Archimandrite later that he did not wish to be a witness at such discussions because he 'feared the possible consequences'.

In addition to clashes with other hierarchs, Justinian was seen as an authoritarian person. An incident which took place on New Year's Eve 1953 seems to illustrate the way in which the hierarchs behaved towards each other. Justinian, who was seated in the altar, assigned a priest, Babuşa, the role of reading a letter to the congregation on the propagandistic message of peace. The letter was signed by all religious leaders in Romania; however, the priest omitted the names of the others ending only with 'the Patriarch Justinian and his colleagues'. Hearing this, the report states that

> Justinian purpled with fury and began to hammer on the altar with his fists, meanwhile pouring forth a torrent of words at the attending priests who were trying to persuade him to be seated in his armchair. With his meagre knowledge of Romanian, the Archimandrite was mystified by the scene and he asked a deacon on the altar, called Duţă, what the Patriarch was saying. The terrified deacon whispered 'He is cursing

like a coachman'. The priests finally succeeded in inducing Justinian to leave the altar and a physician was summoned to the Cathedral to calm him. The next day, the offending priest, Babuşa disappeared from his room in the Patriarchate building and did not return for a month; other priests told the Archimandrite that he had been arrested by the *Securitate*.

The 'Archimandrite' wanted to know more about this event and to understand why Justinian reacted in this manner. He found out that Babu?a was regarded as 'bourgeois' by the communists and his colleagues were jealous of him especially because he inherited some expensive furniture and a radio. He was listening to foreign radio broadcasts at night together with some of his friends. Someone informed Bishop Teoctist of his habits, who one evening came unexpectedly to his room, but at that time the radio was not switched on. The 'Archimandrite' recalled, 'After sardonically congratulating Babuşa on his luxuriously-furnished room, Teoctist told him that if he ever caught him listening to foreign broadcasts he would break "every tube in that radio set."' After this incident Babuşa sent the radio to a friend. In addition to the jealousy of his colleagues, the 'Archimandrite' discovered that Babuşa was practicing spiritualism. On the day before New Year's Eve he mentioned to a friend that Patriarch Nicodim appeared to him in one of his sessions telling him that he died because he was poisoned by Justinian. The other priests' jealousy ensured that the patriarch heard about this and the report claims that this was what infuriated Justinian.

The 'Archimandrite' stated that Justinian was permanently under the surveillance of Bishop Teoctist and Bishop Zaharia and that every time he left Bucharest he was accompanied by a member of the *Securitate*. The 'Archimandrite' remembered an example when Justinian invited him to travel to Curtea de Argeş. The patriarch was accompanied by his daughter, Silvica, her husband Dr Solomon, who was also one of his physicians, the chauffeur and the *Securitate* man. When the car left Bucharest 'the Patriarch asked the man ingratiatingly "What's your name, comrade?" The man replied 'Vasile'. That was the only word the agent spoke during the entire journey; no one addressed him and he addressed no one.'

The example offered by the 'Archimandrite' suggests that despite officially having good relations, the hierarchs were not trusted by the communists who wanted to be informed of all of their activities and discussions. In order to maintain cordial contact with the communists, Justinian organised lavish parties to which top officials were invited. In December 1952 the 'Archimandrite' attended one of these parties at which Groza and Constantinescu-Iaşi were present. When the 'Archimandrite' was introduced to Groza as the student who came from Beirut to learn about the church under communism, Groza allegedly said 'Moi je suis un croyant' and made the sign of the cross.

In addition to close surveillance, some rumours inside the Patriarchate were fabricated in order to discredit Justinian and some other hierarchs.[214] As the report mentions

> according to the gossip at the Patriarchate, he [Justinian] has several mistresses – difficult as this is to believe of a man of his age [. . .] Justinian is also supposed to be in amorous relations with a nun named Anatolia who lives in the Patriarchal Palace in the guise of a 'housekeeper'. The Archimandrite said that Anatolia is very beautiful; when he first met her he asked her if she spoke French. A priest standing within hearing remarked 'She knows only the language of love'. [. . .] It is also gossiped among the priests that Justinian's daughter, Silvica, is the mistress of the Bishop Teoctist.

In the opinion of the 'Archimandrite', 90 per cent of the clergy were anti-communists, five per cent were communists while the other five per cent pretended to be communist only to foster their personal careers. Although he did not witness the arrests of any priests, some clergy inside the Patriarchate suggested to him that around 1500 priests were arrested at that time mainly because they told their parishioners that they did not believe in communist propaganda or because they refused to read the Pastoral letters and sermons in which the regime was praised.

In March 1953 the 'Archimandrite' wanted to leave Romania and went to the Alien Control Office to apply for an exit visa where he encountered another incident. The clerk wanted to hold his passport but he was afraid that the only document which proved his identity could be lost and refused to leave it there. The clerk started to shout that 'We are in a communist state and the functionaries are honourable men. With us the documents do not disappear in government offices, as they do in your capitalist states'. He mentioned that he was a guest of Justinian and that he could be contacted at the Patriarchate when his visa was ready. He obtained his exit visa just 18 days later, thus illustrating the influence of the patriarch outside his religious domain.

The night before his departure, the 'Archimandrite' had a final meeting with Justinian who asked him to convey only a verbal, rather than a written, communication to Patriarch Alexander. As the report points out 'Justinian's last words to the Archimandrite were somewhat Delphic: "With you in Syria and the Lebanon the situation is different. Here we are closer to God."' The Archimandrite left Bucharest on 2 April 1953, being terrified up to the last moment that he would not be allowed to leave the country. The report concludes that 'I could hardly believe that I was in a free country again. Now I know the meaning of that phrase I had heard so often in Bucharest '*Nu mai există primăvară în România, totul e propagandă*' (There is no longer any spring in Romania, it is all propaganda)'. After his return to the Middle East he received death threats from pro-communists who accused him of receiving money from 'the imperialists' and went to Paris to continue his studies.

The Archimandrite's account of his experiences in Romania in 1952–3 is extremely evocative. On the one hand, it shows that, behind the scenes, the communists encouraged the clash for ecclesiastical power among members of the hierarchy; on the other hand, they also extensively controlled all activities of the church. As the report suggests, the actions of the church hierarchy were limited and their positions depended entirely on their good relations with the communist authorities. Despite the incident with Father Babuşa, the Archimandrite presented Justinian in a favourable light as being the only hierarch who was genuinely interested in preserving the church from communist control. The Archimandrite pointed out that the other clergy had communist sympathies because they were only interested in acquiring better positions in the church.

Conclusion

The installation of communism in Romania brought dramatic changes to all aspects of public and private life. Soviet pressure on Romanian politics and adherence to Stalinist policy affected large numbers, as both former political leaders and ordinary members of the population were imprisoned or deported.

Religion was one of the most important aspects of Romanian life which suffered during communism, and the communist leadership acted differently towards the various religions in the country according to their political agenda. Thus, while the Roman Catholic Church hierarchy was in prison or repatriated and its church was transformed into a national body, other churches and religious groups such as the Uniate Church and the Lord's Army were completely disbanded and integrated into the Orthodox Church. As the institution of the predominant religious confession, the Orthodox Church had advantages. However, these came at the cost of collaborating with the regime and suffering major transformations in its structure and organisation.

The communists used the Orthodox Church as a political tool, following the pattern of previous regimes which had used the nationalist discourse of the church in order to impose their own political dominance. The atheist communists in Romania used the church according to the Soviet perception of religion, but they also took into account the role that the church had during the interwar period and previously. They imposed their own people in the hierarchy and propagated communist ideology to the masses through the channels of the church. The church leaders encouraged the faithful to support the communist reconstruction of the country. For their obedience, the clergy continued to receive salaries from the state and subsidies in rebuilding churches and publishing journals.

The relationship between the church and the state during this period was twofold: on the one hand, the church was employed by the communists in order to guarantee support in elections; while on the other hand, the church was engaged in the ideological propaganda of peace. The communists attributed to themselves the image of promoters of peace while Western leaders were presented as the warmongers. The church widely supported this ideology, from the involvement of priests in national committees for the defence of peace to international protests addressed to the United Nations against the American war in Korea and the use of bacteriological weapons.

The church was accepted by the communists in their ideological battle because the Orthodox commonwealth was seen as a possible means of spreading political influence and control outside the Eastern bloc. Orthodox clergy in the Middle East and Northern Africa enjoyed good relations with the communists in the Soviet Union and other communist countries like Romania. The establishment of autocephalous churches in the communist bloc in countries where Orthodoxy was insignificant reinforced the concept of Moscow as the third Rome. The Romanian Orthodox Church supported these Soviet claims and Romanian delegations frequently visited Moscow and ensured the implementation of its policies. The regime encouraged connections with the Orthodox diaspora in attempting to spread communism in Western countries and to inculcate a sense of superiority of communist ideology. The communists extensively used Orthodoxy rather than other religions in the first years of communism because it was considered the best political tool through which they could both assert religious connections with the Soviet Union and influence Orthodox communities abroad.

As in other countries under the influence of the Soviet Union, a false perception of religious liberty was promoted in Romania. Religious purges and transformations were presented as popular decisions and the state imposed its authority in rewriting the history of the country in order to protect the 'interests' of its people. In this way the communists created an institutional machine to propagate state terror reaching its climax with the establishment of the Ministry of State Security which controlled all aspects of social life. A simple denunciation meant the loss of individual liberties and even physical elimination. The church was affected by state terror, especially those members who opposed collectivisation, who had supported previous regimes or who publicly condemned the communists. The general perception among the lower clergy during this period was that collaboration with the communists was necessary in order to survive. However, at the same time, there was a belief among both ordinary people and some church hierarchs that a new war would soon start, led by the Americans, which would change the atheist regime in all communist countries. Thus some members, on certain occasions, were cautious in not extensively supporting the regime too strongly, in case of possible future reprisals should the communists fall from power.

The Orthodox Church attempted to resist communist dominance by promoting a discourse which included references to the national past. The idea of creating national saints who did not have a connection with the Russian Church, the reestablishment of a bishopric which had an anti-Russian message in its name and the commemoration of Romanian patriarchs from previous 'bourgeoisie' regimes were all gestures which reinforced the position of the Romanian Orthodox Church as not completely under Moscow's control even if it shared the same religious doctrine. The ambivalent role of the patriarch who praised the communists but surrounded himself with people from the previous regime was a way of increasing both his political influence in society and gaining control of the hierarchy. For this reason the church could be seen as supporting the Romanian communist nationalist faction over the Soviet-trained Romanian communists who would gain complete political power in 1952. In the next chapters the relationship between the communists and the church hierarchy would become stronger. In attempting to distance the country from the Soviet Union, the church would promote a stronger form of nationalism.

5

Orthodoxy and the Romanian Road to Communism, 1953–5

The death of Stalin

The beginning of 1953 seemed to be like any other year since the inception of communism. Following the promulgation of a new constitution, in the general election for seats in the GNA on 30 November 1952, the communists obtained a majority. On 24 January 1953, the GNA elected a new Presidium which was headed by Petru Groza and a new Council of Ministers headed by Gheorghiu-Dej.

At the time of these political changes, church life continued to follow the same path supporting the regime. Promoting the idea that all religious confessions enjoyed freedom under the new constitution, on 25 February Justinian received 106 students and six professors from the Protestant Theological Institute in Cluj, who had come to visit an exhibition, the 'Scheduling of the Romanian People's Republic's Economy', in Bucharest.[1] Dialogue between the Orthodox Church and other religious bodies in Romania suggested that all religious confessions were involved in the economic evolution of the country.

On 5 March 1953 a shock affected the political and religious course of Romania: Stalin, 'the great and peace-maker leader of the people', died. Responses to his death show how well organised religious confessions in Romania were after only five years of communist rule. On the following day Patriarch Justinian summoned the leaders of all religious confessions in the country to a meeting at the Patriarchal Palace in Bucharest. The leaders decided to act together to preserve the memory of the Soviet leader and agreed that from 6 March until the funeral day all churches would ring their bells four times a day for half an hour. Clerics of all religious confessions would hold special religious services and sermons, and on the burial day would preach a special necrology dedicated to the life of Stalin. After their meeting, the leaders sent a telegram to Gheorghiu-Dej expressing their sorrow on hearing the tragic news. Justinian went even further and sent a telegram of condolences to his Russian counterpart, Patriarch Alexius in Moscow,[2] and visited the Soviet Embassy to sign a special register.[3] Even if the religious leaders encouraged their faithful to participate in these ceremonies, the regime feared a rise of religious and political opposition. The official communiqué of the meeting did not mention the names of those who attended the meeting in Bucharest, inducing the idea that, although some religious confessions were persecuted, all people conformed to these religious measures.[4]

On the funeral day, 9 March, Patriarch Justinian and Bishop Teoctist, together with a group of priests from Bucharest laid flowers at Stalin's statue and attended a liturgy at the Patriarchal Cathedral. Furthermore, in the afternoon the patriarch gave a lecture to 300 priests at the Theological Institute, in which he emphasised the country's connections with the Soviet Union, attributing to Stalin the 'liberty and independence' of Romania. In his speech he emphasised the propagandistic message of the regime claiming that

> We shall never forget that to the name of the great Stalin is bound the free life of our people today; we shall never forget that without the heroic sacrifices of Stalin's soldiers, our country would not have been able to tear apart the chains of slavery in which the fascists enchained her and we would not have been able to reunite with our dear Transylvania; we shall never forget that the Great Soviet Union, built through the work and wisdom of Stalin, is the guarantee of liberty and independence of our dear Fatherland, is the guarantee of peace in the whole world.[5]

While official propaganda commemorated Stalin's death, ordinary people in Romania did not devotedly express their sorrow as in Moscow, where hundreds of people died queuing to see the body of their leader. An eyewitness in Bucharest, a member of the British Legation, reported that religious confessions followed the official line in decorating their churches for mourning. In his report he presented his own experiences:

> On the morning of Saturday, the 7th March, I prowled about the town to look for signs of mourning and other reactions of possible interests on the occasion of Stalin's death. The Orthodox churches were if anything less frequented than usual on a busy working day; at about 11 a.m. I decided to drop in the Roman Catholic Cathedral to study the form there, and noticed the Romanian flag decked with the customary black ribbon flying above the gate.[6]

In addition to this account, the Radio Free Europe report of the 'Archimandrite' offers more details on Romanian perception of the Soviet leader's death and the private reactions of church hierarchs. The 'Archimandrite' stated that

> Bucharest took on a special aspect on that first night when Stalin's critical illness was made public. The downtown districts were almost deserted

and official automobiles roared through the streets at high speed. From 21.00 hours on the *Securitate* began to check the identity cards of the few pedestrians.[7]

The regime's apprehension that news of Stalin's imminent death would trigger anti-communist manifestations was borne out the following day when some papers were distributed around the University of Bucharest and the Dunărea Restaurant welcoming the death, stating that this represented 'the beginning of the end of communism' and proclaiming 'Long Live Free Romania'.[8] In addition, despite the official decision of the religious leaders in Romania to organise special prayers, the clergy were generally reluctant 'to pray for the soul of the one who denied the existence of God'.[9] The priests conducted their services for fear that *Securitate* agents had infiltrated the crowds and would later report if they had not obeyed orders.

The 'Archimandrite' claimed that the patriarch was aware of the propagandistic message which the church was forced to follow, and that he was not personally mourning Stalin's death. This was in direct contrast with the speech he gave at the Theological Institute. In his report, the 'Archimandrite' describes an incident during the liturgy dedicated to the commemoration of Stalin:

When the service ended the Patriarch emerged from the cathedral's entrance attended by his acolytes, with the laity bringing up the rear. On seeing the waiting throng, Justinian raised his hands to the sky as if to say 'God has taken him'. A priest at our source's side commented in a low voice 'The Patriarch seems to say by that gesture, Thank God for ridding us of him'. Some of the people in the cathedral's entrance overheard the remark and burst out laughing.[10]

In contrast with the official propaganda of Stalin's funeral was the death shortly after of the former Romanian King Carol II on 4 April at Estoril, Portugal. Carol II was buried in the Royal Pantheon at Sao Vicente Cathedral in Lisbon, the burial place of the kings of Portugal.[11] The death of the former king was officially unobserved in Romania and the Romanian Orthodox Church did not perform any official ceremonies in his honour. The church paid tribute only to the Soviet leader and not to the former royal political ruler of its own country.

Responses to the death of Stalin show that despite collaboration between church and regime and the special position of the Orthodox Church in Romania, the clergy were reluctant to pursue the party line. Publicly, they conformed to the regime's policy of engaging the faithful in mourning the death of the Soviet leader but privately they differed from the official line.

Orthodoxy and the new course of Romanian politics

The communists were interested in developing contact with other countries which might embrace communism and the church was a factor which could influence the perception of foreign leaders. Thus, on 22 March 1953 delegates from various Latin American countries visited the Romanian Patriarchate headquarters.[12] The group listened to the Orthodox liturgy and Bishop Teoctist showed them around the Patriarchal Palace. In line with official discourse, he indicated that their presence there was an undisputed sign that Romania enjoyed complete religious freedom.

The communists continued their policy of controlling all aspects of church life and on 5 March 1953, the Ministry of Religious Confessions registered the regulations for the organisation of monasteries, endorsed by the Holy Synod on 26 March 1953.[13] The Holy Synod was asked to propose new regulations for monasteries at its meeting on 25 February 1950; however, it took three years until they were officially approved by the ministry.[14] The main purpose of decreeing these regulations was to have stricter control of monastic life and to benefit from the production of traditional ecclesiastical crafts. Most products made in monasteries were exported and the state gained important revenue.[15] It was also hoped that by imposing strict rules for monasteries the number of religious would diminish in the future. However, instead, more people embraced monastic life, increasing the number of those who wanted to live in monasteries. In the following years some monasteries would even be accused of protecting outlaws and becoming centres of resistance against the regime. The regulations for monasteries lasted until 1959 when the regime would dramatically change its policy and impose harder conditions for those who wanted to embrace monastic life.

The collaboration of the church hierarchy with the regime was visible again at Easter, when Justinian sent a Pastoral Letter combining religious teachings with practical advice on collecting the harvest in Romania.[16] In addition, preserving the ecclesiastical tradition of international Orthodox relations, on 19 April 1953 he sent an Easter telegram to all other Orthodox patriarchs. Contact between Justinian and the other patriarchs, especially the Ecumenical Patriarch Athenagoras, indicated that although he had good relations with the regime, he was recognised by the broader Orthodox commonwealth as the Romanian leader and that his position was not threatened.

Imposing his spiritual authority abroad and supporting the Soviet policy of granting autocephaly to other Orthodox countries, on 9 May 1953, a Romanian delegation led by Justinian went to Sofia to attend the enthronement of Patriarch Kiril. The Bulgarian Orthodox Church was in conflict with the Ecumenical Patriarchate which refused to grant it the status of Patriarchate. In spite of this dispute, with the political support of the Bulgarian communist regime and the spiritual help of the Russian Orthodox Church, on 10 May Patriarch Kiril was enthroned in Saint Alexander Nevski

Cathedral. Justinian was the first among the foreign delegations to congratulate him publicly and to offer him religious gifts suggesting good connections between their churches and recognising the authority of the new Patriarchate.[17] In fostering close relations between the leaders of the Orthodox churches, on 16 May all members of foreign delegations attended the meeting of the Bulgarian Holy Synod. The leaders emphasised the desire to advance peace in the world and that the Orthodox churches should maintain permanent contact between them. Following religious ceremonials, Justinian celebrated the liturgy in the Romanian Orthodox Church in Sofia and, until 21 May, visited various Bulgarian cities which had an historical resonance in Romanian history.

On returning home, Justinian reported on his Bulgarian visit at meetings of the Holy Synod and National Clerical Assembly in Bucharest between 6 and 8 June 1953. Sharing his impressions, Justinian emphasised the propagandistic need to support the defence of peace and, following his advice, the Holy Synod sent a telegram to the World Council of Peace which showed the favourable attitude of the Romanian faithful towards the peace movement. For his ongoing support of the regime, the communist authorities decorated him with the order 'The Star of the Romanian People's Republic', first class. These meetings also marked the symbolic moment of five years under Justinian's leadership. In a special religious ceremony, *Te Deum*, he was congratulated for leading the church by Bishop Teoctist in the name of the church hierarchy while the government awarded medals to other clerics indicating their role in promoting the defence of peace.[18]

The theme of discussions in the National Clerical Assembly on 8 June 1953 showed the level of obedience of the church hierarchy towards the regime. The main point concerned the attitude of the church towards Romanians living abroad, especially the most important community in the US.[19] Four days before this meeting, on 4 July 1953, the congress of Romanians in the US condemned the appointment of Moldovan by the communist regime and approved an independent constitution organising the episcopate with around 50 parishes. The Holy Synod in Bucharest labelled as non-canonical both the new constitution and the election of Bishop Valerian Trifa as the new leader of the Episcopate,[20] declaring him as a 'traitor of the Church and the Fatherland' and pointing out that the bishops who celebrated his enthronement had been excommunicated by the Russian Patriarchate.[21] The relationship between the Romanian community in the US and the Romanian Patriarchate would be strained throughout the communist period. This was especially because the Romanian communities in both American continents and Western Europe came under the spiritual authority of the bishop in the US.

From the installation of communism, Romania suffered economic deprivation; however, the propagandistic messages of peace were aimed at alleviating these internal problems. In addition to holding meetings which praised their achievements, the communists imitated the international grandiose manifestations of other European countries dedicated to the cause of peace. In Bucharest, from 25 to 30 July 1953, the 3rd World Youth Congress was held with 1500 delegates from 106 countries and from 2 to 14 August, the 4th World Youth and Students Festival was held with 30,000 participants from 111 countries. The church hierarchy supported the regime's policy and encouraged the faithful to help the state and engage in the preparation of these events.

In addition to making public statements supporting the communist agenda, church hierarchs offered interviews to the international mass media which attended the festivities. The main reason was to prove to foreign journalists that the Romanian people enjoyed full religious freedom and that there were no restrictions on foreign representatives in meeting with the hierarchy. An example was the report of the British Legation in Bucharest concerning the experience of Mr Preston, the *Times* correspondent from Vienna. On 17 August 1953, he visited the Patriarchal Palace on a tour especially for foreign journalists and was shown around by Bishop Teoctist. According to the report, he set out a number of examples which proved that Romanians enjoyed full religious liberty. As the report stated, Bishop Teoctist

> alleged that the Church had full liberty and was able to carry out education [. . .] He claimed that there had been no changes in the constitution of the Church since the 1920s when, for the first time, secular elements had been introduced into the assembly. He said that the Church lands had been expropriated since 1864, though the Church had voluntarily handed over 30 per cent of its land to the present regime, while retaining the rest even in the cases where it exceeded 50 hectares [. . .] [In] Romania there were 7000 nuns and monks in 275 monasteries [. . .] [The Church had] 1500 candidates to priesthood and 400 students in two theological institutes [. . .] The salary of a priest was the same of that of a university professor and ranged from 600 to 2,000 lei. A bishop was paid from 1200 lei upwards and Teoctist said that he himself earned from 2000 to 4000 lei.

Further to this official and detailed data, the report describes an incident that occurred during Preston's visit which showed another face of the church, different from that of official propaganda:

> After the harangue by Teoctist a priest who was present succeeded in leading Preston aside from his interpreter on the pretext of showing him some pictures and made the usual vague remarks critical of the regime, regretting the existence of the Iron Curtain and maintaining that the only freedom was within one's self. He also spoke highly of the BBC Romanian Service broadcast, with which he appeared to be well acquainted,

but criticised Father Gâldău's religious talks [. . .][22] On the wall opposite of that carrying religious pictures and icons of the Patriarchate there were two large pictures of Gheorghiu-Dej and Groza.[23]

The fact that a priest at the Patriarchal Palace managed to express privately his opposition to the regime to a foreign journalist indicated that within the church there were divergent opinions between those who supported the regime and those who were discontent with the collaboration. Had the priest's attitude been made public, it probably would have cost him his life. His discreet comments to Mr Preston showed that in the inner circle of priests around the patriarch some hoped for a change of Romanian communism even with possible help from abroad. The priest described Father Gâldău's talks on the BBC as ineffective and worsening the position of the church regarding the regime. The priest's reference to Father Gâldău's talks indicated the lack of religious authority he had in Romania even if he used the forum of a prestigious radio station to criticise religious abuses.[24]

In addition to youth congress and festivals, the communists inaugurated the '23 August' Stadium in Bucharest, the largest sport centre in Romania, with 80,000 seats. The name of the stadium, as with so many other public places, reminded the visitor of the coup d'état or turn of politics when Romania joined the Soviet Union against fascist Germany in 1944. The stadium's opening ceremony offered another example of the regime's use of religious leaders. As a British report points out, various religious leaders were present, among them Monsignor Alexandru Cisar, the Roman Catholic archbishop of Bucharest, who was under house arrest. He was seen in the VIP lounge; however, no public mention was made of his presence there. This suggests that the regime made use of religious leaders as participants in the most important achievements to bring it credibility and also as a means of testing their personal support.[25]

Developments in international relations and the beginning of the process of détente between the great powers had an impact on Romanian politics. Gheorghiu-Dej was not a nationalist leader, interested in promoting the national characteristics of Romanian communism, as would be the case with Nicolae Ceauşescu, the country's next political leader. He was more concerned with maintaining authority and strengthening his own political position in the party while following a Stalinist model.

In August 1953 Gheorghiu-Dej gave a speech to the party on the 'new course' of Romanian politics. The new course represented a more independent Romania in the Eastern bloc. The party wanted to have its own distinct voice within the communist world while asserting close connections with the Soviet Union. In this sense the party reinterpreted Romania's history and emphasised its Slavic elements. From 13 September 1953 'excessive' Latin words were removed from usage and the country's name was change from *România* to *Romînia*. Throughout the language Romanians were forced to change 'â's to 'î's, an orthography which looks more Slavic without changing the sound or meaning of the words.[26]

After the Berlin meeting of foreign ministers from France, Great Britain, the US and the Soviet Union, from 25 January to 18 February 1954, the communist regimes in Eastern Europe began to change its compromised leaders who applied Stalinist measures. However, Romania remained one of the only countries in the region which failed to transform its party leadership and despite his associations with Stalinist politics, Gheorghiu-Dej remained in power. He realised that in order to maintain full control he had to reform the organisation of the party and eliminate any possible enemies who could compromise his position. The trial of Lucreţiu Pătrăşcanu, former communist Minister of Justice who criticised his Stalinist policy, was held between 6 and 13 April 1954. Just a few days before the Soviet Plenum officially declared a new attitude towards the West, Pătrăşcanu was sentenced to death and executed in the night of 16–17 April at Jilava prison.[27] In this way, the only serious threat to Gheorghiu-Dej was removed and Pătrăşcanu's death set an example for anyone who might oppose the leader.[28] Furthermore, a few days after Pătrăşcanu's death, on 19 April, the Plenum of the Central Committee of the RWP decided to abolish the position of general secretary of the party, replacing it with a secretariat composed of three people led by a first secretary. This position was occupied by Gheorghe Apostol who was seen as a reformist leader according to the new Soviet directives. The Organisatoric Bureau of the party was abolished while in the Political Bureau two new names were added, Alexandru Drăghici and Nicolae Ceauşescu.

The new course of Romanian politics was supported by the church. In May 1954 the official journal of the church, *Biserica Ortodoxă Română* (The Romanian Orthodox Church), dedicated a special issue to two Romanian saints, Hierarch Calinic of Cernica and Martyr Ioan Romanul.[29] By asserting the significance of two national saints, the church hierarchy wanted to induce the idea that the Romanians were a people with their own distinct spirituality within the broad Orthodox commonwealth. Although the articles were limited to brief presentations of the lives of these two saints, this publication could be seen as a form of support for a Romanian version of communism. A few years earlier the Holy Synod adopted the decision to canonise some Romanian saints, including Hierarch Calinic and Martyr Ioan, however, the communists had not fully endorsed the idea. This special issue of the most important church journal showed that the communists were aware of how the church could contribute to the new course of Romanian communism.

The Romanian Orthodox Church maintained good relations with the other Orthodox churches. Promoting communist support in its international contacts, the hierarchy indirectly claimed religious freedom. The day after more priests were awarded medals by the regime on 3 September 1954,[30] Justinian offered the title of 'Doctor in Theology' to the Russian Metropolitan Nikolai of Krutic[31] in a ceremony at the Theological Institute in Bucharest. Metropolitan Nikolai was a member of the World Council of Peace and the communist authorities arranged for him to visit the National Committee for the Defence of Peace where he met other Romanian religious leaders. Close relations with other Orthodox churches were not limited to those under communism but also with those in other countries. On 13 September 1954, Justinian welcomed Patriarch Alexander III of Antioch at Bucharest airport. Alexander III was on his way to spend a few days in Moscow on holiday with Patriarch Alexius and before his departure he celebrated the liturgy together with Justinian and offered him the highest distinction of the Antioch Church: 'The Order of Apostles Peter and Paul', first class. Both Russian and Antioch leaders were received by Petru Groza and their visits were widely publicised in the press.[32]

Tensions with the Church of England

While strengthening their relations with other Orthodox leaders, the church hierarchy took a direct stance against accusations from the West that it was infiltrated by communists and was completely subordinate to their power. The hierarchy wanted to enforce the image, particularly abroad, that it was independent of political interference and that its actions came from the popular will of the people. The most significant complaint was the letter from Patriarch Justinian to the archbishop of Canterbury, Geoffrey Francis Fisher, dated 24 September 1954. Justinian protested about the recent publication of a report on 'The Churches of Europe under the Communists Governments' which was written at the request of the archbishop in order to be presented by the Council on Foreign Relations in the Anglican Church Assembly in 1954. The tone of the patriarch's letter was abrupt and lacked basic diplomatic etiquette. The letter was addressed directly 'To His Eminence Geoffrey' rather than with the traditional ecclesiastical formula 'Beloved Brother in Christ'. Justinian objected to the content of the report and demanded that the archbishop rewrite it, threatening with a rupture between their churches if he should fail to do so:

> In case Your Eminence is not willing to take into consideration the points we have raised, we shall with much regret find ourselves forced to consider breaking off brotherly relations between our Churches.[33]

In the 1950s relations between the Orthodox and Anglican churches represented the most significant religious contacts between East and West. The only church that could influence the perception of Romanian Orthodoxy in Western Europe was the Church of England. This was especially due to close connections that had developed during the interwar period and the fact that the Anglican hierarchy was aware of the evolution of Orthodoxy through personal contacts with former and current leaders in the Eastern bloc.

The Anglican publication contained a strong political message, stating the lack of religious freedom and the difficult positions of the Orthodox churches under communism. The report had 33 pages analysing the religious situation in Russia, the Baltic States, Bulgaria, Czechoslovakia, East Germany and Romania. The religious framework in Romania was described in seven pages which condemned the abolition of the Uniate Church and the personal role of Justinian in suppressing this religious confession in 1948:

> The Romanian Patriarch is, of course, entitled to his own point of view, but there is something incongruous in the head of one Christian Church hastening to join avowed atheists and enemies of religion in attacking the head of another.[34]

The report was extremely critical of the election of Justinian, an unknown priest who was promoted with communist help to reach the position of patriarch, stating that

> The Patriarch of Romania, Justinian, was appointed to his present position as a result of the interference of the communist government in the appointments of the Orthodox Church. He was regarded by the government as the man most likely to collaborate with them in their plans.[35]

Concluding the Romanian section, the report suggested economic discrepancies between the church hierarchy and lower clergy and a contrast in their attitudes towards communism:

> Latest reports from Romania state the present position of the Orthodox Church in Romania is not good. While the bishops are said to be living in comfortable, not to say luxurious, conditions, the clergy are suffering from poverty and oppression, an unknown but considerable number being in prison.[36]

Justinian was aware that his close connections with the communists were not favourably regarded in the West. However, there were a few western clerics who supported communist ideology and preached for a possible collaboration between Christianity and communism. In his letter of protest to Geoffrey, Justinian recommended that he consult 'The Church in Rumania', a pamphlet written by Reverends Stanley Evans and John Bliss. Stanley Evans was well known for his left-wing political beliefs. He had easy access to religious leaders in the Eastern bloc and travelled to Bucharest a few times where he met the patriarch. Evans and Bliss's pamphlet was published immediately after the Anglican report, with exactly the same number of pages. It was published by the so-called British-Romanian Friendship Association and focused exclusively on the religious situation in Romania.[37] This publication was a propagandistic description of Romania portraying the patriarch as a hero: '[The] present Patriarch Justinian, then [in the interwar period] a parish priest, [. . .] fought heroically against the fascist version of Christianity and had a large support'.[38] The abolition of the Uniate Church was presented as the will of the people and the authors quoted from various speeches of Justinian. Evans and Bliss claimed that they visited Romania in 1952 without any difficulty. They 'were able to visit Churches and talk to Church people wherever they went and to bring previous impressions up to date'.[39] In conclusion it was stated that

> In general the two writers of this pamphlet would record it as their considered opinion that the Churches in Great Britain at this historic moment of world affairs have much to learn from the Churches in Romania and could only benefit from a widespread resumption of contacts with them.[40] [. . .] The truth is that the Orthodox Church in Romania today is in a flourishing condition, and if the Church of England Council on Foreign Relations doubts this it should communicate directly with the Patriarch of the Romanian Orthodox Church who would, no doubt, be only too pleased to answer their queries and find means of convincing them of the truth.[41]

On 25 November 1954 on the advice of the Foreign Office, the Archbishop of Canterbury wrote a short reply to Justinian. He stated that the responsibility for the comments expressed in the report lay entirely with the Council on Foreign Relations and did not reflect his personal opinion. In order to clarify the situation he asked Justinian to specify exactly which points he thought were incorrect.[42] Justinian received the letter. He neither replied to it nor did he break off official relations between the churches.

   At a time when in Western Europe the Romanian Orthodox Church was accused of collaboration with the communists, the hierarchy continued to develop its relations with the other Orthodox churches under communist regimes. On 11 October 1954 Justinian met Kiril, patriarch of the Bulgarian Orthodox Church, on his first official visit to Romania, thus returning the visit of the Romanian patriarch.[43] The Bulgarian delegation was welcomed at the frontier between the two countries and escorted to Bucharest. On 24 October 1954, Justinian and Kiril celebrated the liturgy together in Saint Spiridon Cathedral in Bucharest.[44] The cathedral was the same place where on 6 June 1948 Justinian was enthroned and was now declared as the Patriarchate Chapel.[45]

   While the church hierarchy enjoyed good relations with the regime, ordinary people suffered from religious oppression. Reports from Radio Free Europe stated that according to existing legislation they were allowed to profess their religion; however, in practice most people conducted religious activities secretly at home. Marriages were usually held outside the couple's hometown in churches or monasteries where they were unknown.[46] Moreover, people started to protest indirectly against the church's stance towards the authorities. A Radio Free Europe report suggests that, in 1955, at Easter people immediately left the church where Justinian was celebrating the liturgy when he began preaching about communism.[47]

Canonisation and communism

Competition for political power in the Soviet Comintern in January 1955 led to the rise of Nikita Khrushchev and the dismissal of the prime minister, Georgi Malenkov, who was replaced by Nikolai Bulganian. The Soviet Union began to adopt a more open attitude towards the West and on 15 January 1955 recognised the independence of West Germany. The integration of West Germany into NATO on 9 May 1955 and the proclamation of Austria as a sovereign state on 27 July 1955 deepened the separation between Eastern and Western blocs. These progressive tensions reached a new stage when the European communist states signed the 'Collective and Mutual Assistance Treaty' in Warsaw on 14 May 1955, what became known as the 'Warsaw Pact', in an alliance aimed at fostering military relations among communist countries vis-à-vis the threat from the West.[48]

Representatives of the US, the Soviet Union, Great Britain and France met in Geneva from 18 to 23 July 1955 marking the start of the official process of détente between the superpowers. The relaxation of international tensions had an impact on Romanian politics which wanted to emphasise that it was pursuing a particular version of communism. In July 1955, the surviving prisoners of the slave labour camp in Sighet were released. Furthermore, on 24 August the government decreed the release of all prisoners who had been sentenced for up to ten years imprisonment and who had participated with the Soviet Union in the war against Germany during the Second World War. However, this decree carefully excluded political opponents of the regime, particularly former members of Romanian governments between 6 September 1940 and 23 August 1944. In addition to these political makeovers, the government decided to reduce its military forces on 30 August 1955. The decree stated that the military forces of Romania would be diminished on 1 December 1955 by 40,000 with a further reduction of 20,000 planned for 1 September 1956.

Officially the party transformed itself; however, in fact, Gheorghiu-Dej obtained even tighter control. At the Party Plenum from 30 September to 1 October 1955, he was elected first secretary of the Central Committee of the RWP while the GNA elected Chivu Stoica as chairman of the Council of Ministers. By reassessing the party structure, Gheorghiu-Dej made sure that his own position was untouched. While the other communist bloc leaders criticised their predecessors who had applied Stalinist measures, Gheorghiu-Dej continued the same course of politics while beginning to affirm the Romanian way of constructing communism. According to his approach, communism had to be adapted for the national interests of the country and take into account its historical evolution.

If before July 1955 church discourse was mainly directed at the propaganda of defending peace, the new course of Romanian politics had an influence on its affairs as the hierarchy started to make more reference to the national past. In July 1955, Justinian laid the foundation stones for the construction of new buildings at Dealu monastery, Ghighiu monastery and Schitul Maicilor in Bucharest. The Dealu monastery had a symbolic meaning for Romanian history being the burial site of the head of Michael the Brave. References to Michael the Brave were previously made only during the consolidation of the Romanian state in the nineteenth century and in the interwar period. The church received financial subsidies for the reconstruction and building of religious building and the sudden nationalist reference to Michael the Brave could be perceived as an attempt to support the Romanian way of combining national elements with communism.[49]

At the time of the Geneva conference between the great powers, Justinian was invited to Moscow and began his journey on 16 July 1955.[50] This was his fifth trip to the Soviet Union since being elected patriarch and was dedicated to the celebration of Saint Sergii Radonezhsky on 18 July by the Russian Patriarchate.[51] On his previous visits and through inter-Orthodox contacts, Justinian had acquired an important position among the other Orthodox leaders. This was recognised on 21 July when his proposal to send a peace appeal to all Christians in the world was unanimously approved by all participants.[52] After religious celebrations Justinian visited Leningrad between 26 and 28 July, Kiev between 29 and 31 and spent a few days as the guest of Patriarch Alexius in his summer residence at the Bolshoi Fontan monastery in Odessa between 1 and 4 August.

The progressive policy of imposing a Romanian road to communism achieved a more decisive dimension in October 1955. Already in 1950 and 1951 the Holy Synod had adopted the decision to canonise some Romanian saints but public celebrations were rejected by the communists. In the new course of political affairs, the party sought further advantages of using the church and granted permission. Firstly, by allowing public manifestations, the Romanian communists attempted to differentiate themselves from the other communist regimes in the region, and especially from the Soviet Union. Secondly, the grandiose nature of these public celebrations were an ideal means to promote the propagandistic message of religious freedom in Romania, at a time when the forced unification of the Greek Catholic Church with the Orthodox Church showed signs of failure. Thirdly, inviting the representatives of foreign churches was a means of showing the openness of the regime and of consolidating Romanian Orthodox prestige on the international scene.

The celebrations canonising Romanian saints were perfectly timed with the commemoration of 70 years of autocephaly of the Romanian Orthodox

Church, 30 years after the proclamation of the Romanian Patriarchate and seven years from the 'reunification' of the Uniate Church with the Orthodox Church.[53] These three dates became slogans indicating the timely moment in canonising the first saints in the history of the Romanian Orthodox Church.[54] On 20 August 1955 Justinian sent letters to major religious leaders inviting them to participate in the festivities.[55] Orthodox churches from Russia[56], the Ecumenical Patriarchate[57], Bulgaria[58] and Greece[59] sent delegations. Yugoslavia was in tense political relations with Romania and the Serbian Orthodox Church did not send any representatives, stating in its official response that the hierarchy was preparing to travel to 'a different country'.[60] The most important delegations were from Bulgaria, which was led by the newly elected Patriarch Kiril, and for the first time after the Second World War, two representatives of Ecumenical Patriarch Athenagoras who had their religious jurisdictions in Western Europe.

The saints to be canonised were the same as those nominated in 1950 and 1951. The first group of saints had relics in Romania and the church decided to extend their veneration from their local region to the whole country.[61] In addition, the church also adopted Saint Ioan Valahul who died in Constantinople and was celebrated in Greece, but not in Romania, as a national saint. The church canonised new saints who had significant and symbolic roles both for the history of the Orthodox Church and for the country's political development after the installation of communism. Two saints, Joseph the New of Partoş, Metropolitan of Timişoara (who died on 15 August 1656) and Calinic from Cernica, Bishop of Râmnic (who died on 11 April 1868), were locally venerated and considered as saints in the tradition of the church. They belonged to the history of the church being both monks and members of the hierarchy. Two saints, Ilie Iorest and Sava Brancovici, were hierarchs from Transylvania and were perceived as church leaders who fought against the Catholic Church in the sixteenth century. Three saints, Monk Visarion Sarai, Monk Sofronie of Cioara and Oprea Miclăuş, an ordinary peasant, were considered martyrs for the Orthodox faith against the Catholic Church in the eighteenth century.[62]

The election of a peasant as a Romanian saint promoted the idea that in Romania, a dominantly rural country, any peasant could identify with the saint. His election also had a propagandistic message directed against the Catholic Church as he died preserving the Orthodox faith. The newly canonised saints were exclusively Romanians who did not have any connections with the Russian Orthodox Church; thus, the church wanted to develop its independent voice within the Orthodox commonwealth and would not be accused of being assimilated with its larger Russian counterpart. The sole foreign exception among the saints was Saint Ioan Valahul who was celebrated in Greece and not in Romania. By adopting this saint, the church showed that it was open to other Orthodox churches even if Greece was not under communism. The regime used this moment politically and the propagandistic religious literature showed a parallel between the lives of these people and 'religious freedom' achieved under communist rule. Thus, the state and the church presented the aspiration for religious rights in Transylvania in previous centuries as similar to the decision of people to reunite the Uniate Church with the Orthodox: 'There was no force that could stop this popular movement which gained its religious freedom through fight'.[63]

Public religious celebrations started on 9 October 1955 with the liturgy in Saint Spiridon Cathedral at which all foreign guests gave short speeches. After the liturgy the delegates were invited to an official lunch together with communist officials. In the afternoon the National Clerical Assembly held a special meeting attended by the church hierarchy, foreign delegates and communists who stated support for the regime's policy on religious freedom. The assembly recognised the role of the state in organising these events by sending telegrams to Groza, president of the GNA, Stoica, president of the Council of Ministers, and to the National Committee for the Defence of Peace.[64]

On 10 October, the foreign guests attended the meeting of the Holy Synod which officially proclaimed the canonisation of the Romanian saints. At the end of their assembly, Justinian gave brief biographies of the new saints. Promoting a nationalist discourse in his speech, Justinian showed the positions of these saints for the evolution of the Romanian state:

> The Orthodox faith of our people was defended by blood and sacrifices against all enemies. The nation and the Romanian Orthodox Church formed the barrier against the invasion of Turks towards the West and of Catholics towards the East. The Orthodox faith survived battles and sacrifices. Thus, our Romanian nation added its heroes, martyrs and confessors to those of Orthodoxy from all times.[65]

The celebrations continued on the following days with grandiose public manifestations in each city where the relics of Romanian saints were located or where they were active during their lives.[66]

The participation of some Orthodox hierarchs from Western Europe marked a significant moment of contact with other churches. Foreign bishops were surprised to find that Romania was the only communist country which had a Ministry of Religious Confessions and that the church was heavily subsided by the state. The regime's special interest in the church was evident from the fact that the Patriarchate had six Cadillacs at its disposal with petrol and chauffeurs' salaries paid by the state while the bishoprics were allowed three cars.[67]

On their return the clergymen described their experiences in Romania. In an official report to the Ecumenical Patriarch and the Holy Synod in Istanbul, Archbishop Athenagoras of Thyteira and Bishop Jacob of Malta presented their favourable impressions of the Romanian Orthodox Church: 'We cannot describe in words the deep and sincere respect and unbelievable love and affection of the Patriarch and the clergy as well as the most pious (without exaggeration) people of Romania toward the centre of Orthodoxy'.[68] The two clergymen did not venture to offer an explanation of the reason for canonising Romanian saints but they stated that they were overwhelmed by the piety of the people. This was especially proven on 23 October 1955 at the celebration canonising Saint Calinic, whose relics were held in Cernica monastery, when around 10,000 to 15,000 people attended the ceremony, most of whom walked 17 kilometres from Bucharest to the monastery.[69]

As their report stated, communist influence was present in every aspect of church life. Every room, corridor and workshop of every monastery had the inscription 'work is prayer', while all monasteries worked for the government producing goods for export. They had a very good impression of Justinian and pointed out that

> He is a really select prelate with humility, feeling, sweetness, infinite respect toward the Mother Church, the Ecumenical Patriarchate, and the person of the Ecumenical Patriarch, comprehension, administrative ability, strong influence on the people and the government and a really admirable organising and creative spirit. To these virtues of his are due the whole development, progress and prosperity of the Romanian Church which is really incomparable.[70]

Moreover, they found that around 500 priests were still in prison but every month five to ten priests were released on the patriarch's intervention.[71] The report had a very good impact on the Ecumenical Patriarchate who would strengthen contacts with the Romanian Orthodox Church. After Bishop James of Malta returned from the celebrations, he shared his impressions with his colleagues from the World Council of Churches in Geneva, one of whom recorded them in a note. The note reveals the discrepancies between the lower clergy and the hierarchy and the piety of ordinary people:

> The clergy are all immensely rich whereas the ordinary people are very poor indeed. The government itself produces some money for the bishops and clergy and this is made up by Church sources. The bishops get about 10,000 lei a month and the clergy between 600 and 1,000. They are amply supplied with large cars, lorries and other equipment. There are 7000 monks and nuns, including 5000 nuns nearly three times as many as before the war. These nuns are engaged in embroidery and produce articles of great value to Romanian exports. They therefore earn a great deal of money. The monks and nuns are consequently well clothed, well housed and well looked after and are better off than a great many of the ordinary people. There has in consequence been a substantial increase in the numbers of 'religious' between 25 and 35 years of age.

Figure 5.1   The canonisation of Romanian saints. The ceremony at the Patriarchal Cathedral, 1955 (courtesy of the Romanian National News Agency ROMPRES).

[. . .] The attendance of people at Church is enormous and huge crowds gathered in the square to say good-bye to the visiting delegates. Bishop James was tremendously moved, he found people kissing him all over, kissing his boots and every particle of clothing and so on. The attendance at Church is very high and indeed the religious situation in Romania is on the surface extremely healthy.[72]

Romania was the only communist state which sanctified national saints in the 1950s. The Ecumenical Patriarchate canonised only one saint in May 1955 while the Bulgarian Orthodox Church and the Russian Orthodox Church sanctified only three saints in the 1960s. The number of Romanian saints in 1955 exceeded all other canonised saints in the entire Orthodox commonwealth in the 1950s and 1960s.[73]

Despite the public ceremonies, the communist press did not comment greatly on the canonisations mainly because the regime feared a possible growth of resistance.[74] Allowing these religious manifestations and mentioning them in the main party newspapers show that the regime was interested in using the church in its claim of Romanian communism within the Eastern bloc. That the regime was aware of the impact of religion and wanted to maintain control is evident from the fact that at the time of allowing religious manifestations it was careful to eliminate any possible religious opposition. In 1955, Father Vasile Leu, who had escaped from Romania together with Father Florian Gâldău in 1948 and who criticised the regime on BBC broadcasts, was kidnapped in Austria by Soviet Intelligence and, after being imprisoned in Moscow, was transferred to Romania.[75] Radio Free Europe

Figure 5.2 The public procession during the canonisation of Romanian saints in Bucharest, 1955 (courtesy of the Romanian National News Agency ROMPRES).

reported that while in prison he was asked to collaborate with the regime and that he was even offered the Bishopric of Argeş but he refused.[76]

The communist regimes of Eastern Europe were internationally recognised on 14 December 1955 when Romania and 16 other states were offered membership of the United Nations Organisation. With this recognition and looking for a more independent voice within the communist world, the Second Congress of the RWP from 23 to 28 December 1955 in Bucharest publicly stated for the first time the Romanian path of constructing communism and the adaptation of Marxism to national tradition.[77] The congress acknowledged party control by Gheorghiu-Dej, who was re-elected first secretary of the Central Committee, while three new members, Petre Borilă, Alexandru Drăghici and Nicolae Ceauşescu, were introduced into the Political Bureau.[78]

Khrushchev's condemnation of Stalin's crimes in a secret speech on 24 February 1956 would have a domino effect on the other Eastern communist countries. The demonstrations in Poznan, Poland, in June 1956 and the riots in Budapest in October 1956 had not only political content but also religious substratum. The people in Poland asked for an end to persecution of the Roman Catholic Church and in Hungary there were demands for the release from prison of Cardinal Mindszenty.

Conclusion

The death of Stalin in March 1953 had important social and political consequences in Romania. Political uncertainties within the Soviet Union and the transformation of communist parties in the Eastern bloc led to the removal of those leaders associated with Stalinist methods. Aiming to maintain control of his party, Gheorghiu-Dej launched the concept of a Romanian road to communism. This was not intended to be a complete rupture from Moscow's hegemony but the assertion of a politics which assured continuity in Romanian leadership. In addition, the Romanian road represented the inclusion of societal factors which had an impact on how the communist party ruled; it turned more attention to improving its image in society, as otherwise it would not survive.

A significant and unprecedented gesture in the communist world was the regime's consent to celebrate the canonisation of the first Romanian saints, two months before Gheorghiu-Dej publicly asserted the Romanian road to communism at the Second Congress of the RWP in December 1955. By allowing canonisation of the first Romanian saints in the history of the church, the communists controlled the possibility of religious opposition. Although the decision to sanctify Romanian saints was purely ecclesiastical, public festivities would not have been possible without state support. After demonstrations against Soviet occupation in Poznan and Budapest in 1956, Gheorghiu-Dej feared that the Romanian road to communism would weaken his political control of the party and he returned to Stalinist methods. However, elements of it were reintroduced at the beginning of the 1960s when the regime embarked on a more distinct form of national communism.

6
Religious Diplomacy and Socialism, 1956–9

Orthodoxy and the construction of Romanian socialism

On 24 February 1956, at the Twentieth Congress of the CPSU, Nikita Khrushchev denounced Stalin's crimes in a 'secret speech' criticising the rise of 'the cult of personality' and deviation from the doctrine of 'collective leadership'.[1] Khrushchev's words came as a shock to Gheorghiu-Dej who felt that he was under attack for his authoritarian style. Gheorghiu-Dej was perceived as a Romanian Stalin and, returning from Moscow, took almost a month until he presented the directives of the Soviet Congress to the RWP. In his report to the Plenum of the Central Committee of the RWP, held from 20 to 23 March 1956, Gheorghiu-Dej omitted Khrushchev's 'secret speech' but criticised some of Stalin's political measures without fundamentally condemning his legacy. He suggested that Romanian communists had managed to remove those associated with the Stalinist cult of personality in 1952, referring indirectly to the elimination of Ana Pauker. According to his opinion, Romania witnessed a de-Stalinisation process even in 1952, before the death of Stalin, and there was no need for further political transformations. Gheorghiu-Dej's report faced public criticism from Miron Constantinescu and Iosif Chişinevschi, who had attended the Moscow Congress with him. However, their comments did not have any impact on Gheorghiu-Dej's position in the party.

The communists' political objective of promoting a Romanian road to communism led to an increase of religious references particularly during the informal visits of Western politicians to Romania on economics and sports occasions. An example was provided by the visit to Romania of Sir Wavell Wakefield, a British MP, in May 1956, following a tour of the Romanian rugby team of England and Wales in September 1955. Wakefield met top communist ministers and was invited to spend the last day of his visit, 13 May 1956, in Sinaia, an exclusive resort in the Carpathian Mountains. There he had informal discussions with various officials, including Grigore Preoteasa, minister of

Foreign Affairs, who was an old Rugby Football player and president of the Romanian Rugby Football Federation. During this meeting, the communists showed their intention of increasing trade agreements with England, and of encouraging mutual exchanges of scientists, tourists and sports events. A particular point of discussion was their sudden interest in fostering relations between the Romanian Orthodox Church and the Church of England. Showing knowledge of basic theology, Preoteasa pointed out that the dogma of their churches was similar and that they would welcome visits of Anglican clergymen to Romania.[2] As a result of these informal meetings, relations between the Orthodox and Anglican churches would substantially improve, and, a few months later, for the first time after 1948, a bishop from the Church of England travelled to Bucharest.

Discussions with Western politicians show that the regime was aware of the impact of religion and was interested in further using the church's foreign contacts in order to benefit the position of Romanian communism.[3] The regime was afraid that internal political opposition could develop with a religious dimension, similar to that in Poland and Hungary. In May 1956, leaflets were distributed in Constanţa which stated that the priests were with the people and that the end of communism was near.[4]

In order to apply socialism to the national interests of the country, the regime carried out a new census on 21 February 1956.[5] Furthermore, forging international support for Romanian communism, relations were strengthened with other communist countries which were seen as possible competitors to Moscow's hegemony. After establishing close contacts with the Chinese government in previous years, the Romanian communists sought an improvement of relations with Yugoslavia. Tito visited Romania from 23 to 26 June 1956 and Gheorghiu-Dej returned the visit from 20 to 29 October 1956. At their final meeting, the main topic of discussion was the effect of Khrushchev's speech which had led to riots in Poznan in June 1956 and in Budapest in October 1956.

Gheorghiu-Dej's visit to Belgrade was cut short by the outbreak of student and workers' manifestations in Bucharest, Cluj and Iaşi. Students were demanding the improvement of education facilities and the abolition of mandatory Russian classes from university and high school curricula, while workers asked for an increase of their monthly salaries.[6] The main difference between the Romanian demonstrations and those in the other communist states was that the Romanian students and workers did not have a defined political agenda. On the contrary, improvements to education and to the economy indirectly supported Gheorghiu-Dej's position in the party and the particular stance of Romanian communism. The regime ensured that the demonstrations would not spread and Emil Bodnăraş, minister of Military Forces, ordered the arrest of around 1200 people. In addition, foreigners were prevented from entering Romania from those countries neighbouring the regions of Moldavia and Transylvania. The number of police patrols increased and on 5 November the Central Committee of the RWP re-established the workers' guards for the 'defence' of factories. In addition, at Khrushchev's request, Gheorghiu-Dej, Emil Bodnăraş, Valter Roman and Mihai Beniuc visited Budapest, from 22 to 25 November 1956, with around 200 political activists of Hungarian origin from Transylvania in order to consolidate the reorganisation of the Hungarian communist party and of the security forces.[7]

On 6 October 1956, at the time of the Hungarian revolution, Patriarch Justinian presided over a meeting of the Holy Synod in Timişoara, the largest town near the Hungarian border, at which Saint Hierarch Iosif the New was canonised.[8] There is no direct connection between the patriarch's presence and the student demonstrations in Timişoara. However, preaching the veneration of national saints and celebrating the canonisation of a new saint near the border with Hungary could be seen as direct support for the new path of Romanian communism. On 30 October 1956, Justinian met Patriarch Vikentije of the Serbian Orthodox Church who was returning home from a trip to Moscow. Vikentije stopped off for a day in Bucharest and their short meeting would form the basis of future contacts between their churches paralleling the political improvement of relations between the Romanian and Yugoslav governments.

The brutal suppression of the Hungarian revolution had an impact on the Romanian church hierarchy. On the one hand, the use of Soviet military force and the active engagement of Romanian communists in rebuilding the Hungarian Communist Party and Security Intelligence gave signals to the clergy that their positions depended on collaboration with the regime. On the other hand, the communists wanted to assure the West that after these events, Romania continued the same policy towards the church. On 4 December 1956, Justinian was allowed to meet Frederick William Thomas Craske, the Anglican Bishop of Gibraltar, who had jurisdiction over British citizens in the Balkans. The Right Reverend Cecil Douglas Horsley, who had been the last Anglican bishop to visit Justinian in 1948, had been constantly refused a Romanian visa; the presence of the new bishop was aimed at proving that the regime had a more open attitude towards the church.

Bishop Craske visited Sofia and Bucharest and in both places was able to meet church leaders without communist escorts.[9] On his return home he wrote a report to Geoffrey Francis Fisher, the archbishop of Canterbury, detailing his discussions. The account of his meeting with Justinian offers an eyewitness image of how the church hierarchs feared pressure after the Hungarian events and the attitude of the Orthodox Church towards the Church of England.

The report stated that Bishop Craske met Justinian, accompanied by two advisors of the Patriarchate, Bishop Antim Nica and Bishop Teoctist, as well as Deacon Vintilă Popescu, professor at the Theological Institute in Bucharest, as a translator. At the start of their conversation, the patriarch declared that he wanted to resume official discussions between their churches according to the agreement reached in 1935. By directly mentioning this, Justinian showed that the Romanian Orthodox Church was the same ecclesiastical body as in the interwar period even if it was now under a different political regime. The patriarch expressed criticism of the report written by the Church of England Council on Foreign Relations in 1954, about which he had written to the Archbishop, in which the Romanian hierarchy was accused of collaboration with the communists and suppression of the Uniate Church. Justinian highlighted pages 18 and 19 of this report complaining that the quotations used were extracted from Roman Catholic sources. This claim would alleviate the harsh tone of his accusations, suggesting that responsibility for these quotations lay with Catholic hierarchs and not the Anglicans. Justinian stated that the regime did not allow further visits to Romania by Bishop Horsley of Gibraltar because they believed that, after the last time, he was responsible for securing a job for an opponent of the regime, Father Florian Gâldău, at the BBC. Justinian complained that Gâldău's talks on the BBC and on the Voice of America produced a negative image of Romania in the West which did not conform to reality.[10]

Justinian showed Bishop Craske that the church was publishing four theological journals and stated that he would welcome more publications from London. On his departure, accompanying the bishop to his car, Vintilă Popescu and Antim Nica sent their greetings to various clergymen whom they had met on their visits to England in 1935 and 1937, 20 years earlier.[11] Returning to London, Bishop Craske shared his personal impressions on the patriarch with his colleagues. A note written by one of his colleagues after his return gave his personal impression of the patriarch:

> Justinian was a person of very considerable character but not altogether a desirable one. He [the Bishop] thought that very likely during his interview there was a microphone under the table, and even if there were not, the Patriarch was deliberately talking as if there were. He [Justinian] had been seen the worse for drink at a party a week before the Bishop arrived.[12]

Bishop Craske's report conformed to the negative image of Justinian abroad, due to his enthronement with the help of the communists in 1948. This image was supported by the fact that Western religious leaders condemned the collaboration between church and regime and believed that the church hierarchy had a more political than spiritual mission. The report showed that in these initial Anglican-Orthodox contacts after 1948 both sides were affected by recent political developments in the Eastern bloc. By mentioning the speeches of Father Gâldău, Justinian wanted to be officially recognised as the leading hierarch of the Romanian Orthodox Church by other churches in the West. However, religious leaders in the West were cautious in engaging in dialogue with the Romanian hierarchs. As the British Minister in Bucharest reported

> exchanges between Churches [. . .] seemed to offer the minimum opportunity of useful observation and the maximum risk of misunderstanding and misinterpretation [. . .] The whole line of their propaganda is to represent us as evil imperialists [. . .] I think the Church of England are mistaken in thinking that the Romanian Church which recognised the Anglican Church in 1935 is in anything but an empty formal sense the Church which now exists here. The Orthodox Church seems to be the most reliable of the regime's religious instruments. But I must confess that my judgment may have been warped by Justinian hobnobbing and drinking with (and drinking more than) the Communist leaders. He has a particularly repellent aspect when a little dishevelled. I have never seen the Protestant leaders in such a situation.[13]

After the visit of Bishop Craske, Professor Theodor M. Popescu, a prominent Romanian theologian, published an article in *Ortodoxia*, one of the major church journals, on 'The Union of the Churches as a Christian Problem of Special Interest at Present'. Popescu claimed that the church was a 'body of Christian congregations which have the same organisation and which are freely connected in same faith, in love, hope and worship'.[14] The main message of the article was that Orthodox leaders were keen to engage more deeply in dialogue with the Church of England. In addition, on 26 January 1957 Patriarch Justinian sent a letter to the archbishop of Canterbury in which he emphasised that the visit of the Anglican prelate 'has opened the perspective of a closer relationship between our Churches in the future'.[15] In his letter Justinian complained that his church was sending theological journals but did not receive any from England in the last years and would be interested in continuing the exchange.[16] Justinian's comment was a sign that the Romanian clergymen wanted to be informed about the latest religious events in Western Europe directly from the source rather than via Moscow.

The church hierarchy and the communists

The composition of the church hierarchy suffered a few significant changes. Nicolae Bălan, metropolitan of Transylvania, died in 1955 and in his place the Electoral Collegium elected Professor Justin Moisescu of the Theological Institute in Bucharest on 26 February 1956.[17] The patriarch enthroned Metropolitan Justin on 15 March 1956 in Bucharest[18] and three days later offered him the pastoral staff in Sibiu. Archival sources which could shed more light on personal contacts between Justin and the communist leadership before he became metropolitan are scarce. *Securitate* reports during this period state that, as a professor, he expressed dissatisfaction with various actions of the patriarch. In Justin's opinion, the patriarch did not completely conform to the social and political reality of communism.[19] Most probably the regime thought that he could be the best possible opponent of the patriarch and that he was the hierarch who could better implement the regime's policies. The fact that Metropolitan Justin had the support of the communists to succeed Justinian became visible within one year.

On 15 September 1956 Sebastian Ruşan, metropolitan of Moldavia and Suceava, died.[20] Although Ruşan, who was a spiritual relative of Petru Groza, president of the GNA, was well known as a communist supporter, in the last years he seemed to have adopted a more anti-communist position. Radio Free Europe reported that when he was in Damascus in 1954, after reading an article from the Romanian diaspora newspaper, *Cronica Românească* (Romanian Chronicle), in which he was criticised for collaborating with the communists, Ruşan, with tears in eyes, said to the reporter, 'It is easy to judge from abroad; why not try to be in our position to see what conditions we live in. I am over 70 years old and I think it is my duty to do everything I can, even if I do not like something, to save whatever can be saved.'[21] According to church tradition, the metropolitan of Moldavia held the position of *locum tenens*, and if the patriarch died or was deposed, he would most likely be elected to this position. For this reason, Metropolitan Ruşan was considered likely to have been the next patriarch; however, his anti-communist stance made him a danger for the regime. Ruşan died in suspicious circumstances and his death opened the way for the rapid ascension of Metropolitan Justin.

After a short time as metropolitan of Transylvania, Justin was appointed by the Electoral Collegium as metropolitan of Moldavia and Suceava on 10 January 1957.[22] The Electoral Collegium was composed of 81 members of whom 80 were present on that day. It included representatives of the regime, such as Petru Groza; Gheorghe Hossu, vice-president of the Council of Ministers who represented Chivu Stoica, president of the Council of Ministers; Mihail Mujic, vice-president of the Presidium of the GNA; and Petre Constantinescu-Iaşi, minister of Religious Confessions. In the secret ballot, Justin received the maximum number of votes, 80.[23]

On 12 January 1957, the patriarch offered the pastoral staff to Metropolitan Justin in Bucharest and on the following day they went to Iaşi, where he was enthroned. The presence of communists showed the official support of the government. In his speech, Constantinescu-Iaşi referred to major national historical figures, from monk Teoctist, advisor of Stephen the Great, to Metropolitans Varlaam, Dosoftei, Iacob Stamate, Veniamin Costache and Iosif Naniescu of Moldavia. Furthermore, Constantinescu-Iaşi mentioned the former Patriarch Nicodim Munteanu who had been metropolitan of Moldavia, and other hierarchs who had held this position, such as Irineu

Mihălcescu and Justinian.[24] A few days after his enthronement, Justin showed his public allegiance to the regime, indicating the relationship between church and state in Romania at the time, in a speech to the GNA:

> The teachings of our Orthodox Church, which in its structure is favourable to progress, show us the profound moral basis on which the high principles of an advanced social life have their source. We therefore strongly support all actions undertaken by the state government for the construction of the new life of our people.[25]

The regime's support for Justin to become the successor of the patriarch was evident even from his appointment as metropolitan of Transylvania. Radio Free Europe reported that the communists offered him a villa in Filipescu Park in Bucharest and a Zim car, while the lavish party after his enthronement in Sibiu cost the extravagant sum of 50,000 lei which was paid by the state.[26] Moreover, an article titled 'Metropolitan Justin Moisescu' was published with his photograph in the newspaper *Romînia Liberă* (Free Romania) on 27 December 1956. The article gave some biographical details and stated that Justin was proposed as a deputy in the GNA for the next elections.

The fast promotion of Metropolitan Justin from professor to metropolitan of Transylvania and later of Moldavia and Suceava reflected political will through which the communists continued to impose their people in the hierarchy.[27] In the following years Metropolitan Justin clashed with the patriarch over the relationship between church and state and his attitude caused tensions within the hierarchy. In addition, Reverend Richard Wurmbrand accused Metropolitan Justin of being a member of the *Securitate*. In his statement to the US Senate, Wurmbrand stated that 'Justin Moisescu personally has denounced a group of 400 Christian peasants from his own bishopric. He has denounced the priests whom he ordained when he knew that they said some counterrevolutionary words'.[28]

The election of Justin to metropolitan of Moldavia and Suceava was perceived with suspicion in the West. In a letter from the Foreign Office to the British Legation in Bucharest, the authorities asked for continuous observation of religious affairs in Romania: 'It occurs to us, that if Justinian were to die, from excess of drink or other cause, Moisescu would become *locum tenens*'.[29] In his response, Alan Dudley, the British minister reported on 13 March 1957:

> The fact that he [Metropolitan Justin] was allowed to occupy the Chair of Theology in Bucharest for ten years shows that the regime must regard him as a sound man, with sufficient theological background to implement their religious programme convincingly. He has written articles in support of the Peace campaign and at the time of his consecration last year he promised that the clergy and faithful would 'support with all

their might the construction of the new life in our beloved Fatherland'. While he may not be personally as unprepossessing as Justinian, I think we must assume that he is a faithful servant of the regime and that he would not be where he is otherwise.[30]

Despite formal good relations between the communists and the church hierarchy, the regime wanted to diminish the role of religion in society. The new government of Chivu Stoica on 19 March 1957 reduced the Ministry of Religious Confessions to a simple Department of Religious Confessions. Petre Constantinescu-Iaşi was released from his position, and Dumitru Dogaru was appointed General Secretary, serving until 1975.

On 25 March 1957 the church was allowed to publish 'the Prayer Book' and Justinian continued to show his public support for the regime. The following day he attended the ceremonies on the anniversary of the peasants' revolt in 1907 at the Theological Institute in Bucharest and on 5 April 1957 gave a speech to the National Committee for the Defence of Peace. Moreover, on 22 April 1957, a fragment of his Pastoral Letter was read on the radio.[31]

After Justin's election to metropolitan of Moldavia and Suceava his former position became vacant. On 23 May 1957, the Electoral Collegium appointed Bishop Nicolae Colan of Cluj as the next metropolitan of Transylvania.[32] Colan had been a bishop since 1936 and was seen as a moderate figure without being actively engaged in the campaign for peace.[33] On 19 December 1957 his vacant position in the Bishopric of Cluj was filled by Teofil Herineanu, bishop of Roman and Huşi, a former Uniate priest who was promoted to the church hierarchy due to his collaboration with the regime.[34]

Looking for a way to strengthen Romanian communism, on 15 April 1957, Gheorghiu-Dej signed an agreement in Bucharest on the juridical status of the Soviet troops in Romania. The agreement stated that the troops would remain on Romanian territory until a further decision was taken, and this action was considered an important step in asserting military autonomy from Moscow. In addition, on 28 and 29 June 1957 at the Plenum of the Central Committee of the RWP, Gheorghiu-Dej acquired complete control of the party when Miron Constantinescu and Iosif Chişinevschi were excluded from their official positions and accused of conspiring against the communist leadership. From that moment until his death in 1965 Gheorghiu-Dej remained the uncontested leader of Romanian politics.

Following the party leadership, the church hierarchy sought to strengthen its relations with other communist countries within the Eastern bloc. On 2 June 1957, a church delegation led by Justinian began a visit to his counterpart in Belgrade.[35] The Romanian delegation was received on 3 June by the Federal Executive Council, the Executive Council of Serbia and the Popular Committee Presidency of Belgrade and on 5 June, Justinian had an audience with Tito. Both regimes made use of their national pasts and Justinian, together with Patriarch Viketije, visited Ljubostinja monastery built by Milica, the wife of Prince Lazar who died at the battle of Kosovo in 1389. According to communist propaganda, both Serbian and Romanian heroes who died in the Kosovo battle for preserving the Orthodox faith were buried at the monastery. Vikentije held a special religious ceremony at Ljubostinja, celebrating nine years since Justinian's enthronement as patriarch, and both of them signed a special communiqué strengthening relations between their churches.[36] Justinian visited places of historical importance and, before returning home, was received by Alexander Rankovic, chief of the Security Police, who awarded him the 'Order of the Yugoslav Flag', first class, while the other bishops received the same medal, second class.[37]

The Romanian version of communism did not lead to a rupture in relations between the church hierarchies in Bucharest and Moscow. On the contrary, religious leaders from the Soviet Union and Romania maintained contact and ensured that their churches were actively engaged in supporting the socialist development of their countries. Thus, on 10 September 1957 Patriarch Alexius of Moscow, who was going on a trip to Bulgaria, stopped for a few hours in Bucharest where he met Justinian[38] while, a few days later, on 27 September, Justinian became one of the organisers of the month long festival celebrating Romanian-Soviet friendship.

The funeral of Petru Groza

On 7 January 1958 the Romanian political leadership suffered an important loss; Petru Groza, president of the GNA, died and his funeral proved to be an occasion combining religion with communism on an unprecedented scale. Although Groza had never joined the communist party, he was one of the most important artisans of the installation of communism in Romania. He was imposed by the Soviet Union as prime minister from 1946 to 1952, when the communists appointed him as president of the GNA, a position without great political influence. Groza was a personal friend of the religious hierarchy and also masterminded political control of the Orthodox Church. He did not make secret his connections with the clergy; as a report from Radio Free Europe stated, Groza often claimed in conversations with Western officials: 'Imagine it, all my ancestors were Orthodox priests. My father was a chief priest and he sent me to study theology in Berlin and Leipzig. I later switched over to law and economics.'[39] Due to his involvement in religious affairs, Groza's death was an opportunity for the regime to promote special relations between the church and the state.

On 10 January top communist officials and foreign diplomatic representatives gathered at the Palace of the Republic, the former Royal Palace. Groza's coffin lay in what was formerly the throne room and, at 8 o'clock in the morning, Patriarch Justinian, seconded by the metropolitans of Transylvania and Moldavia and the choir of the Romanian Patriarchate, held a full religious service in the presence of Groza's family and the communist leadership.[40]

At the end of the service Romanian religious leaders held prayers remembering the life of Groza while the ceremony was broadcast and reported in detail in the press.[41]

At 10 o'clock, immediately after the religious service, the coffin was taken outside the palace and placed on a special tribune. For almost an hour Gheorghiu-Dej and officials of other communist countries used the occasion to address the crowds, praising the political achievements of Groza. An eyewitness, the British minister in Bucharest, who attended the funeral, reported his impressions on the use of religion by the communists:

> I am therefore not inclined to accept as sufficient the assumption that Dr Groza was given a religious funeral because he was a practicing member of the Romanian Church and a member of its synod, since it would have been quite practicable to have had a private religious ceremony before the public ceremonials took place. The fact that the leaders of the Romanian party and state were prepared to associate themselves closely and publicly with a service of this kind suggests that they have an attitude towards the relationship between religion and the state somewhat different from that adopted in the Soviet Union and in other satellite countries. [. . .]
>
> In the square itself, and along the route to the cemetery, the crowds were immense. Even after some two and a half hours, having gone from the centre of the town to the cemetery at the very edge of the city (where there was another short religious ceremony before the interment) the cortege was still passing along streets lined on either side by at least three or four rows of people, and every roof and gallery and tree held its spectators. Although the majority were probably by curiosity, I am inclined to believe that the size of the crowds was to some extent a testimonial to the fact that Dr Groza had held the interest and attention of the people and had attained a certain degree of personal popularity, which distinguished him from the rest of the ruling group.[42]

Groza was allowed a religious funeral mainly at the request of his family and perhaps of his own wish. However, it was unusual for a member of the Romanian political leadership to be offered a religious burial as most were cremated without a religious service. Since the start of the communist regime in Romania, many communists were expelled from the party simply for being seen attending religious ceremonies. In addition, the usual funeral procedure was that the deceased was transported in a red truck and cremated without any religious reference. Groza would seem to have been permitted a special procession mainly because the regime was interested in presenting a different image of Romania abroad, while internally the regime used Groza's popularity to assert strongly the Romanian version of socialism. The communists made sure that his absence would not affect the Romanian political leadership and on 11 January 1958, the day after his burial, the GNA elected Ion Gheorghe Maurer as president of the GNA Presidium.

Anglican-Orthodox contacts

Following the propagandistic message of 'religious freedom' in Romania, church hierarchs continued to meet their foreign counterparts, being particularly interested in the development of relations with the Church of England. On 8 April 1958, Justinian received Canon Herbert M. Waddams from the Foreign Affairs Service of the Anglican Church who brought an Easter message from the archbishop of Canterbury.[43] After his return from Bucharest, Waddams wrote a report detailing his meeting with Justinian. The report is extremely suggestive, as after his visit, the perception of Justinian in the West would start to change. The report shows the degree of relative independence which the regime allowed the church and the position of Justinian within the hierarchy.

The report mentions that Waddams was welcomed at the Patriarchal Palace by Father Ioan Cazacu, Patriarchate secretary, and he met Deacon Vintilă Popescu by chance. Popescu was a professor at the Theological Institute in Bucharest and a close friend of Justinian, having studied together at the same theological seminary; he was married to an Englishwoman and his son was in England. Justinian was accompanied at the interview by Father Gheorghe Soare and Father Traian Ghica, advisors in the Romanian Patriarchate; Father Ioan Gagiu, director of Patriarchate Administration; Father Cazacu and Deacon Popescu. Father Cazacu was the only one who took notes and only the patriarch participated in discussions.

Every ten years the Church of England organised a conference at Lambeth Palace analysing the evolution of the church. An important point on the conference agenda was relations with other churches, especially with the Orthodox. The archbishop was not personally very knowledgeable about Orthodoxy; however, some members of the Council of Foreign Relations were particularly interested in monitoring the Eastern churches and sought an extension of relations between their churches.[44] For this reason, the Church of England sent invitations to all Orthodox churches under communism to attend the Lambeth Conference mainly because it was the only church at that time which could oppose Moscow's influence. The Church of England was the only church that could summon religious leaders from East and West, as the Ecumenical Patriarchate still lacked authority among all Orthodox leaders. The decision to invite Orthodox leaders was also encouraged by the fact that the patriarchs of Alexandria and Serbia indicated that they would like to send representatives if they were invited.[45] In addition, the regime in Romania was eager to use the Orthodox Church as a means of promoting closer contacts with Britain and the Romanian Minister in London suggested to some Anglican hierarchs as early as 1957 that Patriarch Justinian should be invited.[46]

At the start of his conversation with the patriarch, Canon Waddams stated that the archbishop hoped that Justinian would come to London to attend the Lambeth Conference scheduled for June 1958. Justinian replied that the

Holy Synod had already discussed the matter and that he accepted the invitation. He pointed out that in case he would not be able to attend he would appoint someone else. An Anglican-Orthodox conference[47] had been held in Moscow in July 1956 and Waddams brought a copy of the resolutions of the meeting. Justinian thanked him for the copy and emphasised the benefit of ecumenical contacts between East and West. He spoke favourably of the position of the Orthodox Church of Greece in the World Council of Churches and suggested that religious leaders should follow the example of political leaders and meet more often in order to understand each other better. Justinian pointed out that the position of the Ecumenical Patriarch was extremely difficult as the other Orthodox leaders did not trust the Turks. He thought that further dialogue should be held at Mount Athos which was under the spiritual jurisdiction of the Ecumenical Patriarchate rather than in Turkey. At the end of their discussions, Waddams suggested an exchange of students. Justinian welcomed the proposal and pointed out that they already sent a student to India, Andrei Scrima, and they were expecting someone to arrive in Romania. Justinian thanked Canon Waddams for the publications that he brought and asked him to send more books and journals.

After meeting Justinian, Canon Waddams was shown around the Patriarchal Palace by Popescu who mentioned that he translated into Romanian all documents regarding official dialogue between the Orthodox and the Anglican churches in the 1930s. Waddams offered to drive Popescu home where he met his British wife and they had a private discussion on the regime in Romania. According to Waddams's report, Popescu stated that

> the Church was paid by the state but that the people recognised that the priests were the only people who could not be removed at the whim of the regime, and that although there were a number of things which the Church could not do, yet on the whole they were left alone if they did not have any connection with politics. He said that the people had come closer to the Church under this regime and there were more people going to Church than before. The economic situation of the clergy was not too bad as their salaries were supplemented by payment from their parishioners although there were some very poor parishes.

Concluding his report, Canon Waddams presented his personal opinion of the patriarch, stating that

> My impressions of the Patriarch were more favourable than I had expected. He reminded me somewhat of the Patriarch of Moscow and had a certain crafty element about his look. But he appeared more at ease than the Patriarch of Moscow has ever appeared in interviews when I have been present, and he was also less equivocal and more outspoken. He did not think it necessary to mention the subject of peace during the

interview, though he gave me a book in Romanian about it, and in general he gave the impression of being in command. Whatever his attitude to the regime I have the impression from this interview, and from previous information, that he is genuinely interested in promoting the good of the Church so far as its contacts with Christians outside the country are concerned. The whole atmosphere of the Patriarchate struck me as less furtive than in other Patriarchates which I have visited, and the other Churchmen besides the Patriarch seemed more sure of themselves.[48]

After meeting Waddams, Justinian's image started to improve among religious leaders in the West who believed that he was genuine in fighting for the rights of his church under communism. In the following years sympathy and support from the West towards the church hierarchy would increase. In addition, Justinian would begin to turn more to the West both as a means of asserting the independent politics of the Romanian regime and of strengthening his personal position in the church. Closer contact between their churches resulted in sending more publications from London[49] while the church journal *Ortodoxia* published a special issue titled 'The Anglican-Orthodox Relationship, Introduction to the Unity of Churches'.[50]

The dialogue between the Romanian Orthodox Church and the Church of England had an impact on a broader scale in the Orthodox commonwealth. On 9 May 1958, Justinian went on his seventh trip since his enthronement as patriarch to Moscow attending the religious celebrations of 40 years since the re-establishment of the Russian Patriarchate[51] and ten years since the 1948 Pan-Orthodox Conference at which the Orthodox churches condemned the West.[52] Representatives of all Orthodox churches with the exception of Cyprus and Jerusalem were present. On 13 May all of the religious leaders met at Troitse-Sergiyeva Lavra and on this occasion publicly declared that their churches were ready to engage in more contact with other churches in the ecumenical movement, thus implying that the attitude of the Romanian Orthodox Church was in line with the general perception in the Eastern bloc. At the end of their meeting, on 19 May, the leaders signed an appeal for peace addressed to all Christians in the world.

The church hierarchy deepened its contact with other Orthodox leaders and after the religious festivities in Moscow, Justinian, together with Patriarch Kiril of the Bulgarian Orthodox Church, went for a short visit to Yerevan, from 25 May to 29 May 1958, and met the Armenian Catholicos-Patriarch Vaszgen I who was born and studied in Romania. The Romanian Orthodox Church took advantage of the Moscow meeting to invite Patriarch Christopher of Alexandria to Bucharest on his way home. The last visit to Romania of the patriarch of Alexandria had taken place two centuries earlier[53] and the regime reported the visit in its newspapers because it was interested in deepening collaboration with other countries in the Middle East.[54] Patriarch Christopher arrived in Bucharest on 30 May and his presence was another sign of 'religious freedom', as, according to the propaganda, only during communism had the church acquired the privilege of such a visit.[55]

The visit coincided with the celebration of ten years since Justinian's election and, on 6 June 1958, both patriarchs attended special ceremonies marking the end of renovations of the Patriarchal Palace. The palace had been significantly transformed and now had decorations that incorporated themes of Romanian history such as sculptures in wood representing monks working in agriculture and priests who had played a role in the national independence of Romania. On this special occasion, Justinian offered the medal 'The Jubilee Patriarchal Cross' (*Crucea Patriarhală Jubiliară*) to members of Christopher's delegation and to some Romanian clergy. In the following days, Justinian opened new workshops and a new press of the Biblical Institute of Orthodox Mission, showing how the church benefited from its collaboration with the communist regime. A few days later, on 8 June, Justinian celebrated ten years of his enthronement and 75 years of the Theological Institute in Bucharest at which Patriarch Christopher was joined by a Russian delegation led by Bishop Pavel of Perm.[56] The presence of Russian and Egyptian delegations showed that Justinian had an important position in the Orthodox commonwealth.

The use of religion in communist propaganda extended to all countries in the communist bloc, including not only Orthodox churches but also other confessions. Thus, leaders of the Evangelical churches of Czechoslovakia, Hungary, Romania, Poland, the Soviet Union, the German Democratic Republic, the Federal Republic of Germany, the Union of South Africa and representatives of the Orthodox churches of the Soviet Union, Bulgaria and Czechoslovakia established the so-called 'Prague Conference' in their meeting from 1 to 3 June 1958. The Conference represented the communist interpretation of ecumenism according to the decisions made in Moscow a few months earlier. The establishment of the Prague Conference was meant to indicate the collaboration between churches in an ecumenical movement within the communist countries, opposed to other Christian movements in Western Europe. At first the Romanian Orthodox Church sent only an observer, Father Milan Seşan, from the Theological Institute in Bucharest who attended a session of the Conference in Debrecen, Hungary, between 30 October and 2 November 1958.[57] In the following years, the Prague Conference would become the most important ecumenical movement in the Eastern bloc with regular meetings between religious leaders mainly dedicated to the theme of peace.

After Canon Waddams met Patriarch Justinian in April 1958, Archbishop Fischer sent a letter of invitation to Justinian in May 1958 inviting him to come personally or to send a representative to the Lambeth Conference. In June 1958 Justinian replied that he would send Metropolitan Justin of Moldavia and Suceava and Father Alexandru Ionescu, vicar of the Archbishopric of Bucharest. Justinian did not go to London in person because he would have been the only head of an Orthodox Church at the conference as all other patriarchs sent only representatives. The regime ensured that the church was represented by clergy that it trusted, as Metropolitan Justin was one of the most influential hierarchs and Father Ionescu was well known for his active engagement in the peace campaign movement. Justin and Ionescu arrived in London on 27 June 1958; on 30 June they met Archbishop Fischer and gave him a letter from Justinian.[58] The Lambeth Conference gathered Orthodox clergy from Constantinople, Russia, Romania and Bulgaria who met representatives of Anglican, Old Catholic and Protestant churches from Western Europe.

Metropolitan Justin's involvement in the debates was limited. He was not fluent in English, speaking only in French, and generally he did not play a significant role in the discussions. After informal meetings with various church leaders, he gave his acknowledgement for being invited and representing the Romanian Orthodox Church, stating in one of his few public interventions:

> Your Grace, my Lords, my dear brethren in Jesus Christ. Our souls are filled with a holy joy that we have the opportunity of being here among you [. . .] The Orthodox Church of Romania had walked along this road of good relations with the Anglican Church now for several decades. We are sure that this road of good relations is the way of love and the way of love is the way of God.[59]

The Romanian delegation returned on 15 July 1958. On 10 August, the archbishop sent a letter and the report of the conference to Justinian, suggesting closer relations between their churches.

Persecution and survival

The course of Romanian politics within the communist bloc reached a new stage in June and July 1958 when the Soviet troops, which had been stationed in Romania for 14 years since the end of the Second World War, were withdrawn. The departure of the Soviet troops led to fears among the communists that they would face a growth of anti-communist sentiment. In order to prevent resistance, intellectuals and anyone suspected of opposition were arrested. On 21 July 1958 the Penal Code was amended to include punishment by imprisonment for those who insulted insignia, the national symbols of other countries, the government or the Army. It also decreed capital punishment for those who attempted to change the communist system.

Communist pressure increased on the church. While officially continuing to support the regime, an example of which was publication of the fourth volume of 'Social Apostolate. Serving the Church and the Fatherland', the church witnessed one of its most important arrests. In August 1958, a number of intellectuals and theologians, who were part of the so-called group *Rugul aprins* (the Burning Bush), were arrested, accused of having mystical meetings at the Antim monastery.[60] The communists stated that behind these mystical discussions they were promoting a dangerous attitude towards the state order. In the following months, important theologians, such as Dumitru Stăniloae[61] and Theodor M. Popescu[62] were arrested and sentenced, accused of mysticism. The regime even arrested Dr Solomon, Justinian's son-in-law, but on his intervention he was released before the trial.[63] The arrests showed the fear of the regime that the youth could be influenced by monastic life which would eventually lead to oppositional activity.[64]

While increasing pressure on the hierarchy, the regime continued its propagandistic control of the church. On 21 October 1958, at the commemoration of ten years of the unification of the Uniate Church with the Orthodox Church, Justinian gave a speech in Alba Iulia. The liturgy was celebrated with great pomp by Metropolitan Nicolae of Transylvania, Bishops Teofil of Cluj, Valerian of Oradea and six clerics. The celebrations were orchestrated to combine politics with religion. At the end of the religious service, Dr Vasile Aştileanu, a well known Uniate priest who had refused unification since 1948, publicly addressed the crowds in the name of himself and 16 other Uniate priests. He declared that they had decided to 'return' to the Romanian Orthodox Church and asked for forgiveness from the hierarchy and the faithful for opposing unification until then. Aştileanu's gesture showed that the Orthodox Church still faced great problems in assimilating the Uniate Church and that this type of obedience mainly came through the support of the *Securitate*. After the religious service, Justinian distinguished some clergy with the Patriarchal Cross and visited the Museum of Reintegration where he signed the Golden Book.[65]

General relations between church and state in 1958 have been summarised by Alan Dudley, the British Minister, at the end of his term in Bucharest, in a confidential report addressed to the Foreign Office on 1 January 1959:

> During a large part of my time here I have had the feeling that the regime was fairly satisfied with the present state of religious affairs and with its relationship with Churches, which on the whole have acquiesced in the distribution of its propaganda, not only in connexion with the peace campaign, but also in support of other state activities. The large part played by the leaders of the Orthodox Church and the leaders of the other sects in Groza's funeral a year ago seemed to confirm this opinion, and to suggest, by the presence of the whole Politburo at the religious service, that the situation of the Churches here was very different from that in the Soviet Union and in some of the other communist countries. Within the last six months I have had the impression that this situation was beginning gradually to change, and that Romanian communism was beginning to take on a more active anti-religious aspect.[66]

Dudley suggested a new face of Justinian, depicting him as a defender of the church in front of the communists and pointing out that he was facing great difficulties within the hierarchy. Being vulnerable when his son-in-law was arrested and facing an internecine ecclesiastical struggle for power, Justinian's position was threatened. As the minister stated,

I wrote [. . .] about rumours of trouble inside the Orthodox Church. Since then the rumours have continued to develop. The Italian Minister has heard from a source he considers to be reliable that during the recent meeting of the Synod [. . .] Justin Moisescu, the Metropolitan of Iaşi, made a violent attack on the Patriarch for his wrong conception of the relationship between the Church and the state and it is also reported that he has succeeded in arranging the transfer of some important part of the theological seminary from Bucharest to Iaşi. Whether this merely represents a struggle for power within the Church or is part of the development imposed by the regime I do not know, but since it is connected with increasing rumours of arrests of priests the tendency is to believe that the communists are behind it. There are also stories that the workshops run by the convents are to be closed down; my impression is that it was Justinian who arranged for their continuance on the ground that they not only allowed the convents to make a living but that they were a definite economic asset to the state.[67]

The increase of communist interference in the church reached such a level that at the end of 1958 more rumours circulated in the West about the start of a systemised persecution of the Orthodox Church and even the house arrest of Justinian.[68] On 13 January 1959, Visser't Hooft, general secretary of the World Council of Churches, sent a confidential memorandum to religious leaders in the West, sharing with them what he had been informed by a source in Romania. According to his memorandum, 145 Orthodox priests and monks were arrested; the patriarch was under house arrest and was not allowed to communicate with anybody; Justinian's archdeacon, physician and son-in-law were also under house arrest; Justinian was publicly accused by Metropolitan Justin of being non-patriotic and of spending two million lei on 'self-advertisement'; the police claimed that they found a radio transmitter in the cellar of the Patriarchate; all workshops of the church were closed in an attempt to cut off the church from any financial subsidies and discourage people from entering monasteries. It was also predicted that Justinian would be soon dethroned and Metropolitan Justin would be raised as the new patriarch.[69]

Rumours of the patriarch's arrest were published in the Romanian diaspora newspaper *Lumea noastră* (Our World) on 9 December 1958 in a short article stating that 'Patriarch Justinian and 50 priests were arrested'. Moreover, a pamphlet of the World Council of Churches published in January 1959 asserted that Justinian was under house arrest and that he was taken to an unknown destination outside Bucharest where he was kept under police surveillance.[70] The regime aimed to disprove the rumours and published a telegram from Justinian to Chivu Stoica in *Romînia Liberă* of 14 December 1958 suggesting that it remained on good official terms with the hierarchy.

The rumours in the West have never been confirmed. However, their main basis was due to increasing pressure of the regime on the church and the regime's intention to reduce the number of religious in monasteries and of students in theological seminaries. Justinian's opposition to these measures presented a good opportunity for other hierarchs to secure better positions for themselves in the eyes of the regime. The regime planned to reduce the number of priests from 12,000 to 7000. The patriarch suggested retaining the same number of priests but reducing financial expenditure so that the state would only pay for 7000. His proposal was rejected.[71] Radio Free Europe reported that the regime was not happy with Justinian because, after their release, priests who had been imprisoned for anti-communist activities spent a short period of time in monasteries and were re-employed in their former parishes. In addition, the *Securitate* complained that monasteries accepted many former political detainees as monks and could thus become centres of resistance.[72]

The government modified the decree law of 4 August 1948 issuing the famous Decree 410 on 28 October 1959 which now stated that monks entering monasteries should be at least 55 years old and nuns 50 years old. They should renounce any pension or salary from the state and should not be married. Theological students were allowed to enter monasteries at any age after completing military conscription. The decree was the official means of closing monasteries and expelling religious on the largest scale in Romania after the Second World War.[73]

A Radio Free Europe report details the procedure used to persuade monks to leave monasteries. The person in charge was Vasilescu, a former officer of the *Securitate* who worked as General Inspector at the Department of Religious Confessions. With the support of the party secretary of each particular county, he went to the monastery which was supposed to be closed or have its religious reduced. Vasilescu read the names of the monks who were selected to leave and demanded to see them the following day. As the report states, his usual argument was, 'What are you doing with these dirty beards and hair? In these modern and progressive times are you still living in monasteries when you should work for the construction of socialism? By tomorrow morning I want to see that all of you have cut your hair and beards and then come to receive civil clothes'.[74]

The new law affected the constitution of the church. Many monasteries were closed and almost one third of monks and nuns were expelled. Empty monasteries were given different functions, most of them being transformed into museums. A calendar published in 1959 by the Orthodox Church of Greece stated that the composition of the Romanian Orthodox Church, according to official data received from the hierarchy, was as follows: 8326 parishes, 10,153 priests, 182 monasteries, 1657 monks, 4440 nuns and 11,506,217 faithful; a Theological Institute in Bucharest had 290 students and one in Sibiu 338 students; 6 seminaries for monks had 759 pupils and 2 seminaries 138 nuns. The same type of calendar for 1961 did not give any such information as the church officials did not respond to their request.[75]

State control of religion spanned not only to the church's activities but also to education. At Easter, school teachers and university professors were warned not to attend the religious ceremonials.[76] The regime organised dance evenings and excursions exactly over the Easter period and expelled 350 students from universities and high schools who were identified as churchgoers. The clergy's sermons were censored to refer only to morality and not to 'belief', which was described as superstition.[77] At the same time as these concrete measures, propaganda targeted young people. The Romanian youth newspaper, *Scînteia Tineretului* (The Star of Youth), mocked the lives of saints and claimed that those who continued to believe in religion were living in the Middle Ages.[78] In addition, the communist Society for the Dissemination of Science and Culture began to publish brochures with strong anti-religious contents: 'The Universe as seen by Science and Culture', 'The Origins and Evolution of Man', 'Religion or Science?', 'Superstitions are Harmful for the Health',[79] while the Sahia Film Studio made the first film aimed at combating mysticism titled *Din trecutul pămîntului* (About the Past of the Land), written and directed by Ion Bostan, a member of the Romanian Academy.[80]

While closing a number of monasteries down, the regime wanted to assure the West that its policy did not represent a general persecution of the church but more the implementation of socialisation in the country. For this reason, the regime allowed the visit of Reverend John R. Satterthwaite, Assistant Secretary of the Church of England Council on Foreign Relations, to Romania for one week from 3 March 1959.[81] He received his visa very quickly, surprising the British authorities. His visit was the most significant religious encounter between a Western religious leader and the Orthodox hierarchy as he was allowed to meet Justinian and personally witness the fact that the patriarch was not under house arrest. On 7 March, for one hour and a half, Satterthwaite had a meeting with Justinian at which they discussed the relationship between the Orthodox and Anglican churches in the light of the previous Lambeth Conference. After his return from Bucharest, Satterthwaite reported in detail his interview and his personal impression of Justinian's position in the church.

The report states that Satterthwaite presented an Easter letter from the archbishop of Canterbury, and greetings from the bishop of Gibraltar and Canon Waddams whom Justinian met a few years earlier. The patriarch pointed out that he had wanted to participate in person in the Lambeth

Conference. He was interested in attending with a professor, but in the end he decided to send some representatives because he would have been the only patriarch. Justinian emphasised that their churches should continue their discussions on the same line as in 1935, indicating that more contacts in the future would lead to better understanding between them. Satterthwaite pointed out that he hoped to renew Joint Doctrinal Discussions between Anglican and other Orthodox churches, giving as examples the meetings between the Ecumenical Patriarchate and the Church of England. Justinian showed impatience and argued that the Ecumenical Patriarch was at that time 'a prisoner of the Turks, without any effective say in Orthodoxy'.[82] Asked what he thought about the pope's decision to summon the Second Vatican Council, Justinian 'brought forth a tirade against the Vatican and all things Roman. He said the Roman Church was not interested in the real message of the Gospel but was in fact political menace with its evil machinations stretching into every avenue.'[83] Justinian stated that he would like to receive the archbishop of Canterbury in Romania. As Satterthwaite recounts, 'I thought at first he was joking, but he said a return visit was really necessary as Patriarch Miron Cristea had already visited England'.[84] When Satterthwaite began to present the Anglican position in the World Council of Churches, the patriarch became more relaxed and made an unusual comment. As Satterthwaite reports:

> I explained that Anglicans hoped for a much greater contribution from Orthodoxy as without the Eastern Churches, the WCC would be hopelessly one-sided. This in turn brought forth from Justinian a torrent of abuse about the WCC and how it had allowed itself to be caught with American imperialism. To my astonishment he then said, 'I suppose you did not expect to find me here, and thought that I had been arrested!' I replied immediately that that was impossible when I had applied for an audience on my arrival in Bucharest, and added that I had not come for the purpose of ecclesiastical espionage, but to increase Anglo-Orthodox relations. This produced a great laugh from Justinian but he went on to say some very strong words about Visser't Hooft. He said that his former admiration for him had waned and that he was allowing himself to be the 'foolish centre' of intrigues and was passing on rumours which did nothing but discredit to the Orthodox [. . .]
>
> Before taking my leave, the Patriarch gave me a large book, and an inscribed photograph of himself, with two pieces of tapestry. He said these would show that he had actually met me, and that he was not under arrest! He became very boisterous and good humoured, and insisted on my seeing the Patriarchate. This I was shown in great detail, including his own private apartment.[85]

In Bucharest Satterthwaite also met Bishop Teoctist, Metropolitan Justin, the Armenian bishop, and visited the Theological Institute and two convents outside the capital.[86] On 8 March 1959, on the recommendation of Canon Waddams he met Deacon Popescu, and after he returned to London, Satterthwaite transcribed their conversations. Popescu was seen as an intimate friend of Justinian and Satterthwaite was interested in knowing more about the position of the patriarch in recent months. As Satterthwaite states:

> I asked first about the rumours that Justinian had been under house arrest and that a large number (in the hundreds) of priests had also been imprisoned. Popescu said that the Patriarch had certainly not been under arrest. He sees him frequently and hears of things which are much less important than arrests! Popescu did say that it was obvious several months ago that the Patriarch was having trouble with the government, and that Moisescu was 'cashing in' on it. Now, however, he says that 'Justin's shares have fallen very low'. A number of priests have without doubt been arrested, but certainly not even a hundred, and not as the result of some change of policy in the last few months. Of those priests most of them have been mixed up to some extent with political activities but one doesn't have to be a priest to get shut away for anything which is even remotely connected with political subversive activity [. . .]
>
> Finally I asked Popescu what was his frank opinion of Justinian. He said that many have been horrified by his early apparent selling out to the communists. He was known and referred to as 'that red priest'. Now, however, the faithful realise that the Church's strength and present position is entirely due to his handling the situation, and wonder how much longer it can continue.[87]

After Satterthwaite's visit, on 9 March 1959, the patriarch sent a letter to the archbishop of Canterbury thanking him for sending his representative. Justinian acknowledged that in the last months he received 51 books from England as a gift after Waddams's meeting the previous year.[88] Satterthwaite maintained contact with the Romanian hierarchy and on 28 July 1959 replied to Justinian informing him that the Church of England in fact sent 109 books and he hoped that the missing ones would arrive soon.[89]

Satterthwaite's account of his experiences in Romania had a positive impact on Justinian's image abroad. Canon Waddams wrote a letter to Visser't Hooft on 16 March 1959 indicating that rumours about Justinian's imprisonment were untrue.[90] After the visit, Anglican leaders and religious leaders from the World Council of Churches would start to have more sympathy for Justinian and the church hierarchy in Romania. In the following years, the Romanian Orthodox Church was encouraged to become a member of the World Council of Churches and the regime permitted more meetings with Western churches. The regime was interested in promoting the idea that religious confessions enjoyed full religious freedom and that the contacts of the Orthodox Church abroad could be useful in fostering relations with governments in the West. The attitude of the regime was visible a few months later when, looking for closer economic contact with the West, a delegation was sent, led by Alexandru Bârlădeanu, from 20 June to 14 August 1959, to Switzerland, France, England, Belgium, Holland and Italy to develop trade agreements with these countries.

Political pressure on the hierarchy did not bring unexpected changes as the church continued to be easily used in the propagandistic theme of peace.[91] On 4 August 1959 Justinian followed the same pattern of meeting other Orthodox leaders and greeted Patriarch Theodosius of Antioch who stopped off briefly in Bucharest on his way home from Moscow.[92] For his support of the regime, on 12 August 1959 Justinian was decorated with the medal 'Order 23 August', second class, a lower ranking than he had previously received. Communist pressure on the church continued and although Justinian retained his position, he continued to face the challenge of other clergymen. A report from the British Legation on 14 December 1959 stated the general situation of the church in Romania during this period and the difficult position of Justinian:

> The Patriarch attended the celebration at the Soviet Embassy of the anniversary of the Russian revolution on November 7, and made a speech about peace and peaceful coexistence to a separate gathering consisting of leading Romanians and the heads of foreign diplomatic missions. Gheorghiu-Dej and President Maurer snubbed the Patriarch in front of these people by interrupting him and correcting what he said, and the Patriarch, who was clearly nervous, soon fell silent. On the other hand, the Metropolitan Justin who accepted an invitation to a reception at HM Minister's house a few days later was in good spirits and appeared full of self-confidence.
>
> Such evidence as is available, which is by no means conclusive, indicates therefore that there is conflict between the Patriarch and the metropolitan and that the latter feels himself well-placed as a result of his greater willingness to fall in with the demands on the Church by the Romanian authorities.[93]

The withdrawal of Soviet troops, economic progress and closer contact with the West were signs that the regime was achieving progress in implementing its political programme. However, while its policies seemed to have a favourable effect, the general perception of those who travelled to Romania at the end of the 1950s was that Romanians were still facing one of the most oppressive regimes in the Eastern bloc. An eyewitness to Romania in those months was Mr Rotunda, Head of the Romanian desk of the Voice of America, who was offered a visa to Romania in October 1959 after three

Romanian journalists were allowed to travel to the US. In a conversation with BBC staff on his way home from Bucharest, Rotunda stated that Romania had made economic progress as new flats were everywhere but people continued to be 'shabbily dressed' and noted what little impact political leaders in exile had on ordinary people after years of communism. When Rotunda left Bucharest he stopped in Belgrade, and even there, in another communist country, saw a difference in people's attitude and felt like 'coming out of darkness into light'.[94]

Conclusion

After the canonisation of Romanian saints in 1955, the communists sought better control of the hierarchy and to use the church more effectively in propaganda. In order to encourage internal clashes for ecclesiastical power, the communists ensured the appointment of people who were communist sympathisers, most notably Metropolitan Justin, who, soon after his enthronement, become a challenge to Patriarch Justinian. In its construction of Romanian socialism, the regime aimed at reducing the influence of the church in society, reaching a climax of religious persecution with the issuing of new regulations for monasteries in 1959. The church hierarchs were encouraged to have more meetings with their foreign counterparts, especially in the Church of England, which, for its part, was interested in supporting the church in Romania and in using its influence in order to diminish religious persecution in the Eastern bloc. However, the Church of England did not have a coherent programme in the 1950s for supporting religious resistance in Romania.

The interest of the Church of England in Romania was especially due to the Romanian stance within the communist bloc and the signs that Romania had begun to pursue an independent position from Moscow. After the visits of Canon Waddam and Reverend Satherthwaite to Bucharest, the perception of the Anglican hierarchy towards Justinian changed and began to be more sympathetic. Furthermore, facing religious persecution, Justinian showed a different image of his church. In his meetings with Anglican hierarchs he mocked the idea that he was under house arrest, thus implying that he was aware of the political pressure of the church. There is still no clear evidence if he was ever under arrest but, by speaking frankly, he gained the sympathy of the Anglican leaders. After the visits of Anglican hierarchs to Romania, the image of Patriarch Justinian would change in the West. He was no longer perceived as the 'Red Patriarch' but as a leader who was genuinely interested in preserving the church from communism. The Church of England was instrumental in promoting this new image; in the following years more clergymen would visit the country and the Romanian Orthodox Church would be invited to join the World Council of Churches.

The regime allowed relations to develop between Orthodox hierarchs and foreign religious leaders because this promoted the image of a Romanian version of communism. Furthermore, through these contacts it hoped for an increase of economic and political contacts between Romania and the West. In the 1960s, the regime would continue to assert a distinctively independent voice in the communist bloc, imposing Romanian national communism.

7

Between Moscow and London: Romanian Orthodoxy and National Communism, 1960–5

Church and communism

The particular position of the Romanian Orthodox Church in the Eastern bloc was depicted by Reverend Francis House in a confidential report written to the World Council of Churches after travelling to Bulgaria, Romania and the Soviet Union in 1961. According to his report, the Orthodox churches performed three similar purposes in these countries: to diminish or eliminate any opposition to the regimes; to preserve 'religious sentiments' of their population; and to support the governmental 'peace' propaganda and political triumph of communism. While the churches had the same roles, their regimes followed different attitudes towards religion. The church in the Soviet Union was financially separated from the state and the salaries of the clergy were only paid by their congregations. In Bulgaria the clergy received salaries from the state but which covered only the subsistence level. Romania enjoyed the best financial position as salaries were equivalent to the average wage of the population. The Romanian Orthodox Church distinguished itself in the communist bloc by receiving financial subsidies for restoring and building around 30 new churches since the end of the Second World War, with the state covering almost one third of the expenses. It was the only church in Eastern Europe which published significant religious and liturgical works, and even translated some books from the West. In addition, compared to other patriarchs, Justinian was a more visible figure in the public arena and a guest at major diplomatic receptions.[1]

The Romanian Orthodox Church was financially supported by the state mainly because by controlling it the regime managed to suppress opposition. If in the 1950s the main concern of the regime was having direct power over the hierarchy, at the beginning of the 1960s the church had been assigned an international mission. By taking advantage of the church's international religious contacts, the regime aimed to foster better relations with the West in order to promote Romania's political and economic interests. The international employment of the church went hand in hand with the rise of national communism in Romania.

Throughout the 1950s the church publicly showed its support for the regime by criticising the West and praising the political achievements of the Soviet Union. An increasingly nationalist tendency of the church became evident in the actions of the hierarchy in the early 1960s. Although the regime continued to arrest any possible dissenters,[2] Patriarch Justinian adopted a different tone in his sermons and in his Pastoral Letters to the faithful. Whereas previously they had a strong political content, they now began focusing more on theological issues combined with nationalist elements. The patriarch continued to mention the main propagandistic themes of the regime, such as the battle for peace, but generally they did not have the previously hostile tone towards Catholics and the West.[3]

Justinian's visibly changing attitude emerged around the time of the Third Congress of the Romanian Workers' Party, held from 20 to 25 June 1960. The congress restated the political position of Gheorghiu-Dej, who was elected first secretary of the Party's Central Committee.[4] In his message, Gheorghiu-Dej emphasised the need for continued industrialisation of the country, implying a direct stance against Soviet directives which wanted to transform Romania into an agricultural satellite.[5] The congress analysed previous achievements of the regime and concluded that 'socialism is victorious in our country. We are starting a new stage of development in our country – the stage of completion of the construction of socialism (*etapa desăvârşirii construcţiei socialismului*).'[6]

The increased signs of independent politics under Gheorghiu-Dej and the efforts of the Romanian government to enter into dialogue with non-communist countries were encouraged by the West. On 27 September 1960, Gheorghiu-Dej was invited to give a speech at the 15th session of the United Nations General Assembly and on 9 December 1960, for the first time after the Second World War, the government signed a convention with the US on cultural, educational and scientific measures.

Changes to the political organisation of the regime created a new form of collective leadership which was intended to look more transparent. On 21 March 1961, the communists established a State Council, the highest political authority in the country. The council was a collective body which ensured the supreme exercise of state power between the sessions of the GNA while its president represented the country in international relations. Gheorghiu-Dej was elected president of the State Council and Ştefan Voitec was elected chairman of the GNA. In addition, the GNA elected Ion Gheorghe Maurer as chairman of the government, a position which he served until he was forcibly pensioned in 1974.

At this time of internal political transformation, the church hierarchy publicly supported the achievements of the regime. After Gheorghiu-Dej's speech, on 7 October 1960, Patriarch Justinian sent a telegram to the United

Nations condemning, in the name of his church, those members who called themselves Christians while 'enslaving' colonies. As the telegram stated:

> The day has come when the United Nations Organisation, this high forum of international collaboration, must respond to the trust peoples have put in it and bring about the general and total disarmament demanded today by all [. . .] Many representatives of member states of the United Nations are, or pretend to be, Christians. A heavy responsibility weighs upon their shoulders: to prove by facts that true Christianity cannot have anything in common with enslaving colonialism and that they must not let the enslaved peoples identify the enemies of their national freedom with believers in Christ.[7]

The telegram was published in the Romanian communist press and made reference to the 'national freedom' of 'believers in Christ'. This type of message, supporting national liberation, would begin to be employed more frequently by the hierarchy. The church's message paralleled the regime's discourse in asserting a more independent course of Romanian politics from Moscow. An example was the Pastoral Letter for Christmas 1960 which clearly indicated to the faithful that 'The time has come when people will build their own destiny in liberty and independence, in national sovereignty and human dignity'.[8] Moreover, the same sentence was used in the letters sent to the other Orthodox patriarchs in the usual exchange of Christmas messages.[9]

Communist control of the church hierarchy was evident when an important position became vacant after the death of Andrei Magier, bishop of Arad. On 15 December 1960, the Electoral Collegium, composed of 73 members out of 85, elected Nicolae Corneanu to this position. At the time of his ordination, Corneanu was a lecturer at the Theological Institute in Sibiu and not a monk as in canonical tradition. He was ordained on 15 January 1961 and four days later in a special ceremony Justinian gave him the pastoral staff. After the fall of communism in 1989, Corneanu was the only hierarch who publicly admitted that he was a collaborator of the *Securitate*. Having their own people in the hierarchy, the communists sought to generate a favourable image in the West of religious freedom and encouraged more contact with Western political and religious authorities.

The World Council of Churches and the Pan-Orthodox Conference

On 21 February 1961, Visser't Hooft, general secretary of the World Council of Churches, wrote to Justinian inviting him to send an observer to the WCC General Assembly in New Delhi.[10] Relations between the WCC and Romanian hierarchs were limited in the 1950s. On 2 May 1956, Hooft had invited Justinian to send an observer to the meeting of the WCC Central Committee in Matrahaza, Hungary. On 29 June 1956, Patriarch Justinian replied that his church was unable to send someone due to the celebrations for the canonisation of Romanian saints. A few days later, on 13 July, Justinian sent a second letter which signalled his church's policy on the ecumenical movement. The letter stated that a Romanian delegation would be part of a meeting between the Russian Orthodox Church and representatives of the World Council of Churches in Moscow, scheduled for 1957 and that there would be other opportunities for future meetings.[11] Therefore, the position of the Romanian Orthodox Church was determined by Moscow.[12]

This lack of contact lasted until early 1960s when the regime pursued a new political path. The first official visit after the Second World War of a member of the WCC took place on 30 March 1961 when Reverend Francis House came to Romania as a representative of the bishop of Gibraltar to celebrate Easter for the Anglican community in Bucharest.[13] He renewed the invitation to the patriarch to send an observer to the WCC General Assembly in New Delhi. At the meeting Justinian was accompanied by Bishop Teoctist, Bishop Antim and Father Cazacu as translator; the patriarch mainly spoke and Bishop Antim made a few remarks. The most significant aspect of the meeting was that Justinian had already decided to send an observer to New Delhi, accompanied by two experts on church-state relations,[14] a fact which took Reverend House by surprise.[15]

Reverend House's report stated that discussions between them were restricted to religious matters. After the closure of some Orthodox monasteries in the previous year, the major concerns of the West were with regard to the situation of those monks and nuns who had been expelled. The patriarch tried to persuade Reverend House that the West paid too much attention to this matter and suggested that he had personally attempted to convince the communists to postpone implementation of the decree. Justinian claimed that the church offered clothes to those expelled and looked after their economic reintegration in society.[16]

Reverend House reported that the Romanian hierarchy replied extremely fast in agreeing to send an observer to New Delhi. This was in contrast to the attitude of the Bulgarian hierarchy which hesitated to give even a verbal answer. Moreover, Patriarch Justinian made clear that he should be informed of the WCC's activity through the Ecumenical Patriarch and not through Moscow. This particular request of Justinian showed that the Romanian hierarchy wanted to avoid the monopoly of the Russian Patriarchate and Soviet channels in disseminating information from the West.[17] Justinian even suggested that the WCC should move its headquarters from Geneva to Mount Athos in Greece where there was 'less political atmosphere'.[18] At the end of the conversation, Justinian attempted to distance his church from association with the Russian Orthodox Church and stated that 'in the past the Romanian Church (unlike the Russian) had taken an active part in the ecumenical movement and had acquired much experience of ecumenical meetings. The time had again come for the Romanian Church to take an active part.'[19] In the following days, Reverend House visited the Theological Institute in Bucharest and some monasteries around the capital. That he was able to do so was intended to suggest the regime's religious tolerance.

The eagerness of the Romanian Orthodox Church to become more active internationally was visible in the swift exchange of letters between the Patriarchate and the WCC. While Reverend House was still in Bucharest, on 1 April 1961 Bishop Antim sent a letter to Visser't Hooft stating that his church would like to send an observer to New Delhi.[20] This was followed by a second letter on 10 June in which the patriarch stated the official intention of his church to join the organisation in the near future.[21]

While Justinian seemed to distance his church from Moscow's influence, the Romanian stance on this matter followed the Russian Orthodox Church's intention of joining the same organisation. The Russian Orthodox Church first indicated its intention to join the WCC during an informal visit of Metropolitan Nikodim, head of the Department of Foreign Church Relations of the Moscow Patriarchate, to the WCC in Geneva on 7 November 1960.[22] He visited again on 10 March 1961 and during his meeting with Visser't Hooft agreed to submit an application before 15 April.[23] On 11 April 1961, without consulting his counterparts in Bucharest, Patriarch Alexius wrote a letter to Justinian in which he stated the decision of the Russian Church to join the WCC. Justinian replied to his letter, but omitted the mention of his church's intention of also joining the ecumenical body.[24]

The political development of Romanian independence within the communist bloc became stronger in 1961 and was reflected in church affairs. The Romanian Orthodox Church found a supporter in the Church of England, especially after the election of Arthur Michael Ramsey as the hundredth archbishop of Canterbury.[25] Archbishop Ramsey was a member of the Central Committee of the WCC and through his theological background was personally interested in Orthodox churches. On 28 April 1961, the archbishop sent invitations to the patriarchs in Moscow, Bucharest, Sofia and Belgrade to attend his enthronement on 27 June 1961.[26] The Romanian hierarchy took advantage of the invitation and eagerly replied that Metropolitan Justin of Moldavia and Suceava would represent the church.[27] In addition, the most important journal of the Romanian Orthodox Church published a short biographical article on Archbishop Ramsey and on Justin's presence at the ceremony.[28] Metropolitan Justin met Visser't Hooft in London and after the enthronement ceremonies, on his way back to Romania, stopped at the WCC headquarters in Geneva for a five-day visit to negotiate the membership of the Romanian Orthodox Church.

Metropolitan Justin came to Geneva with Father Gheorghe I. Moisescu, a priest of the Romanian community in Vienna; their visit coincided with that of two other Orthodox prelates, namely Metropolitan Maximos of Sardis, chairman of the Pan-Christian Committee of the Holy Synod of the Ecumenical Patriarchate and Metropolitan Joseph of Varna and Preslav, Bulgaria. The official purpose of the visit was to get acquainted with the work of the WCC; unofficially the purpose was to negotiate the conditions for the membership of the Romanian and the Bulgarian Orthodox churches.

Justin tried to gain sympathy for his church and entertained 24 people at a lavish dinner, including members of the WCC staff. In discussions he made it clear that his church was very interested in becoming a member. He was also interested in theological exchanges between students, especially in sending mature students for between one semester and a year to study in Geneva. In addition, he ensured that by accepting the Romanian Orthodox Church as a member, the WCC would not also admit the Romanian Orthodox Episcopate from the US as it lacked the recognition of his Church.[29]

The application to join the WCC did not face any resistance in the West with the exception of a letter written by Father Florian Gâldău, who was now living in New York. However, his claim that the ecumenical organisation should not enter into dialogue with churches under communism did not have any effect.[30] On 31 August 1961, the Holy Synod stated that the church should become a member of the WCC and sent an official application signed by Justinian and Bishop Antim on 15 September 1961. In the letter the patriarch indicated that the church numbered 13 million faithful, 9500 priests and deacons, 9200 parishes, 'together with a large number of monks and nuns'.[31]

The main reasons for applying for membership to this organisation were not only to distance the church from the tutelage of Moscow, but to become more involved in religious affairs in the West which would also benefit the regime. The Romanian Orthodox Church and the Russian Orthodox Church applied for membership at the same time at the WCC's assembly in New Delhi from 18 November to 6 December 1961 and both were accepted.[32]

The results of Romanian participation in the WCC were significant. Firstly, Metropolitan Justin was elected to the Central Committee of the WCC where he lobbied extensively for a better image of the Romanian regime within Western religious and political circles. Secondly, a number of mature students in their thirties attended courses at the Theological Institute in Geneva.[33] On their return, some of them acquired higher positions in the church's structures or were promoted to positions abroad. Some were employed in international ecumenical organisations. Thirdly, membership of the organisation led to an exchange of carefully orchestrated visits between prelates and members of the WCC.[34]

The WCC refrained from criticising the Romanian Orthodox Church in its submission to the communist regime.[35] A suggestive example of its approach is a letter from Visser't Hooft to Father Gâldău dated 23 April 1979. Gâldău, whom he had met previously, invited him to contribute to his forthcoming book on Romanian Orthodoxy during communism.[36] Hooft's letter reflected the general position of his ecumenical organisation during the Cold War period and its lack of criticism of the church hierarchs who collaborated with the communist regimes:

> It is true that I had the occasion to get to know the late Patriarch Justinian and also the present leaders of the Romanian Orthodox Church. But it is precisely because of that reason that I do not want to make any public statements about them or to contribute to such statements. The problem with regard to church leaders in communist countries is that they cannot possibly let us know all the facts of the situation. They cannot tell us what goes in the conversations which they have with government officials. So we do not know exactly in how far they give in and how far they resist. I have of course my guesses about the attitude of those I got to know, but this is not solid basis for the forming of a definite opinion.
>
> One can of course say that the relatively speaking situation in Romania is better than in the USSR or Bulgaria. I think of three aspects of the situation: theological education, the monasteries and the construction of new church buildings.
>
> I am sorry that I cannot give more help. You may feel that as I am getting older I become too prudent. But in a world in which so many imprudent opinions are delivered, it may be necessary to go against the current.[37]

In addition to engagement in the activities of the WCC, the church hierarchy agreed to attend a Pan-Orthodox Conference in Rhodes organised by the Ecumenical Patriarchate. The Ecumenical Patriarchate had asked Moscow and other Orthodox churches to participate as early as 1951, but the request was constantly refused.[38] At the beginning of the 1960s, the Ecumenical Patriarch began to regain his influence among his Orthodox counterparts and raised a previously proposed subject for the conference, the possibility of organising the Eighth Ecumenical Synod in the near future. The Orthodox churches had, at their doctrinal and theological core, the teachings of Seven Ecumenical Councils which were held in the first eight centuries. The idea of organising the Eighth Synod had been previously put forward in 1923, during the interwar period, at a conference in Constantinople, and in 1930 at a conference at Vatoped monastery, when the date was even fixed for 1932. However, it was postponed and nothing happened until 1961. The Ecumenical Patriarchate proposed that all Orthodox churches should meet and discuss the main themes which would be covered by the Synod.

This proposal was very sensitive, especially among those Orthodox churches under communism, as, by participating in discussions, their leadership would be recognised by other churches, and future generations would have to obey the Synod's decisions. The Pan-Orthodox Conference was held from 24 September until 30 September 1961 in Rhodes, with Greek, Arabic and Russian as the official languages. The Romanian Orthodox Church sent an impressive delegation, but its participation was mentioned only briefly in the communist newspaper *Romînia Liberă* on 11 October 1961.[39] On the same day, the newspaper *Glasul Patriei* (The Voice of the Fatherland), which was aimed at encouraging Romanians abroad to return to their country, published a detailed programme of the conference.[40]

In the following years, regular meetings in Rhodes remained mainly on the level of religious tourism and the Orthodox churches never agreed on a date for an Eighth Ecumenical Synod.[41] By participating in the conference in a significant number, the church hierarchy showed that it was ready to embark on the construction of a Romanian communism. The regime did not extensively report the conference in the national newspapers, but ensured that the church's actions would be propagandistically presented for the benefit of its image abroad. In this way, the church was employed as a messenger of Romanian communism, while inside the country ordinary people continued to face atheist propaganda.[42]

Between East and West

The increased openness of the Romanian Orthodox Church to the West and developments in Romanian politics led to a more favourable image of Justinian abroad. The suppression of monasteries at the end of the 1950s, communist pressure over religious matters and the determination of the patri-arch to alleviate the impact of communism on the church brought an impor-tant change of image of Justinian among other religious leaders. This image was in direct relation to the continuous piety of the Romanian population. On 25 October 1961, Reverend Francis Bailey of Malta came with Dalton Murray, the new British Minister in Bucharest, to meet Justinian at the Patriarchal Palace. The time of their visit coincided with the religious celebra-tions for Saint Dimitrie, the patron saint of Bucharest, whose relics were held in the Patriarchal Cathedral. British officials were surprised to see huge crowds of people and to witness a strong religious fervour in Romania after almost 14 years of communism. The report from the British Legation questioned the perception of Justinian in the West as a man of the communists stating that

> While the Church accepts the existing political situation, it is genuinely concerned to maintain the standards of the priesthood and of the religious communities, and to encourage the Christian way of life by public witness. [. . .] The American Chargé d'Affaires here told me that he had recently asked the Ecumenical Patriarch, an old friend, what he thought about the Patriarch of Romania. His Holiness replied that 'my son has done very fine work for the Church in Romania under the most difficult conditions'.[43]

The changing attitude of the West towards the church hierarchy, and especially towards Justinian, would have a deep impact on the evolution of relations between church and state in Romania. Justinian was encouraged by the communists to meet various political and religious leaders from abroad in order to indicate the freedom of religion in Romania. Until his death in 1977, the party would not carry out another systematic persecution of the church. Furthermore, the favourable image of the patriarch encouraged the communists to maintain the same church leadership after the death of Gheorghiu-Dej in 1965.

Collaboration between the church and the communists acquired a new dimension after the 22nd Congress of the CPSU held from 17 to 31 October 1961. In his attempt to transform the party, Nikita Khrushchev denounced the crimes committed during Stalin's leadership and the Congress decided to remove Stalin's remains from the specially built mausoleum. The Congress had a significant impact on Romanian politics as the Romanian leadership under Gheorghiu-Dej was associated with Stalinist measures. Attempting to alleviate the situation, at the Plenum of the Central Committee of the RWP from 30 November to 5 December 1961, Gheorghiu-Dej reiterated the idea that the process of de-Stalinisation had already taken place in Romania and that the party did not have to rehabilitate anyone. He criticised Ana Pauker, who died the previous year, for pursuing Stalinist concepts, and blamed Teohari Georgescu, former Minister of the Interior, who was in prison, for arresting more than 80,000 peasants, 30,000 of whom had been put on trial.[44] Gheorghiu-Dej turned to his other opponents and accused Iosif Chişinevschi and Miron Constantinescu of promoting an 'anarchical and petty-bourgeois spirit'. In Gheorghiu-Dej's opinion, Chişinevschi was particularly responsible for extensively publishing Stalin's writings in Romanian, thus diminishing the importance of Lenin.[45] In order to combat the cult of personality, Gheorghiu-Dej proposed that public spaces, such as streets and museums would no longer be named after communist leaders who were alive. Furthermore, statues of Stalin in Romania were removed.

In addition to these changes, the Plenum marked a critical point in relations with Moscow mainly because of economic issues. The Soviet Union wanted to transform Comecon into a supranational organism in which each country would have a specific economic function. According to Moscow's directives, Romania was supposed to increase its agricultural role and not its industrial sectors. This was in direct contrast with the Romanian national interests which saw a stronger economy in industry rather than agriculture. In taking an independent economic stance, Romania increased foreign trade with the West to such a level that in 1960 and 1961 the country had the highest growth of economic development in the communist bloc. The country's particular position was recognised internationally when on 1 December 1961, for the first time, Romania was elected as a non-permanent member of the Security Council for one year.

The Plenum had an impact on the church which would continue to use a distinct nationalist discourse following the country's political trajectory. In addition, in order to ensure control of the hierarchy, the communists deposed an important figure who was suspected of opposition. On 16 December 1961, the Holy Synod voted for Vasile Lăzărescu, archbishop of Timişoara and metropolitan of Banat, to be sent into retirement at Cernica monastery.[46] Officially Metropolitan Lăzărescu was accused by the Holy Synod of not correctly supervising his employees and of embezzlement,[47] using money of the church for personal purposes.[48] In reality he was deposed due to communist pressure because he helped the families of some imprisoned priests financially.[49] The vacant position led to the rise of Bishop Nicolae Corneanu who was elected on 17 February 1962 by the Electoral Collegium.

The Plenum of the Central Committee of the RWP, from 23 to 25 April 1962, declared the completion of the process of agricultural collectivisation in Romania. 11,000 peasants were brought to Bucharest in a gesture to indicate their support for the achievements of the party and the transformation of Romanian agriculture. According to official data, 3,201,000 families were working in collectivised farms representing 93.4 per cent of the agricultural land in Romania. Church hierarchs were present at the Plenum and on 26 April 1962, Justinian was awarded the medal 'In honour of the completion of agricultural collectivisation'. The claim that agriculture had been fully collectivised was intended to be a sign towards Moscow that Romania was ready to embark onto a more industrial stage of its economy.

Support for Romanian political independence was also evident in church publications. An article in *Glasul Bisericii* (The Voice of the Church) launched the idea that, within the current international context, the old Patriarchate Sees could no longer claim religious primacy. The article claimed that 'Constantinople and Rome, can no longer today justify their previous political importance' and that 'only Jerusalem, where Christ worked and suffered, has retained its former importance – which is of a religious and not political honour [. . .] Jerusalem has primacy of suffering rather than primacy of honour and power'.[50] The article represented a change in the church's attitude towards Moscow, as after the Second World War all Orthodox churches recognised the Russian Patriarchate as having the most important position in the Orthodox commonwealth. By asserting Jerusalem, a church which was seen as politically neutral, as the first Patriarchate among Orthodox churches, the Romanian hierarchy wanted to suggest that the role of Moscow was diminishing.

The Romanian Orthodox Church took a more distant position from Moscow during the second official visit of Patriarch Alexius. On 31 May 1962, at Bucharest airport, Justinian welcomed his counterpart who was returning from a trip to Yugoslavia and Bulgaria. After visiting Ghighiu and Curtea de Argeş monasteries, on 2 June Alexius was received by Ion Gheorghe Maurer, president of the Council of Ministers, who was accompanied by Emil Bodnăraş, vice-president of the Council of Ministers, and Dumitru Dogaru, general secretary of the Department of Religious Confessions.[51] For his special contribution to fostering relations between the Soviet Union and Romania, Maurer offered the Russian patriarch the medal 'the Star of the Romanian People's Republic', first class. Patriarch Alexius had known Petru Groza, and after meeting the communist leadership, he went together with Justinian to Ghencea cemetery where they laid flowers at his grave. This personal gesture of Alexius showed the popularity and the position that Groza had enjoyed during his life. In the evening Dogaru hosted a dinner at one of the most exclusive restaurants, the Athenée Palace, to which Romanian religious leaders were invited. The visit of Alexius was relatively short. After celebrating the liturgy on 3 June at Saint Spiridon Cathedral with Justinian, and signing a joint communiqué, he returned home on 4 June.[52]

The *Securitate* reports from this period suggest the beginning of strained relations between the Russian and Romanian patriarchs. At one of their meetings, Justinian did not hide his opinion and voiced his criticism of the Soviet politics of denationalisation of the Romanian population in Bessarabia, a former Romanian territory which was under Soviet occupation.[53] Tensions between the two countries became more visible during Khrushchev's visit to Romania from 18 to 25 June 1962. Khrushchev refused to visit an important plant in Craiova, accusing the Romanian communists of pursuing independent economic activity within the Eastern bloc and thus estranging themselves from the 'true' communist leadership.

The rise of Romanian national communism influenced the church which began to take more independent steps. On 31 August 1962, Justinian unofficially sent Deacon Vintilă Popescu to London in order to meet Canon John R. Satterthwaite, who had visited Romania in 1959. Canon Satterthwaite was not in London at the time but Popescu managed to see Reverend M. A. Halliwell. Popescu brought a private message from Justinian which stated that the patriarch would welcome a visit from the archbishop of Canterbury to Romania either between 6 and 17 October 1962 or between 1 and 10 June 1963. Justinian did not want to send an official invitation to the archbishop until he was sure that it would be accepted. Popescu also pointed out that the patriarch was interested in having a priest in London who would act as a liaison between their churches. Until 1955 Father Gâldău was the only Romanian Orthodox priest in London and, after he moved to the US, the Romanian community no longer had a spiritual leader. Popescu claimed that Justinian wanted to appoint him to this position but they needed a building in which to hold their religious services.[54]

The message from Justinian showed a particular position of the church hierarchy. On the one hand, the meeting showed that Justinian was aware that closer relations would please the communist position of independence; but, on the other hand, by sending an unofficial messenger to London, Justinian made sure that the church would also benefit from collaboration with the West. A visit from the archbishop would represent a gain not only for the regime but also for the church, assuring it a stronger position in

Romanian society and internationally. Any future measures against the church would have an unfavourable echo abroad and affect the image of Romanian politics.

In addition to increasing religious diplomacy with the West, the patriarch ensured that he had people around him with experience in international affairs or who would follow his orders. A significant change in the church hierarchy occurred on 28 July 1962 when Bishop Teoctist, one of the vicars of the Patriarchate, was elected by the Electoral Collegium as bishop of Arad. On 9 September 1962, Archimandrite Visarion Aştileanu was elected as the new vicar. The change from Teoctist to Visarion could be seen as an indication of Justinian's intention to have better control of the affairs of the church. Visarion was ordained only one year previously and, although he was officially in charge of foreign affairs of the church, it was, in fact, Justinian who controlled every aspect of religious diplomacy. The other vicar of the Patriarchate, Antim Nica, remained in his position until 1973 when he was moved to Galaţi.[55] Nica had travelled in Western Europe in the interwar period and maintained contacts thereafter. Sending Teoctist far away from Bucharest while keeping Nica in his position could be seen as a way of strengthening the group of people around the patriarch who had contacts abroad rather than those associated with the regime.

At a time of developing greater contact with Western churches, the Romanian Orthodox Church also deepened its relations with the Serbian Orthodox Church in Yugoslavia, both of which promoted distinct politics within the communist bloc. On 23 October, Justinian welcomed a Serbian delegation led by Patriarch German at the North Railway Station in Bucharest.[56] The Serbian delegation had a lengthy visit to Romania. On 27 October 1962, both patriarchs celebrated the renovation of Bălaşa Church in Bucharest and until 1 November they visited a number of other cities and monasteries. On 2 November, on their last day in Bucharest, they signed a joint communiqué. Returning to his country by train, Justinian accompanied German for most of his journey, only saying farewell at the Timişoara railway station on 5 November.[57]

In addition to close relations with the Serbian Church, the hierarchy also welcomed delegates of other religious confessions who might give a favourable image of the country abroad. Thus, on 8 November, a delegation of the World Council of Churches visited for one week.[58] However, Romania continued to maintain a critical position towards the Catholic Church. When the Pope summoned the Second Vatican Council, the regime refused to send any representatives to Rome and the church declared that the Vatican lacked the spiritual authority to establish the Council.[59] Romania was the only communist country which failed to send any delegation to Rome, as even Moscow sent observers.[60]

In October 1962, relations between Bucharest and Moscow reached another difficult point. On their way home from a trip to Indonesia, India and Burma,

Gheorghiu-Dej and Maurer were invited to stop in Moscow, where Khrushchev revealed the existence of Soviet missiles in Cuba. The Soviet Union's decision to install the missiles provoked great concern as the party leadership thought that Romania could become engaged in a nuclear war without being consulted. For this reason, Romania would begin moving towards a greater separation from the Soviet Union. In the following year all Romanian institutions which were created between 1946 and 1948 and contributed to the process of Sovietization were closed. The Maxim Gorki Institute, the Romanian-Russian Museum and the Romanian-Russian Institute were closed while numerous institutions and public spaces which had Russian names were changed. A new Institute of South-Eastern European Studies was established in Bucharest, while the Polytechnic, the most important complex of industrial education in Romania, was built from 1963 to 1966.

The church hierarchy continued to support the regime without engaging extensively in political disputes. The Pastoral Letter for Easter on 1 April 1963 focused on the theme of peace[61] and an eyewitness from the British Legation in Bucharest reported a significant increase in the number of faithful at the Easter services. Justinian's prayers to the congregation did not mention political leaders but only the battle for peace. Nevertheless, as the report states, young people were giggling when Justinian gave the blessing for the People's Republic.[62]

Despite good official contacts between the hierarchy and the regime, communist propaganda continued its anti-religious indoctrination especially targeting the youth. Students were forced to attend special pioneers' assemblies that were timed to coincide with religious festivals while, in 1963, the atheist Society for the Dissemination of Science and Culture published more brochures against religion: 'Adam and Eve our Ancestors?', 'When and Why Did Religion Appear?', 'The Origin of Christianity', 'Anthology of Atheism in Romania', and 'The Bible in Pictures'.[63] These publications followed the translation into Romanian of the satirical book by Leo Taxil, 'La Bible amusante' in 1962, in which the main Bible characters and events were mocked.[64]

Justinian welcomed contact not only with foreign religious leaders but also with Western academics who were working on Romanian history. On 5 April 1963, the patriarch received Eric Tappe, lecturer at the School of Slavonic Studies, University of London, who was fluent in Romanian.[65] After his return from Bucharest he gave an account of his conversations at the Patriarchal Palace in a meeting with the Anglican hierarchy. He was received by Justinian in the presence of Deacon Vintilă Popescu and two other ecclesiastical representatives. Tappe noted the close friendship between Popescu and the patriarch. Justinian emphasised that the archbishop of Canterbury should visit Romania especially because the church hierarchy sent a representative to his enthronement.[66]

A significant change in the church came about with the death of Andrei Moldovan, bishop of Romanians in the US, on 14 March 1963. Moldovan had under his jurisdiction those Romanian communities in the US which recognised the Holy Synod in Romania, as opposed to those Romanians who separated from the Orthodox Church and installed their own spiritual leader. The appointment of a bishop who was under the jurisdiction of the Romanian Orthodox Church was a means through which the regime sought to control political opposition in the Romanian diaspora. The significance of Moldovan's position was evident from the prompt response of the church hierarchy to his death. On 17 March 1963, Justinian led a remembrance service in Bucharest and sent Metropolitan Justin of Moldavia and Suceava and Deacon Nicolae Nicolaescu, rector of the Theological Institute in Bucharest, to attend the funeral on 19 March in Detroit.[67] The delegation was asked to remain in the US until the Romanian communities had a new bishop.[68] On 7 May 1963, in the meeting of the Holy Synod, Justinian claimed that the Council of the Episcopate had elected Bishop Teoctist of Arad to this position. The election of Teoctist raised questions over his real activities in the US. A report from the British Embassy in Washington showed the problems of Teoctist's nomination:

> the visa application submitted on behalf of Bishop Teoctist caused great difficulty. Because of his connection with the Peace Movement he was not eligible for an immigration visa; yet because he had indicated his wish to take up residence in the United States, he could not be granted a visitor's visa.[69]

Teoctist's application was withdrawn from the US Legation and the American authorities were instructed to say that 'the visa application is still under consideration'.[70] The main reason for rejecting a visa for the Romanian hierarch was the fact that the US law did not allow members of a 'communist mass organisation' to receive immigration visas. Teoctist was well known for his engagement in the communist propagandistic battle for peace and American authorities made sure that he would not be permitted residence.[71] A few months later Teoctist received a visitor's visa which allowed him to travel together with Metropolitan Justin, Deacon Nicolaescu and Professor Nicolae Chiţescu to Canada, the US and France, visiting the Romanian communities from 26 July to 7 September 1963 and encouraging them to remain under the jurisdiction of the Romanian Orthodox Church.[72]

The importance of ecclesiastical matters in communist countries was acknowledged not only in the US but also in London. The Anglican hierarchy sought greater contact with Orthodox leaders in order to benefit the position of religious communities under atheist regimes. On returning home from a visit to Moscow, Archbishop Ramsey declared his disappointment with his trip, but was impressed with the 'independence of spirit of the Romanian Orthodox Church' which, by now was clearly pursuing its own path in the communist bloc. Despite political control of the church leadership, the archbishop particularly remarked on the involvement of the Romanian Orthodox Church in the ecumenical movement in recent years. A report from the Foreign Office stated, he 'was inclined to believe that he [Justinian] cooperated with the regime very largely so as to draw upon himself the odium for this and thus protect his clergy'.[73] The Church of England was also approached by some Romanian émigrés in London, who, refusing to accept the jurisdiction of the Holy Synod, wanted to have their own priest and a church in which to conduct their religious services. The Anglican hierarchy rejected their request and instead considered the possibility of offering a place of worship to Deacon Popescu in case he was appointed to London.[74]

The stance of Romanian communism was visible again when Justinian strengthened his relations with the Ecumenical Patriarch by personally attending the celebration of one millennium from the foundation of Great Lavra monastery on Mount Athos from 20 June to 5 July 1963.[75] The religious celebrations for this event brought together for the first time the most important Orthodox patriarchs outside Moscow's supervision. On 23 June, Ecumenical Patriarch Athenagoras, Patriarch Justinian, Patriarch Benedict of Jerusalem, Patriarch German of Serbia, Patriarch Kiril of Bulgaria and eleven metropolitans and bishops celebrated the liturgy. The Romanian delegation visited various monasteries, Thessaloniki and Athens and, before returning to Bucharest, the patriarch stopped off in Sofia.

Close contact with other Orthodox leaders and attendance at the Mount Athos festivities were not directed against the Russian Church. On the contrary, they were part of a dialogue among Orthodox churches and a few weeks later, from 12 to 22 July 1963, Justinian went on his eighth trip to Moscow since his election as patriarch, to celebrate 50 years of the pastoral leadership of Patriarch Alexius.[76] The church hierarchy continued to enjoy good relations with Moscow but their position changed from total obedience as in the 1950s to the assertion of an independent position in the Orthodox commonwealth.

Opposition to Moscow's supremacy reached another level on 17 September 1963. For the first time after the Second World War, the Romanian representative at the General Assembly of the United Nations voted differently from the other communist countries regarding the establishment of a nuclear-free zone in South America.[77] This action was seen as another political measure of pursuing a different politics behind the Iron Curtain and was reinforced by the decision to stop blocking Romanian language radio broadcasts from the West in August 1963, a measure which was welcomed by the population.[78]

The claimed openness of the regime allowed the hierarchy to continue its dialogue with the other churches in Europe. Relations between the Romanian Orthodox Church and the Church of England reached a new level with the visit of Canon John R. Satterthwaite, general secretary of the Church of England Council on Foreign Relations, to Romania from 17 to 24 October

1963. His visit represented the most important meeting before a decision was made as to whether the archbishop of Canterbury would visit Romania. Canon Satterthwaite was received by Justinian on 18 October and on 19 October they had a private dinner together. Discussions between these two religious leaders revealed a different face of the patriarch. As a man who was considered to support the communists, Justinian made surprising statements, presenting a new image of his church, which would influence the Anglican leadership's decision.

A report from the British Legation mentioned that Satterthwaite was accommodated in luxury at the Patriarchal Palace, being offered caviar and rum for breakfast. Canon Satterthwaite noted that public attendance at churches in Romania was higher than in Moscow and in private meetings he discovered that priests were going to party officials to carry out baptisms and funerals. In addition, during a tour of Bucharest, ordinary people piously greeted him, spontaneously kissing his hand as a sign of reverence.[79] The British Legation reported Satterthwaite's own impression of Justinian in relation to the other hierarchs and the communists:

> Satterthwaite said that when the Patriarch was surrounded by his own staff, and particularly when a representative of the Ministry of Religious Confessions was present, he tended to be extremely cagey and even foxy, but during the tête-à-tête his attitude changed completely and he felt free to discuss his own position, that of the Church.[80]

After his return to London, Canon Satterthwaite wrote a report detailing his meetings with Justinian. Their first meeting, on 18 October, was formal, lasting only half an hour, and they did not broach any delicate subjects. The next day Satterthwaite was invited for a private dinner with Justinian, accompanied only by Deacon Popescu who acted as translator. Canon Satterthwaite brought the patriarch good news. The Church of England was willing to offer a building for Romanians in London although their community was small, comprising around only 200 émigré members. Canon Satterthwaite was happy to offer his own church, Saint Dustan-in-the-West on Fleet Street in the centre of London, for Sundays and if the project seemed to be a success the Anglican Church could look for a permanent chapel for the future. As Father Gâldău had moved to New York in 1955, the Romanian community attended a Roman Catholic church where a Uniate priest, Father Goya, led the religious services in Romanian. In his report Satterthwaite gave Justinian's reaction on hearing the decision:

> I told the Patriarch that I was pleased to see so much vitality in the Church here, and that from my own observation the Church appeared to be in a stronger position than in other Iron Curtain countries. He attributed this to his own friendship with Gheorghiu-Dej. He said he had

known him since he was a parish priest, and that he had been able to help him a little in difficult days [. . .] He could not make any forecast for the future, but he thought it unlikely that Gheorghiu-Dej would make any drastic alterations. His own policy therefore was to strengthen the Church wherever possible, and encourage the clergy to ally themselves with the people. In his opinion if the clergy in the countryside became centres of resistance against the collectivisation, they would merely cause the government to alter its present lenient policy towards the Church.[81]

The patriarch stated that he was a socialist and not a communist and that he was very disturbed by the fact that in the West people thought of him as a communist.[82] After the meeting with Justinian, Satterthwaite spoke to Deacon Popescu who, referring to the missionary courses imposed by the regime, stated that 'indoctrination has now ended, as it had been bitterly resented by the clergy. They did not want to come to courses to hear the same things as the local leaders of the party were always trying to propagate'.[83]

Canon Satterthwaite's recollection of Justinian showed a different image of the patriarch who was pleased that the Church of England offered support for a church in London. On the one hand, Justinian acknowledged that he was a personal friend of Gheorghiu-Dej, suggesting that he reached the position of patriarch due to this connection. On the other hand, he expressed his own position that the church would face persecution if it opposed collectivisation and other communist measures and that, from this perspective, collaboration was necessary. After Canon Satterthwaite returned to England, Justinian sent a letter to Archbishop Ramsey thanking him for sending his representative and for their fruitful discussions. The patriarch mentioned the intention of sending a Romanian priest to London, who would represent a direct liaison between their churches.[84] The archbishop replied on 7 November 1963, officially offering a church to the Romanian community and considering the future priest as a special connection between them.[85]

The meeting between Canon Satterthwaite and the patriarch had a tremendous impact on the Church of England leadership, convincing the archbishop to visit Romania if this would not cause any political problems.[86] In a letter from 21 November 1963 the British Minister in Bucharest agreed with a visit to Romania[87] for the following year pointing out that most probably 'the Romanian State authorities would stand to benefit from a visit, but I feel that the encouragement which it would give to the Romanian Orthodox Church would outweigh any disadvantages on the first count'.[88]

The visit of Archbishop Arthur Michael Ramsey

In constructing Romanian national communism the political leaders strengthened their relations with other communists who had an independent stance, and from 22 to 30 November 1963, Gheorghiu-Dej visited Yugoslavia.

Relations between their countries were extremely positive and Gheorghiu-Dej had the privilege of addressing the Yugoslav Parliamentary Assembly, the first foreign communist leader who was awarded such a special honour. Yugoslavia was not a member of Comecon and this gesture was seen as a possible indication that Romania was interested in withdrawing from this organisation. Furthermore, closer relations with the West led to the Romanian Legations being raised to embassy level, in London on 1 December and in Paris on 17 December 1963.

Despite improved international relations, Romanian communism retained its Stalinist character as party control remained strong. This political position affected the church; while entering dialogue with religious confessions from the West, the church hierarchy remained reluctant to engage in interfaith dialogue which had direct political connotations. Thus, when Pope Paul VI indicated his intention to visit the Holy Places and meet the Orthodox patriarchs, Justinian was asked by the Ecumenical Patriarch Athenagoras for the response of his church. On 16 December 1963, the Holy Synod replied that any meeting between Patriarch Athenagoras and the pope would only be personal and would not involve the Romanian Orthodox Church.[89]

After Canon Satterthwaite's visit to Romania, the patriarch was convinced that Archbishop Ramsey would not refuse an invitation. On 15 February 1964, without having any confirmation from the Anglican hierarchy, Justinian sent an official letter of invitation asking if he could come to Romania for 'at least 10 days on 25 May this year'.[90] The tone of Justinian's letter produced great surprise in London and made the archbishop remark that 'this was the most encouraging letter he had received so far from an Orthodox Patriarch' and that it 'would be extremely difficult to refuse the invitation' without at least suggesting another date.[91] Because the archbishop had planned to see the Ecumenical Patriarch in May, he proposed coming to Romania from 1 to 5 December 1964.[92] Patriarch Justinian replied that December was not the best month for him as the church held various financial meetings then and proposed postponing the visit to any week in June 1965 that was suitable for the archbishop.[93]

Justinian's confidence in inviting the archbishop to Romania came at the time when Romania publicly asserted its political distance from Moscow's supremacy in the communist bloc. The Plenum of the Central Committee of the Romanian Workers' Party, from 15 to 22 April 1964, inscribed the new position of Romanian communism and declared national communism as state politics. Gheorghiu-Dej publicly proclaimed the refusal of Romania to unconditionally follow the Soviet Union, claiming the right of every communist party 'to elaborate, choose or change the forms and methods of socialist construction'.[94] The Plenum excluded the idea of a 'parent' and 'son' party relationship between Moscow and Bucharest marking the beginning of strained relations between the two countries.[95]

The combination of nationalism promoted by the church and Romanian communism was witnessed at that time by Doreen Berry, a journalist working in Romanian Department of the BBC in London who had a special interest in the affairs of the Orthodox Church.[96] She was in contact with the Anglican hierarchy and especially Canon Satterthwaite, with whom she discussed the evolution of the Romanian Orthodox Church. She wanted to travel to Romania and after overcoming difficulties in obtaining a visa,[97] she arrived in Bucharest in June 1964. Returning to London, she presented a report of her visit to the Anglican hierarchy supporting Canon Satterthwaite's experiences regarding the Church in Romania.

Doreen Berry met Justinian on 7 June 1964 and her first impression was that 'he seemed to have a certain humility and simplicity, a steady eye, and a strong and pleasing personality'.[98] In their conversation, the patriarch confirmed his intention of sending Deacon Popescu to become the priest of the Romanian community in London. Justinian intervened to persuade the communist authorities to accept Popescu as the church's representative and in the end he managed to acquire their approval. Justinian complained about Father Gâldău's addresses on the BBC in the 1950s and said that he was pleased that Gâldău had recently managed to take his wife and children out of Romania.

Berry had various conversations with Deacon Popescu who stated that there was strong pressure from the *Securitate* to take another person with him to London. The patriarch rejected this idea especially because the community in London was small. Popescu would have problems in integrating and a second person would make this even harder. Popescu claimed that he would not want to have anything to do with the Romanian Embassy in London, especially as his salary would be paid by the patriarch and not by the state.[99] Asked about church policy under communism and the image of Justinian in the West, Popescu replied that Justinian used to tell him in the past that he was doing everything he could and that he had to give in sometimes to communist pressure. In his opinion, Popescu thought that the Romanian Orthodox Church had flourished only with God's grace through which such a patriarch was appointed. Popescu was later ordained a priest and went to London in December 1964.[100]

Berry met Justinian again on 11 June when she attended the liturgy celebrated by Justinian at the religious service for Ascension Day. The service was held at the recently restored Bălaşa Church, attended by Prince Constantin Brâncoveanu. The presence of Brâncoveanu, an 89-year-old member of the family of the former ruler of Wallachia, and the financial sponsor of the church whose son was in prison, was a sign that the church continued to attract people despite their political activities.[101]

Berry describes her emotions in witnessing how the faithful welcomed Brâncoveanu and Justinian's reactions. Her account shows how the church leaders used nationalist elements in their discourses and their impact on the masses:

> I was very happy to see that this man [Justinian] – whose spirituality had so long been doubted by those who deplored his past political speeches – officiated the most be[autiful?][102] holy parts of the liturgy with a reverence, simplicity, and total absence of extravagant gesture, that convinced me of his genuine spirituality. The choir sang beautifully, the congregation fully participated in the responses, the parish priest preached an entirely religious, non-political sermon on the meaning of the Ascension [. . .] At the end of the liturgy, the Patriarch spoke briefly about the Ascension and Princely founders, and then conducted the requiem for them – Prince Brâncoveanu standing near him. Names of prince after prince were read out (It was hard to remember I was in a communist country!) [. . .] During the liturgy, the Patriarch announced that the famous Prince Brâncoveanu (martyred by the Turks) was in process of being considered for beatification.[103]

Later that day Berry met Justinian once more before her departure to London. Justinian was informal and made jokes about the Anglican Church and Romanian history. They both laughed at the idea that it was good in the end that Archbishop Ramsey did not come to Romania in December, especially because 'cold winds then blew from the Soviet Union'.[104]

The nationalist discourse of church hierarchy had a double effect. On the one hand, the church became stronger, while at the same time it helped to oppose persecution and to acquire a more favourable standing in the West. The changing image of Justinian, from the 'Red Patriarch' who helped the communists to the 'humble' spiritual leader who protected the church, was in fact influenced by the development of Romanian politics. Justinian was genuinely interested in the survival of the church and his actions to this end were appreciated and encouraged in the West. After Berry's visit, the archbishop wrote a letter to Justinian that he would come to Romania from 2 to 8 June 1965. In his reply, Justinian expressed his pleasure to receive the archbishop and encouraged him to stay for a longer period of time.[105]

The international political contacts and the independent stance of the Romanian government seemed to bring good results. From 27 July to 3 August 1964, a Romanian delegation, led by Prime Minister Maurer, held talks with the French President Charles de Gaulle. Romania was looking for support in case of a possible embargo from other Comecon countries due to its politics. In addition, collaboration with Yugoslavia benefited the Romanian economy. On 7 September 1964, at Gura Văii and Sip on the Danube River, in Romania and Yugoslavia respectively, Gheorghiu-Dej and Tito officially inaugurated the hydro-electrical works Iron Gates I, which represented an independent source of energy for their countries' industries.

Economic development was also supported by the church. In its expansion to new markets Romania extended its collaboration with other countries, whatever their form of government. From 26 to 29 September 1964, Emperor Haile Selassie I of Ethiopia visited Romania and on 27 September he met Justinian. Emperor Selassie was accompanied by high-ranking Romanian communists and attended a special religious service, *Te Deum*, in his honour.[106] Ethiopia was the only Orthodox country in Africa and the communists encouraged the delegation to deepen their contacts with the church in Romania. Although before this visit the Orthodox churches in Romania and Ethiopia did not have strong official relations, and although there was no religious leader in the emperor's delegation, Justinian pointed out that both churches had the same apostolic origin and that they shared the same attitude towards achieving national independence.[107]

The political distance from the Soviet Union did not fundamentally change the regime and Romanian communist leaders continued to enjoy official contact with their Soviet counterparts even after the removal of Khrushchev. A Romanian delegation led by Maurer met Leonid Brezhnev, the new Soviet leader, on 7 November 1964 and assured him that the regime remained faithful to communism. It is still unclear if the change of leadership in the Soviet Union was associated with dissatisfaction regarding Romania's position in the communist bloc. However, a few months after Khrushchev's removal, on 19 March 1965 the Romanian political leadership suffered a major transformation when Gheorghiu-Dej died unexpectedly. His death occurred only one day after he was automatically elected by the GNA as president of the State Council, while Maurer and Voitec retained their previous positions. Three days after Gheorghiu-Dej's death, on 22 March, Nicolae Ceauşescu was elected as first secretary of the Central Committee of the RWP and on 24 March the GNA elected Chivu Stoica as president of the State Council.[108]

The sudden death of Gheorghiu-Dej was attributed by the Romanian communists to differences between Romania and the Soviet Union, and rumours were circulated that there had been three Soviet attempts on his life in 1964. The disappearance from the political scene of the Romanian political leader did not immediately have an effect on the country's politics as Ceauşescu continued on the same course. Gheorghiu-Dej did not have a religious funeral and Patriarch Justinian only sent a telegram of condolences to the Council of Ministers. In addition, the main church journal published a special article entitled 'The memory of President Gheorghe Gheorghiu-Dej will live forever in the heart of Romanian people'.[109]

On Maundy Thursday, 22 April 1965, the church celebrated the consecration of the Holy Chrism, commemorating 80 years of autocephaly and a few days later, on Easter Sunday, in a public speech Justinian declared the support of the church for the new political leadership. The new regime continued the same policy towards the church and even presented itself as more open to visits of representatives from other religious confessions. In April 1965, in an unprecedented gesture, the regime allowed the visit of an important member of the Catholic Church, Father Bonaventure Schweizer, superior general of the Salvatorian Fathers. He was the first superior of any Catholic order to travel to Romania since the end of the Second World War. Schweizer travelled freely for one month in Romania and had a meeting with Justinian.[110]

Continuity in the regime's policy towards religion allowed negotiations for the visit of Archbishop Ramsey to Bucharest to proceed according to the agreed scheduled.[111] Justinian had a few meetings with Reverend David Tustin who conducted Easter services in Bucharest for the Anglican community.[112] At the same time, the Romanian Ambassador in London invited the archbishop to attend a special party before travelling to Romania, but he declined as he was ill after a trip to Australia. The archbishop was invited to bring his wife to Romania, but he rejected the proposal as not in keeping with religious protocol.[113] Before the archbishop's departure to Romania, Ion Rațiu, president of the British-Romanian Association in London and a well known member of the Romanian émigré community, expressed his objection to the scheduled visit; however, his intervention had no effect.[114]

The archbishop arrived in Bucharest on 2 June 1965 and his small delegation was accommodated at the Patriarchal Palace.[115] The archbishop initially hinted that he would not like to have meetings with political leaders but on 3 June he met Chivu Stoica, president of the State Council and Ion Gheorghe Maurer, chairman of the Council of Ministers.[116] In the afternoon he gave a speech at the Theological Institute in Bucharest and later attended a special dinner hosted by Dumitru Dogaru from the Department of Religious Confessions. Over the next two days the archbishop and the patriarch visited some monasteries;[117] on 6 June Justinian celebrated the liturgy at Saint Spiridon Cathedral offering the archbishop the highest ecclesiastical distinction, the Patriarchal Cross, while in the evening the archbishop celebrated Vespers at the Anglican Church in Bucharest, accompanied by Justinian; on 7 June Archbishop Ramsey and Patriarch Justinian signed a joint communiqué and had dinner at the British Ambassador's residence.[118] The following day, the Anglican delegation returned to England, accompanied to the airport by the patriarch.[119]

After their return to London, Canon Satterthwaite asked John Andrew, a member of the delegation and the archbishop's Domestic Chaplain, to share with him his impressions of the visit. Andrew stated that the archbishop was impressed by the patriarch and that 'the visit had been a spectacular success'.[120] Justinian was regarded with sympathy; however, the Anglican delegation had an unfavourable impression of Metropolitan Justin with whom they were already familiar from the Lambeth Conference in 1958 and the World Council of Churches. Andrew mentioned that when he asked Justin if he had recently visited his diocese, the metropolitan replied

Figure 7.1 Patriarch Justinian and Archbishop Arthur Michael Ramsey at the Patriarchal Cathedral, 1965 (courtesy of the Romanian National News Agency ROMPRES).

that 'Ils n'aiment pas me voir'.[121] This internal account shows that the Anglican hierarchs regarded only the patriarch as the person who was genuinely interested in the survival of his church.

After the archbishop's visit, L. C. Glass, the British Ambassador to Bucharest, reported that in his opinion the new regime allowed the visit because it was interested in presenting a civilised image of Romania to the West. By showing respect for religion, the regime was looking to increase Anglo-Romanian economic relations. For the church, having one of the most respectable religious figures from the West indicated the importance of religion in Romanian society and after the visit the regime would find it more difficult to persecute the church. Glass claimed that Gheorghiu-Dej allegedly stated in a meeting with the Austrian Ambassador in 1964 that 'as long as the Church has no political power and the state has full control of the education of the young I am not against religion', and the same attitude seemed to be followed by his successor.[122]

The visit had a favourable impact on the public as gestures of the patriarch were interpreted as against the regime's atheist policy. The climax of the visit had been the celebration of the liturgy by Justinian in the presence of the archbishop. In his sermon, the patriarch referred to the national past, emphasising that the church survived 500 years of Turkish occupation. The ambassador reported that those present at the ceremony felt that Justinian was hinting at the communist regime and that, just as the church had

Figure 7.2 Archbishop Arthur Michael Ramsey meeting Ion Gheorghe Maurer, Chairman of the Council of Ministers, 1965 (courtesy of the Romanian National News Agency ROMPRES).

survived the Turks, it would survive communism. Glass claimed that hearing Justinian's sermon, an old bishop with tears in his eyes exclaimed that 'the Church had survived'.[123] The report stated that the Romanian Orthodox Church was in the best position in the communist bloc. For example, Bucharest with a population of one and a half million had around 430 priests while Moscow which numbered seven million had only 43 priests.

The ambassador described a significant moment during the meeting with the party leaders when the archbishop asked Maurer about the government's attitude towards religious minorities. Maurer thought that the archbishop referred to the situation of the Uniate and Roman Catholic churches which were persecuted and replied that England had set a precedent for establishing a national church in order 'to prevent a foreign ecclesiastical ruler interfering in the national affairs'.[124] In aiming to please the regime, the archbishop expressed admiration for the country's economic achievements and suggested further relations between their countries.[125] Concluding his report, the ambassador stated that the visit was a general success and there were no attempts to include political propaganda in clerical speeches or in the final communiqué, as occurred during the archbishop's visit to Moscow.

The Romanian press did not extensively report on the visit.[126] The most important references were factual and without comment. Only one weekly journal, *Lumea* (The World), published a propagandistic article signed by Mircea Ivănescu titled 'In the spirit of peace and friendship' which included photographs of Archbishop Ramsey with Maurer and Stoica.[127] As the British Ambassador suggested, the regime thought that the visit would have a wider impact in the Western mass media and Romanian officials seemed disappointed that the visit was not greatly publicised in England. The most significant coverage of the visit was an interview conducted by the BBC when the archbishop arrived at the airport in London.[128]

In the following months, Archbishop Ramsey remained in contact with Justinian and invited him to return his visit by coming to London the following year. From 21 to 28 June 1966, Justinian, accompanied by a small delegation,[129] visited London where they were received by the Foreign Secretary and Queen Elizabeth II.[130]

The construction of Romanian national communism reached a new level in the following months. The internal clash for political power was visible at the Ninth Congress of the RWP from 19 to 24 July 1965 when the name of the party was changed from the Romanian Workers' Party to the Romanian Communist Party. Consolidating his political position, Nicolae Ceaușescu was elected general secretary of the Central Committee of the party. Political transformations led to a change in the Constitution in a special session of the GNA on 21 August 1965 which proclaimed the renaming of the country as the Socialist Republic of Romania. The GNA re-elected Chivu Stoica as president of the State Council, a position which he occupied only until 9 December 1967 when Ceaușescu took his place. The new leaders retained the same attitude as their predecessors towards the church. In 1966, Ceaușescu attended folk festivals organised by the church, and visited a large number of churches and monasteries around the country, signing the golden book of some monasteries. In this way, by publicly cooperating with the church, the communists ensured that it would remain an important ally.[131]

The visit of Archbishop Ramsey did not have an immediate impact on the development of internal politics but influenced the evolution of church-state relations in Romania. In the following years, with the growth of national communism and the dictatorship of Ceaușescu, the regime would benefit from the nationalist discourse of the church. Until the death of Justinian the regime did not systematically persecute the church and continued to encourage the hierarchy to develop contact with Western religious leaders. The fact that Justinian managed to bring to Romania one of the most respectable religious figures in the West strengthened his position and benefited the church in the long term.

Conclusion

At the beginning of the 1960s the Romanian communist regime aimed to distance itself from Soviet hegemony and began to promote national communism, supported by the Romanian Orthodox Church. The new attitude of the regime was evident in sermons of the church hierarchy which lacked the fierce attacks on the West of the 1950s, and focused more on theological and nationalist issues. The regime wanted to induce the idea that religious freedom was respected while the church was used to help control possible opposition both within the country and from Romanian communities abroad.

Analysing the relationship between church and regime brings to light significant aspects of the evolution of national communism in Romania. While there were some earlier signs, it was in 1960 that elements of national communism could be more visibly discerned in the speech of Gheorghiu-Dej at the United Nations and in the increase of international relations between the Romanian government and the West. These signs became more clearly developed after 1962 when Romania officially started to distance itself from the Soviet Union. The regime allowed and encouraged the church hierarchy to enter into contact with Western churches in its attempt to promote Romanian national communism.

The hierarchy strengthened its contacts with other foreign churchmen, and had a particular interest in the Church of England. While, on the one hand, participation in international religious dialogue represented indirect support for the construction of Romanian national communism, on the other hand, the church became stronger as any overt religious persecution would have a negative impact on the regime's image abroad. The Anglican leadership did not condemn religious abuses in Romania or the country's atheist propaganda. It was criticised in the West for failing to do so and for having contact with the Romanian Church hierarchy which was compromised by its collaboration with the communists. However, the main reason for the Anglican leadership wishing to strengthen relations with the Romanian hierarchy was the particular stance of Romanian communists in the Eastern bloc. By meeting church officials, the Anglican hierarchs offered indirect support for religion in the communist world.

Conclusion

In Orthodoxy, the relationship between church and state is bound on the concept of *symphonia*, which dates back to the Byzantine Empire. The church and the state are exponents of religious and political power and their influences overcome their strict spheres. For this reason, religious leaders seek a position in the political realm while political leaders involve themselves in the life of the church. Orthodoxy plays an active role in political life by claiming to be the protector of the nation.

Analysis of the historical trajectory of the Romanian Orthodox Church has shown that religion has been intrinsically linked to successive political regimes. After the establishment of the state, Prince Cuza founded a Holy Synod which would follow his policy. The ecclesiastical organisation of the church indicated that religion was moulded on nationality while collaboration between the religious and political spheres was seen as necessary for the evolution of the state.

During the communist period the church was used as an instrument to control the population through which the regime extended its propaganda and identified its possible dissidents. The regime enjoyed good relations with the hierarchy and ensured that those who were suspected of conducting anti-communist activities were replaced. At the same time, the state pursued strong anti-religious propaganda aimed at reducing the influence of the church in society. In the People's Republic the regime allowed the church to continue its activity mainly because the hierarchy was politically controlled. By having their own people in the hierarchy and fostering internal clerical clashes for ecclesiastical power, the communists sought domination of the entire church. The church was actively engaged in the propagandistic message of the battle for peace and the Holy Synod adopted regulations and elected hierarchs imposed by the regime as representing the will of the church.

There are two dominant positions in the literature of church-state relations in communist Romania. Firstly, some scholars have argued that the Romanian Orthodox Church established an ideological pact with the communists, in a way comparable to the Sergianism pact between the Moscow

Patriarchate and the Soviet Comintern. Secondly, some scholars have maintained that the church's adaptation to communism should be regarded as a form of resistance. This position exonerates the collaboration between the hierarchy and the communist leadership, suggesting that the hierarchy opposed communism.

This research has offered a third perspective. While some actions of the patriarch and other church hierarchs could be interpreted as resistance towards communism, analysis of the church should be related to the wider use of religion in the communist block as a form of ideological propaganda. This book has shown that nationalism was at the core of the church's actions since the establishment of the Romanian state. The use of the national past and the nationalist discourse of the hierarchy helped the church to survive and adapt to the atheist regime. The doctrine of social apostolate promoted by the church hierarchy represented the application of communism to the particularities of Romanian Orthodoxy combining nationalism with the concept of *symphonia*.

By accepting collaboration with the regime, the church hierarchy acted according to *symphonia*, proclaiming that every regime was the will of God. The use of nationalism was evident at a time of fierce Sovietization of the country when the church hierarchy made references to the national past. For example, the renaming of the Bishopric of Galați back to its previous historical name of the Bishopric of the Lower Danube and the decision to canonise the first Romanian saints in 1950 showed that, by preserving references to the national past, the church was interested in promoting a position which would make it stronger. The regime initially opposed the celebration of canonisation, but after the rise of Gheorghiu-Dej and the implementation of state policy on constructing a Romanian road to communism, public festivities were permitted. The October 1955 canonisation proved that the church continued to have a tremendous impact on the people, a sign which was clearly indicated by the attendance of around 10–15,000 people at Cernica monastery, most of whom had walked the 17 kilometres there from Bucharest. While the church restructured its ecclesiastical organisation and important hierarchical figures were deposed or suddenly died, Orthodoxy remained strongly rooted in the lives of ordinary people. Even the communist activists, who travelled to villages to convince people to join collectivised farms and of the political benefits of the new regime, declared themselves religious. Despite political interference in the nomination of the hierarchy, the Orthodox faith remained strong, and throughout the communist period people remained attached to their church.

The Orthodox Church suffered its greatest persecution in 1959 when the regime decreed the modification of the regulations regarding monastic life. The Orthodox monasteries were seen as centres of resistance sheltering former political dissidents and as places through which the church attained its influence in society. The closure of monasteries and the reduction of the number of clergy showed a different image of Patriarch Justinian. He was personally involved in attempting to change the unfavourable legislation of the church and to postpone its implementation for as long as possible. The actions of the patriarch against the regime's policy attracted sympathy in the West. If in the first years of communism, Justinian was labelled the 'Red Patriarch', at the beginning of the 1960s Western religious and political leaders perceived him as the hierarch who was genuinely interested in the survival of the church under communism.

A major part in changing the patriarch's image abroad was played by the Church of England which developed close contact with the Romanian Orthodox Church in the interwar period. From the start of the Second World War until 1956, contact was reduced. However, when the communists began promoting a Romanian version of communism, more Anglican leaders were allowed to travel to the country and meet church hierarchs. These ecclesiastical meetings formed the basis for inviting the Romanian Orthodox Church to join the World Council of Churches in 1961, an ecumenical organisation which gathered leaders from churches across the world in dialogue.

With the rise of national communism in the 1960s, the church promoted more assertive theological and nationalist discourses. The engagement of the Romanian church with churches abroad and the nationalist message inside the country helped the regime to promote an independent position within the communist bloc. Romania officially declared an independent path of communism at the Party Plenum in 1964; this position was long anticipated and the church was one of the institutions which supported this type of message as early as the 1950s. For this reason, after the sudden death of Gheorghiu-Dej in 1965, the new political leadership under Nicolae Ceaușescu continued the same attitude towards the church. Previous contacts of the church were allowed to develop and the church received the archbishop of Canterbury, Arthur Michael Ramsey, in 1965, the most important ecclesiastical visit to Romania since the Second World War. The archbishop's visit was a sign of the important place of religion in Romanian society and the church benefited as the regime would find it more difficult to easily assert its anti-religious policy, even if atheist propaganda against religion continued throughout the Cold War period.

Reports from Romania of this period show that the majority of ordinary clergy remained anti-communist, but some of them declared themselves to be communist or behaved as communists in order to further their careers. The clergy were forced to attend missionary courses at which they were instructed on religious and social issues, combining religion with communist propaganda. Those who refused to read sermons written by hierarchs who supported the regime, or those who publicly opposed the regime, were arrested and imprisoned.[1] One of the most common accusations was collaboration with the previous fascist regime and many priests were removed from their parishes for this reason. Anti-communist resistance and opposition against collectivisation were limited in Romania; however, some priests continued to oppose the regime and most of them died in prison. Despite the general official collaboration of the church, many priests listened to foreign broadcasts in Romanian and praised the religious programmes of Father Florian Gâldău on the BBC and the Voice of America in which the church hierarchs were condemned for their collaboration with the regime.

The communists were aware of the impact of religion on Romanian society. Despite their anti-religious propaganda, some members of the government declared themselves to be Orthodox believers, such as Prime Minister Petru Groza, who enjoyed good relations with the church hierarchy in the interwar period and after the installation of communism. In addition, some party members were interested in and informed about theological matters. In 1956 when a British MP visited Romania, he was surprised to hear Grigore Preoteasa, the Minister of Foreign Affairs, making theological comparisons between the doctrines of the Romanian Orthodox Church and the Church of England. According to Preoteasa, their churches should bring their countries closer together which would entail an increase in economic exchanges and tourism. The combination of religion and economics was also seen in the support of the regime for the Orthodox monasteries which produced goods for exports and were an important contributor for the Romanian economy.

The communists allowed the church to continue its spiritual activity mainly because they saw that using the church was more profitable than persecuting it. Firstly, the monasteries' output brought valuable financial revenue to the state. Secondly, the church continued to preserve the religious sentiments of the people, thereby 'supporting' the claim that the regime allowed complete religious freedom. Thirdly, the church had important connections abroad and the regime encouraged religious tourism to Romania, promoting the idea of a civilised country. Fourthly, the church helped the regime to foster a Romanian road to communism and consequently the rise of Romanian national communism in the 1960s. The sanctification of the first Romanian saints, subsidies for rebuilding churches and paying the salaries of the clergy and personal contacts between the church hierarchy and top government officials contributed to specific Romanian church-state relations. For these reasons, the party gained enormously through the subservience of the church as public religious opposition was minimal in Romania.

The church hierarchy was not critical of the regime and did not openly condemn the imprisonment of clergy, the closure of monasteries and churches or even the ousting of those hierarchs who were accused of anti-regime activities. Reports from this period note how the hierarchy was living in relative luxury while ordinary people were deprived of elementary economic conditions.

The Orthodox Church was actively engaged in the suppression of the Greek Catholic Uniate Church in 1948, which, due to its connection with the Vatican, was seen as a place of resistance against communism. The incorporation of the Uniate Church, despite sporadic opposition, benefited the regime in the long term, but not the Orthodox Church because people remained attached to their religion. This was evident immediately after the fall of communism when the Uniate Church was recognised as an independent church again. However, relations between the Orthodox and Uniate churches remain strained today as there are still conflicts regarding the restitution of property.

This study of church-state relations has revealed that the Romanian communists promoted a Romanian road to communism in the 1950s. Stalin's death in 1953, and Gheorghiu-Dej's intention to retain Stalinist methods, encouraged the party leadership to look for a more independent position in the communist bloc. After the canonisations and the development of closer contacts with the West after 1955, Romania embarked on a more distinctive form of constructing communism. With positive developments of the economy and growing political interest in Romania in the West, the regime showed a distinctly anti-Moscow attitude in the early 1960s. This stance was clearly reflected in the church, as the hierarchs began using more nationalist terms in their sermons.

The position of the church in Romanian society after Nicolae Ceaușescu came to power in 1965 has to take into account that of Gheorghiu-Dej period. Romanian national communism had its roots during this period and the church would continue to have a strong nationalist message which would help the regime in promoting its own agenda. References to the national past and the rise of the cult of personality, which placed Ceaușescu as the leader who acquired mythical features, were elements which had been used to some extent in the nationalist discourse of the previous regime. In the 1980s Ceaușescu was presented as a semi-divine leader whose place in Romanian history would be equal to the rulers of the Middle Ages and the founders of the Romanian state.[2] The church continued to praise the political leadership, combining national communism with its own theological discourse even at the time of public demonstrations against the regime.[3]

One of the most controversial questions regarding the trajectory of church during the communist regime was the collaboration of the hierarchy with the *Securitate*. This study has shown that by obeying the regime most hierarchs benefited during this period and that many of them lived at the intersection of the spiritual and police realms. At a time when obtaining a visa was extremely difficult for ordinary citizens, some hierarchs were allowed to travel extensively outside the country. It is important to make a distinction between those clergy who were mere informers and those of high rank who were involved in specific state police activities. To the present day, the degree and extent of the involvement of top hierarchs and their relations with the *Securitate* remain unclear. Only Bishop Nicolae Corneanu has admitted his involvement with the *Securitate* and in October 2007 the National Council for the Study of the *Securitate* Archives declared that he had been an agent.

What exactly did it mean to be a member of the *Securitate*? The answer is connected with the concept of power during communism; the higher the rank, the higher the social and political power of an individual. The Cold War proved to be the period when being a member of the *Securitate*'s structure was the condition for being promoted to a hierarchical position. It was not sufficient just to be a member of the Communist Party. The most important distinction between members of the *Securitate* ultimately lies in their actions. Some clergy did so for their own benefit while others were forced to join. Some clergy broke the confessional vows of secrecy and their reports led to the imprisonment of ordinary people, while others wrote reports in such a way that they would be of no benefit to the regime.

Contact between Romanian clergy and their foreign counterparts were closely scrutinised by the *Securitate*. All of the most important visitors to the country were followed at a distance by an agent. Thus, during the visit of Canon John R. Satterthwaite, every street he walked on and every time he stopped to look in a shop window, for example, were noted in his file.[4] The *Securitate* recorded not only the activity of the visitors but also their intimate conversations.

A controversial position remains that of Patriarch Justinian. His personality was complex. He had an unusual career in the church hierarchy, rising, with the support of the communist leadership, from a mere parish priest to the highest ecclesiastical rank in a short period of time. Some reports presented him in awkward positions, such as drinking and partying with communist officials, while others reported how people who witnessed his prayers at religious ceremonies believed he had a most revered presence and that he was a providential man for those times. His influence extended far beyond that of the religious realm. The report of the Arab priest, who was close to him in 1952 and 1953, shows that mentioning his name to the Alien Control Office in Bucharest was enough for him to receive special treatment when he applied for an exit visa.

Justinian's strong personality and his personal contacts with the top communist officials managed to gain for the church an important position in society. In his meetings with foreign religious leaders Justinian claimed to be a socialist and not a communist and that he was doing everything in his position to preserve the church. The results of his policy were clearly indicated by the fact that, despite persecution and anti-religious propaganda, at the time of Justinian's death in 1977, the church still had around 8500 priests, a similar number to that of the first years after the Second World War.[5]

The Orthodox Church survived this difficult period because of Justinian's personal role in the evolution of church-state relations. His close contact with top communist leaders and other Orthodox and Western religious leaders represented key elements in his personal survival as the leader of the church and in the trajectory of the church during communism. Although he publicly collaborated with the regime, in the early 1950s when he travelled outside Bucharest he was escorted by a *Securitate* officer and his correspondence was checked by the Ministry of Religious Confessions before he was allowed to read it. That he was genuinely interested in the survival of Orthodoxy as a faith was proved by many accounts of those around him during this period. At the time of the death of Stalin, when he officially engaged the church in celebrations commemorating the loss of the Soviet leader, his gestures were interpreted as opposing communism. During the visit of the archbishop of Canterbury to Bucharest, his words against the Ottoman occupation of Romanian territory were regarded as support for resistance against the communist regime.

Many *Securitate* reports complained that Justinian did not completely follow party directives and he had an ambivalent position, showing that he was not completely trusted by the authorities. He was accused by the *Securitate* of employing clergy who were politically active during the fascist regime. Although some of them were imprisoned for political reasons, in many cases they were reintegrated and offered employment within the church structures.

Analysis of the religious trajectory of the Romanian Orthodox Church in the Romanian People's Republic has shown that political leaders used the church for their own benefit, while church leaders collaborated with the regime in order to ensure the survival of the Orthodox faith and their personal positions in the hierarchy. Despite religious persecution, the adaptation of the church to communism was comparable to the ways in which it collaborated with previous regimes. Political control of the church dates back to the establishment of the Romanian state when the church was transformed into a state institution serving the political interests of that regime. Close contact between religious and communist leaders was encouraged by the fact that the GNA was placed opposite the Patriarchal Cathedral. Thus, the distance between the religious and political realms was literally only a few metres. The communists saw the cathedral every time they had an assembly while the church hierarchy regularly met the communist leadership. The presence of these buildings facing each other symbolises the entire history of collaboration and survival of the church and the regime.

Annexes

1. Biographies of the Major Figures in the Communist Leadership

2. Biographies of the Main Hierarchs in the Romanian Orthodox Church

3. Tables

Biographies of Major Figures in the Communist Leadership

Gheorghe Apostol (b. 1913). First Secretary of the RWP (1954–55). He was considered one of Gheorghiu-Dej's possible successors after his death but was prevented by Maurer, who supported Ceauşescu. He was removed from the party leadership in 1969. In 1989 he signed the 'Letter of the Six', a document condemning the Ceauşescu regime, which was broadcast in the West. After 1989 he attempted unsuccessfully to enter the political scene with the Socialist Labour Party.

Emil Bodnăraş (1904–76). His father was Ukrainian and his mother German. He carried out a number of missions in Romania in the 1930s as a Soviet spy and was imprisoned for ten years by the Romanian authorities. He became a friend of Gheorghiu-Dej while serving in prison. He led the communist party together with Constantin Pârvulescu and Iosif Rangheţ until Gheorghiu-Dej escaped from prison. After 1944 he became head of the *Securitate*. He was later appointed to other positions including Minister of Defence. Under Ceauşescu he became vice-president of the State Council.

Nicolae Ceauşescu (1918–89). In the 1930s Ceauşescu was a member of the underground communists. He met Gheorghiu-Dej in prison and became one of his closest allies. After 1944 he held various positions in the Ministries of Defence and of Agriculture. In 1965 he was supported by Maurer and Bodnăraş to succeed Gheorghiu-Dej. In 1967 he became the head of state and, together with his wife, Elena, progressively imposed his dictatorship. He was executed on 25 December 1989.

Iosif Chişinevschi (1905–63). He met Gheorghiu-Dej in prison in the 1940s and supported him in the conflict with Ana Pauker's group in 1952. Because of his criticism of Gheorghiu-Dej's policy he was removed from the politburo in 1957.

Miron Constantinescu (1917–74). One of the few intellectuals in the communist leadership. After 1944, he was director of *Scânteia* and Minister of Education. In 1956 he criticised Gheorghiu-Dej and was excluded from the politburo together with Chişinevschi. During the Ceauşescu regime he became Minister of Education, President of the Academy of Political and Social Sciences, Rector of the Ştefan Gheorghiu Party Academy and President of the GNA.

Gheorghe Gheorghiu-Dej (1901–65). A railway worker; for his participation in the 1933 strikes, he was sentenced to ten years imprisonment. He escaped from prison in August 1944 and led the communist party together with Ana Pauker, Vasile Luca and Teohari Georgescu. From 1953, after the removal of his opponents he initiated the Romanian road to communism. In the 1960s he launched national communism. Gheorghiu-Dej remained the sole leader of the party as first secretary and president of the State Council until his death in 1965.

Petru Groza (1884–1958). A lawyer; six times member of parliament in the interwar period and served in various ministries in 1920–1 and 1926–7. In 1933 he founded the Ploughmen Front which became linked with the Communist Party in 1935 and absorbed by it after 1944. He was Vice Prime Minister (1944–5), Prime Minister (1945–52) and President of the GNA (1952–8).

Vasile Luca (1898–1960). Of Hungarian origin (born Luka László). He joined the communist movement in 1919 and was imprisoned in 1940. He became a deputy in the Ukrainian Soviet Republic and was one of the closest allies of Ana Pauker. After 1947 he was Vice Prime Minister and Minister of Finance. In 1952 he was accused of sabotaging economic reform and was imprisoned.

Ion Gheorghe Maurer (1902–2000). A lawyer; joined the communist party in the 1930s. In 1957 he was Minister of External Affairs for two years; President of the GNA (1958–61) and Prime Minister until his forced retirement (1961–74).

Ana Pauker (1893–1960). Born into a Jewish family (Ana Robinsohn); a schoolteacher. She was involved with the communist movement in the 1920s and left Romania for Switzerland. She became a Comintern instructor in France. In 1935 she was arrested in Romania and in 1940 she was exchanged for another prisoner with the Soviet Union. The Romanian battalion in the International Brigades in Spain was named after her. In Moscow she was the uncontested leader of the Romanian communist in exile. In 1945 she proposed Gheorghiu-Dej, whom she met in prison in 1940, to become General Secretary. After 1947 she was Minister of Foreign Affairs. In 1952 she was removed from the party by Gheorghiu-Dej and placed under continuous surveillance. During the last years of her life she translated books from French and Russian for the Political Publishing House.

Lucreţiu Pătrăşcanu (1900–54). A lawyer, educated in Leipzig. He joined the communist movement in his youth and was the representative of the communist party at the Comintern in 1934–5. He met Gheorghiu-Dej in prison. After 1944 he was Minister of Justice. Pătrăşcanu was arrested in 1948 and executed in 1954 on Gheorghiu-Dej's orders.

Constantin Pârvulescu (1895–1992). Founder member of the communist party in Romania in 1929. He was arrested in 1934 and escaped to the Soviet Union. Pârvulescu led the party together with Ana Pauker while Gheorghiu-Dej was in prison. From 1945–60 he was chairman of the party control commission. In 1961 he was removed from the politburo for his criticism together with Iosif Chişinevschi and Miron Constantinescu. In 1979 at the Twelfth Party Congress, Pârvulescu publicly accused Ceauşescu of implementing a personal dictatorship. He signed the 'Letter of the Six' in 1989.

Grigore Preoteasa (1915–57). From 1944 to 1946 he was the editor of the newspaper *România Liberă* and Minister of External Affairs (1956–57). He died in a plane crash on his way to Moscow. At the time of his death he was considered Gheorghiu-Dej's successor.

Chivu Stoica (1908–75). A member of the Politburo. Prime Minister (1955–61) and President of the State Council (1965–7).

Biographies of the Main Hierarchs in the Romanian Orthodox Church

Bishop Teoctist Arăpaşu (1915–2007). Studied at the Theological Seminary at Cernica monastery (1932–40); ordained monk (1935); studied in the Faculty of Theology in Bucharest (1940–5) and Faculty of Letters and Philosophy in Iaşi (1945–7); vicar bishop of the Metropolitanate of Moldavia (1948–50); vicar bishop of the Romanian Patriarchate (1950); rector of the Theological Institute in Bucharest (1950–4); bishop of Arad (1962–72); metropolitan of Oltenia (1973); metropolitan of Moldavia (1977); patriarch (1986–2007). As patriarch he followed the same line as his predecessor, Justin, in allowing the regime to carry out its programme against churches in Romania. He resigned in December 1989 but returned in April 1990.

Metropolitan Nicolae Bălan (1882–1955). Studied in the Faculty of Theology in Cernăuţi (1900–4) where he obtained his doctorate (1905); postgraduate research in Wroclaw; editor of 'Theological Journal' in Sibiu (1907–16). In 1918 he was sent to Iaşi as representative of the Romanian National Council asking for unification of Transylvania with Romania. Ordained priest (1919) and metropolitan of Transylvania (1920).

Bishop Nicolae Corneanu (b. 1923). Studied at the Faculty of Theology in Bucharest (1942–6) where he obtained his doctorate (1949); ordained deacon (1948); administrative advisor at the Metropolitanate of Banat (1952–6); professor at the Theological Seminary in Caransebeş (1956–9); ordained priest (1960); bishop of Arad (1961) and metropolitan of Banat (1962). After 1989 he was the only hierarch of the Romanian Orthodox Church who publicly recognised that he had been an informer for the *Securitate*.

Patriarch Miron Cristea (1868–1939). Studied at the Theological-Pedagogic Institute in Sibiu (1887–90); obtained a doctorate from the Faculty of Letters and Philosophy, Budapest University (1895); ordained monk (1902) and elected bishop of Caransebeş (1909); member of the Grand National Assembly at Alba Iulia in 1918 which decided the union of Transylvania with Romania; primate metropolitan (1919) and patriarch (1925).

Bishop Teofil Herineanu (1909–92). Studied at the Uniate Theological Academy in Gherla-Cluj and Faculty of Roman Catholic Theology in Paris (1928–32); ordained and made parish priest in the Greek Catholic Uniate Church (1931–49); joined the Orthodox Church and was elected bishop of Roman and Huşi (1949); bishop of Vad, Feleac and Cluj (1957) and archbishop of Cluj (1973).

Metropolitan Vasile Lăzărescu (1894–1969). Studied at the Faculty of Theology in Cernăuţi from where he obtained his doctorate (1919); postgraduate research in Budapest and Vienna (1920); professor at the Theological Institutes in Sibiu (1920–4), Oradea (1924–33); bishop of Caransebeş (1933); bishop of Timişoara (1940) and metropolitan of Banat (1947). At the regime's request he was pensioned in 1961 and remained at Cernica monastery.

Patriarch Justinian Marina (1901–77). Studied at the Faculty of Theology in Bucharest (1925–9); director of the Theological Seminary in Râmnicu-Vâlcea (1932–3); parish priest (1924–45); vicar bishop of the Metropolitanate of Moldavia (1945–7); metropolitan of Moldavia (1947–8) and patriarch (1948–77).

Metropolitan Justin Moisescu (1910–86). Studied at the Faculty of Theology in Athens (1930–4); postdoctoral studies at the Faculty of Roman Catholic Theology in Strasbourg (1934–6); obtained his doctorate at the Faculty of Theology in Athens in 1937; professor at the Theological Seminary in Bucharest (1937–8); professor at the Faculty of Orthodox Theology at Warsaw University (1938–9); professor at the Faculty of Theology in Suceava and after 1946 in Bucharest; metropolitan of Transylvania (1956–7); metropolitan of Moldavia and Suceava (1957–77); patriarch (1977–86). During his office as patriarch he accepted the communist programme of removing and closing churches in Bucharest, which drew public condemnation from the West.

Patriarch Nicodim Munteanu (1864–1948). Studied at the Theological Academy in Kiev (1890–5); ordained monk (1894); vicar bishop in Iași (1898–1902) and Galați (1902–9); bishop of Huși (1912); abbot of Neamț monastery (1924–35); metropolitan of Moldavia (1935–9) and patriarch (1939–48).

Bishop Antim Nica (1908–94). Studied at the Faculty of Theology in Chișinău (1928–32) where he obtained his doctorate (1940); postgraduate research in Paris, Strasbourg, Oxford and Beirut; ordained monk (1935); chief of the Romanian religious mission in Transnistria (1941–4); bishop of Cetatea Albă-Ismail (1944–7); bishop deputy of Lower Danube (1947–50); vicar bishop of the Romanian Patriarchate (1950–73); bishop of Lower Danube (1973–5); archbishop of Lower Danube (1975–94).

Father Vintilă Popescu (1895–1971). Studied at the Faculty of Theology in Bucharest (1925–9); postgraduate research in Vienna, Graz, Munich and Oxford; professor at the Theological Academy in Arad (1926–48); ordained deacon (1941); lecturer (1948–56) and professor (1956–8) at the Faculty of Theology in Bucharest; ordained priest of the Romanian community in London (1964–9).

Bishop Nicolae Popovici (1903–60). Studied at the Theological Academy in Sibiu (1923–7); obtained a doctorate at the Faculty of Theology in Cernăuți (1934); postgraduate research in Athens, Munich, Tubingen, Leipzig and Breslau (1927–32); professor at the Theological Academy in Sibiu (1932–6); ordained deacon (1929) and priest (1934); bishop of Oradea (1936). He was pensioned in 1950 by the communists and sent into forced domicile at Cheia monastery until his death.

Table 1   Religion in Romania, 1859–1930[1]

Census	1859	1899	1930
Population	3,864,848	5,956,690	18,057,028
Rural	3,174,352	4,836,904	14,405,989
Urban	551,975	1,119,786	3,651,039
Illiterate Population		3,649,473	6,029,136
Literate Rural		571,302; 15.2%	5,937,488; 52.7%
Literate Urban		461,307; 49.4%	2,462,195; 78%
Orthodox Churches	6858	6678 (in 1904)	8279
Orthodox Priests	9702	4998 (in 1904)	8257
Orthodox Monasteries	173	113 (in 1904)	75
Orthodox Monks/Nuns	4672 monks/ 4078 nuns	861 monks/2,220 nuns (in 1904)	2842 (monks and nuns)

[1] The data was collected from: Commission Princière de la Roumanie a l'Exposition universelle de Paris en 1867, *Notice sur la Roumanie principalement au point de vue de son économie rurale industrielle et commerciale avec une carte de la Principauté de Roumanie*, Paris: Librairie A. Franck, 1867; Ministère de l'Agriculture, de l'Industrie, du Commerce et des Domaines, *La Roumanie. 1866–1906*, Bucharest: Imprimerie Socec, 1907; *Recensământul Populațiunei din Decembrie 1899* [December 1899 Census], Bucharest: Eminescu, 1905; Institutul Central de Statistică, *Recensământul General al Populației României din 29 decembrie 1930* [The General Census of Romanian Population on 29 December 1930], vol. 2–4, Bucharest: Editura Institutului Central de Statistică, 1938; Academia Română, *Istoria Românilor* [The History of Romanians], vol. 8, Bucharest: Editura Enciclopedică, 2003.

Table 2   Religious composition of Romania, 1859–1930[2]

Census	1859	%	1899	%	1930	%
Orthodox Church	3,638,749	94.2	5,451,787	91.5	13,108,227	72.6
Greek-Catholics					1,427,391	7.9
Roman Catholics	45,152	1.2	149,667	2.5	1,234,151	6.8
Protestants (census 1859, 1899)	28,903	0.7	22,749	0.4		
Reformed, Calvins					710,706	3.9
Evangelic, Lutherans					398,759	2.2
Unitarians					69,257	0.4
Armenians (census 1859, 1899)	8178	0.2	5787	0.1		
Armeno-Gregorian					10,005	*
Armeno-Catholic					1440	*
Lipovan	8375	0.2	15,094	0.3	57,288	0.3
Adventists					16,102	*
Baptists					60,562	0.3
Jews	134,168	3.5	266,652	4.5	756,930	4.2
Muslims	1323	*	44,732	0.7	185,486	1.0
Other religions			222	*	7434	
Without religion and atheists					6604	
Undeclared					6686	

[2] The data was collected from: Commission Princière de la Roumanie a l'Exposition universelle de Paris en 1867, *Notice sur la Roumanie principalement au point de vue de son économie rurale industrielle et commerciale avec une carte de la Principauté de Roumanie*, Paris: Librairie A. Franck, 1867; Ministère de l'Agriculture, de l'Industrie, du Commerce et des Domaines, *La Roumanie. 1866–1906*, Bucharest: Imprimerie Socec, 1907; *Recensământul Populaţiunei din Decembrie 1899* [December 1899 Census], Bucharest: Eminescu, 1905; Institutul Central de Statistică, *Recensământul General al Populaţiei României din 29 decembrie 1930* [The General Census of Romanian Population on 29 December 1930], vol. 2–4, Bucharest: Editura Institutului Central de Statistică, 1938; Academia Română, *Istoria Românilor* [The History of Romanians], vol. 8, Bucharest: Editura Enciclopedică, 2003. *represents less than 0.1 per cent.

Table 3 Nationalities in Romania, 1930–66[3]

Census	1930	1956	1966
Total Population	18,057,028	17,489,450	19,103,163
Romanians	12,981,324	14,996,114	16,746,510
Hungarians	1,425,507	1,587,675	1,619,592
Germans	745,421	384,708	382,595
Roma	262,501	104,216	64,197
Ukrainians	582,115	60,479	54,705
Serbians, Croats, Slovaks	51,062	46,517	44,236
Russians	409,150	38,731	39,483
Jews	728,115	146,264	42,888
Tatars	22,141	20,469	22,151
Slovaks	51,842 (including Czechs)	23,331	22,221
Turks	154,772	14,329	18,040
Bulgarians	366,384	12,040	11,193
Czechs	51,842 (including Slovaks)	11,821	9978
Greeks	26,495	11,166	9088
Poles	48,310	7627	5860
Armenians	15,544	6441	3436
Hutsullians	12,456		
Albanians	4670		
Gagauzi	105,750		
Other nationalities	56,355	13,357	4681
Undeclared	7114	4165	2309

[3] The data was collected from: Institutul Central de Statistică, *Recensământul General al Populaţiei României din 29 decembrie 1930*. The General Census of Romanian Population on 29 December 1930], vol. 1, Bucharest: Editura Institutului Central de Statistică, 1938, p. 153; Romania. Direcţia Generală de Statistică, *Recensămîntul populaţiei din 21 februarie 1956, Rezultate Generale* [21 February 1956 Census. General Results], Bucharest: Direcţia Generală de Statistică, 1959, p. xix; Romania. Direcţia Centrală de Statistică, *Recensămîntul populaţiei şi locuinţelor din 15 martie 1966, Rezultate generale* [The Census of Population and Buildings on 15 March 1966. General Results], vol. 1, Bucharest: Direcţia Centrală de Statistică, 1969, p. 153.

Table 4   The number of illiterate population in Romania, 1930–56[4]

Census	Total Illiterate	Illiterate Urban Population	Illiterate Rural Population
1930	38.9	20.5	44.4
1948	23.1	10.7	27.0
1956	10.1	6.5	11.8

[4] Romania. Direcţia Generală de Statistică, *Recensămîntul populaţiei din 21 februarie 1956, Rezultate generale* [The Census of the Population on 21 February 1956. General Results], Bucharest: Direcţia Generală de Statistică, 1959, p. xx.

Table 5   The number of Orthodox monks, nuns and monasteries, 1938–57[5]

	1938	1949	1957
Monks	1638	1528	1773
Nuns	2549	3807	4041
Total Religious	4187	5335	5814
Monasteries for monks	119	122	113
Monasteries for nuns	35	56	77
Total Monasteries	154	178	190

[5] ASRI, Fond D, Dossier 7755, vol. 5, f. 97 in Păiuşan and Ciuceanu, p. 315.

Notes

Introduction

1. I use the generic term *Securitate* instead of the various names which the Romanian Information Service bore during the communist period: *Direcția Generală a Securității Poporului* (People's Security Service, 1948–51), *Direcția Generală a Securității Statului* (State Security Service, 1951–6), *Departamentul Securității* (Security Department, 1956–67), *Departamentul Securității Statului* (State Security Department, 1967–8), *Consiliul Securității Statului* (State Security Council, 1968–72) and *Departamentul Securității Statului* (State Security Department, 1972–89).
2. Richard Wurmbrand (24 March 1909–17 February 2001) was sentenced for 'crimes against humanity' from 1948 to 1956 and then again from 1959 to 1964. He described his ordeal in communist prisons in a best seller titled *Tortured for Christ*, London: Lakeland, 1967. See Wurmbrand in AKI.
3. *The Guardian*, 16 March 2001.
4. *Communist Exploitation of Religion: Hearing before the Subcommittee to Investigate the Administration of the Internal Security Act and Other Internal Security Laws of the Committee on the Judiciary United States Senate, Eighty-Ninth Congress, Second Session, Testimony of Rev. Richard Wurmbrand*, London: European Christian Mission, 6 May 1966, p. 24, AKI.
5. For religion and nationalism see Adrian Hastings, *The Construction of Nationhood: Ethnicity, Religion and Nationalism*, Cambridge, New York: Cambridge University Press, 1997; John A. Armstrong, *Nations before Nationalism*, Chapel Hill: The University of North Carolina Press, 1982; John Hutchinson and Anthony Smith (eds), *Nationalism*, Oxford, New York: Oxford University Press, 1994; Anthony D. Smith, *Chosen People: Sacred Sources of National Identity*, Oxford: Oxford University Press, 2003.
6. Olivier Gillet, *Religion et Nationalisme: L'Ideologie de l'Eglise Orthodoxe Roumaine sous le Regime Communiste*, Bruxelles: Universitè de Bruxelles, 1997; Cristian Vasile, *Biserica Ortodoxă Română în primul deceniu communist* [The Romanian Orthodox Church in the First Communist Decade], Bucharest: Curtea Veche, 2005; Sergiu Grossu, *Calvarul României creștine* [The Sufferings of Christian Romania], Chișinău: Convorbiri literare & ABC Dava, 1992; Sergiu Grossu, *Biserica persecutată: Cronica a doi români în exil la Paris* [The Persecuted Church: The Chronicle of Two Romanians in Paris], Bucharest: Compania, București, 2004.
7. George Enache, *Ortodoxie și putere politică în România contemporană* [Orthodoxy and Political Power in Contemporary Romania], Bucharest: Nemira, 2005; Adrian Gabor and Adrian Nicolae Petcu, 'Biserica Ortodoxă Română și puterea comunistă în timpul patriarhului Justinian' [The Romanian Orthodox Church and Communist Power during Patriarch Justinian], in *Anuarul Facultății de Teologie Ortodoxă a Universității București*, Bucharest: Editura Universității din București, 2002; Adrian Gabor, 'Note de lectură asupra Raportului Tismăneanu' [Notes on Tismăneanu's Report] in *Anuarul Facultății de Teologie Ortodoxă "Patriarhul Justinian Marina"*, Bucharest: Editura Universității din București, 2006, pp. 185–208.
8. For analysis of this period see Vladimir Tismaneanu, *Stalinism for All Seasons: A Political History of Romanian Communism*, Berkeley, Los Angeles, London: University

of California Press, 2003; Dennis Deletant, *Communist Terror in Romania: Gheorghiu-Dej and the Police State, 1948–1965*, London: Hurst, 1999; Stephen Fischer-Galaţi, *Twentieth Century Romania*, New York: Columbia University Press, 1991; Ghita Ionescu, *Communism in Rumania, 1944–1962*, London, New York, Toronto: Oxford University Press, 1964; Kenneth Jowitt, *Revolutionary Breakthroughs and National Development: The Case of Romania, 1944–1965*, Berkeley, Los Angeles: University of California Press, 1971; Robert R. King, *A History of the Romanian Communist Party*, Stanford: Hoover Institution Press, 1980.

9. Stephen Fischer-Galaţi, *The New Rumania: From People's Democracy to Socialist Republic*, Cambridge, MA, London: The MIT Press, 1967, pp. vii.

10. David Floyd, *Rumania: Russia's Dissident Ally*, London, Dunmow: Pall Mall Press, 1965, p. 56.

11. Lavinia Betea, *Alexandru Bârlădeanu despre Dej, Ceauşescu şi Iliescu: Convorbiri* [Alexandru Bârlădeanu on Dej, Ceauşescu and Iliescu: Discussions], Bucharest: Editura Evenimentul Românesc, 1998, p. 133; Lavinia Betea, *Maurer şi lumea de ieri: Mărturii despre stalinizarea României* [Maurer and Yesterday's World: Testimonies on the Stalinisation of Romania], Arad: Fundaţia Ioan Slavici, 1995.

12. Kaisamari Hintikka, *The Romanian Orthodox Church and the World Council of Churches, 1961–1977*, Helsinki: Luther-Agricola-Society, 2000.

13. In 2007 the Archives of Keston Institute, Oxford were transferred to Keston Center for Religion, Politics and Society, Baylor University, Waco, Texas.

14. Most documents in these archives have not yet been catalogued. For this reason, I gave my own short description of each document used.

15. Hintikka, p. 17.

16. Vasile, p. 271.

17. Hintikka, p. 17.

18. Marius Oprea, *Moştenitorii Securităţii* [The *Securitate*'s Successors], Bucharest: Humanitas, 2004, pp. 129–30.

19. Cristina Păiuşan and Radu Ciuceanu, *Biserica Ortodoxă Română sub regimul comunist 1945–1958* [The Romanian Orthodox Church under the Communist Regime 1945–1958], vol. 1, Bucharest: Institutul Naţional Pentru Studiul Totalitarismului. Colecţia Documente, 2001.

20. I use the Romanian spelling of the names of people and places except in cases where the English term is well known; for example, Bucharest instead of Bucureşti, Wallachia and Moldavia instead of Valahia and Moldova, Stephen the Great instead of Ştefan cel Mare, King Michael instead of King Mihai, and Metropolitan Justin instead of Metropolitan Iustin. Otherwise, I use the Romanian spelling, such as Petru Groza instead of Peter Groza. With names in languages other than Romanian, I use the English version, for example, Patriarch Alexius, Patriarch Christopher. The translations from Romanian are my own.

21. *Scînteia*, 23 April 1964.

22. Michael Shafir, 'Romanian Foreign Policy under Dej and Ceauşescu' in George Schöpflin (ed.), *The Soviet Union and Eastern Europe: A Handbook*, Oxford, New York: Muller, Blond & White, 1986, pp. 364–77.

1   Orthodoxy, *Symphonia* and Political Power in East European Communism

1. F. L. Cross, *The Oxford Dictionary of the Christian Church*, Oxford, New York: Oxford University Press, 2005, p. 1012.

2. John Meyendorff, *Byzantine Theology: Historical Trends and Doctrinal Themes*, New York: Fordham University Press, 1974, p. 283; John Meyendorff, *The Byzantine Legacy in the Orthodox Christianity*, Crestwood: St Vladimir's Seminary Press, 1982.
3. Grigorios D. Papathomas, *Le Patriarcat Oecumenique de Constantinople (y compris la Politeia monastique du Mont Athos) dans l'Europe unie*, Katerini, Athens: Editions Epektasis, 1998, p. 705.
4. J. M. Hussey, *The Orthodox Church in the Byzantine Empire*, Oxford: Oxford University Press, 2004.
5. Jaroslav Pelikan, *The Christian Tradition: A History of the Development of Doctrine*, vol. 2: *The Spirit of Eastern Christendom (600–1700)*, Chicago, London: The University of Chicago Press, 1974, p. 168.
6. Steven Runciman, *Byzantine Civilisation*, London: Edward Arnold & Co, 1933, p. 108.
7. Ibid., p. 113.
8. Ibid., p. 114.
9. P. N. Ure, *Justinian and His Age*, Westport: Greenwood Press, 1979, p. 122. The English translations of some documents from this period are published in Ernest Barker, *Social and Political Thought in Byzantium from Justinian I to the last Palaeologus*, Oxford: Clarendon Press, 1957.
10. R. Schoel and W. Kroll, *Corpus Juris Civilis*, vol. 3, Berlin: no publisher, 1928, 35ff in Petros Vassiliadis, 'Orthodox Christianity' in Jacob Neusner (ed.), *God's Rule: The Politics of World Religions*, Washington: Georgetown University Press, 2003, p. 99.
11. Ibid.
12. Steven Runciman, *The Orthodox Churches and the Secular State*, Trentham: Auckland University Press & Oxford University Press, 1971, p. 14.
13. A. A. Vasilev, *History of the Byzantine Empire*, Wisconsin, London: University of Wisconsin Press, 1980.
14. Henri Grégoire, 'The Byzantine Church' in Norman H. Bayner and H. St. L. B. Moss (eds), *Byzantium: An Introduction to East Roman Civilization*, Oxford: Clarendon Press, 1948, p. 127.
15. Timothy Ware, *The Orthodox Church*, London: Penguin Books, 1997.
16. Lucian N. Leustean, 'Orthodoxy and Political Myths in Balkan National Identities', *National Identities* (forthcoming, 2008, 10 (4)).
17. Steven Runciman, *The Great Church in Captivity: A Study of the Patriarchate of Constantinople from the Eve of the Turkish Conquest to the Greek War of Independence*, Cambridge: Cambridge University Press, 1968; Paschalis M. Kitromilides, 'Imagined Communities and the Origins of the National Question in the Balkans', *European History Quarterly*, 1989, 19 (2), pp. 149–92; Paschalis M. Kitromilides, *Enlightenment, Nationalism, Orthodoxy: Studies in the Culture and Political Thought of South-East Europe*, Aldershot: Variorum, 1994.
18. Pedro Ramet, *Cross and Commissar: The Politics of Religion in Eastern Europe and the USSR*, Bloomington: Indiana University Press, 1987, p. 24.
19. The 1937 census had one question regarding religious belief but the results were never published. However, the results have been referred to by some scholars who have stated that two-thirds of the rural and one-third of the urban population regarded themselves as religious. William C. Fletcher, *Soviet Believers: The Religious Sector of the Population*, Lawrence: Regents Press of Kansas, 1981. The data supplied in this text is from the Latvian Legation, London, Item 1329/3/51, TNA PRO FO 371/100744.
20. Sabrina Petra Ramet (ed.), *Religious Policy in the Soviet Union*, Cambridge: Cambridge University Press, 1993.

21. John Shelton Curtiss, *The Russian Church and the Soviet State, 1917–1950*, Boston: Little, Brown & Company, 1953, p. 290.
22. On 8 September 1943 Sergii was appointed patriarch. He died on 15 May 1944 and Metropolitan Alexius of Leningrad was elected his successor. Robert Conquest, *Religion in the USSR*, London: The Bodley Head, 1968.
23. Latvian Legation, London, Item 1329/3/51, TNA PRO FO 371/100744; Matthew Spinka, *The Church in Soviet Russia*, New York: Oxford University Press, 1956; Walter Kolarz, *Religion in the Soviet Union*, London, New York: Macmillan, 1961; Dimitri V. Pospielovsky, *A History of Soviet Atheism in Theory and Practice, and the Believer*, London: Macmillan, 1987.
24. Ramet, *Cross and Commissar*, p. 4.
25. A similar attempt was against the Finnish Orthodox Church in 1945.
26. The head of the Polish Orthodox Church, Metropolitan Dionysie, was deposed in May 1948. His successor Metropolitan Timothy asked for a new autocephaly from Moscow on 1 December 1948.
27. Tatiana A. Chumachenko, *Church and State in Soviet Russia: Russian Orthodoxy from World War II to the Khrushchev Years*, trans. by Edward E. Roslof, Armonk, London: M. E. Sharpe, 2002.
28. The first Uniate Church forced to become Orthodox was in Polish Galicia in 1946.
29. William B. Stroyen, *Communist Russia and the Russian Orthodox Church, 1943–1962*, Washington: The Catholic University of America Press, 1967.
30. Ramet, *Cross and Commissar*, p. 5.
31. Ibid., p. 187; Sabrina P. Ramet, *Nihil Obstat: Religion, Politics, and Social Change in East-Central Europe and Russia*, Durham, London: Duke University Press, 1998, pp. 10–50.
32. Lucian N. Leustean (ed.), *Eastern Christianity and the Cold War, 1945–91*, London: Routledge (forthcoming 2009).
33. Daniel Chirot, *Social Change in a Peripheral Society: The Creation of a Balkan Colony*, New York, London: Academic Press, 1976; Gilberg, Trond, 'Religion and Nationalism in Romania' in Pedro Ramet (ed.), *Religion and Nationalism in Soviet and East European Politics*, Durham: Duke University Press, 1984.
34. Keith Hitchins, *Orthodoxy and Nationality: Andreiu Şaguna and the Rumanians of Transylvania, 1846–1873*, Cambridge, MA: Harvard University Press, 1977.
35. Lucian Boia, *History and Myth in Romanian Consciousness*, Budapest: Central European University Press, 2001.
36. Cristian Troncotă, *Istoria securităţii regimului comunist din România, vol. 1, 1948–1964* [The History of the *Securitate* of the Communist Regime in Romania. vol. 1, 1948–64], Bucharest: Institutul Naţional pentru Studiul Totalitarismului, 2003; Krzysztof Persak, Lukasz Kamiński (eds), *A Handbook of the Communist Security Apparatus in East Central Europe, 1944–1989*, Warsaw: Institute of National Remembrance, 2005.
37. Earl A. Pope, 'Protestantism in Romania' in Sabrina Petra Ramet (ed.), *Protestantism and Politics in Eastern Europe and Russia: The Communist and Postcommunist Eras*, Durham, London: Duke University Press, 1992, pp. 191–2.
38. Vincent C. Chrypinski, 'The Catholic Church in Poland, 1944–1989' in Pedro Ramet (ed.), *Catholicism and Politics in Communist Societies*, Durham, London: Duke University Press, 1990, p. 118.
39. Denis R. Janz, *World Christianity and Marxism*, New York, Oxford: Oxford University Press, 1998, p. 72.

40. Jan Kubik, *The Power of Symbols Against the Symbols of Power: The Rise of Solidarity and the Fall of State Socialism in Poland*, University Park: Pennsylvania State University Press, 1994.
41. Spas T. Raikin, 'The Communists and the Bulgarian Orthodox Church, 1944–1948: The Rise and Fall of Exarch Stefan', *Religion in Communist Lands*, 1984, 12, pp. 281–91.
42. Carsten Riis, *Religion, Politics, and Historiography in Bulgaria*, Boulder, CO: East European Monographs, 2002, p. 104.
43. Spas T. Raikin, 'The Bulgarian Orthodox Church' in Pedro Ramet (ed.), *Eastern Christianity and Politics in the Twentieth Century* (Durham, London: Duke University Press), 1988, pp. 160–82.
44. Ibid.
45. Stella Alexander, *Church and State in Yugoslavia since 1945*, Cambridge: Cambridge University Press, 1979, p. 164; Klaus Buchenau, *Orthodoxie und Katholizismus in Jugoslawien 1945–1991: Ein serbisch-kroatischer Vergleich*, Wiesbaden: Harrassowitz 2004.
46. Owen Chadwick, *The Christian Church in the Cold War*, London: Penguin, 1993, p. 32.
47. Pedro Ramet, 'The Serbian Orthodox Church' in Pedro Ramet (ed.), *Eastern Christianity and Politics in the Twentieth Century*, Durham, London: Duke University Press, 1988, pp. 232–48.
48. Sabrina Petra Ramet, 'The Serbian Church and the Serbian Nation' in Sabrina Petra Ramet and Donald W. Treadgold (eds), *Render Unto Caesar: The Religious Sphere in World Politics*, Washington: The American University Press, 1995, pp. 301–24; Vjekoslav Perica, *Balkan Idols: Religion and Nationalism in Yugoslav States*, Oxford: Oxford University Press, 2002; Christos Mylonas, *Serbian Orthodox Fundamentals: The Quest for an Eternal Identity*, Budapest, New York: Central European University Press, 2003.
49. Martin, David, *Does Christianity Cause War?* London: Clarendon Press, 1997. See the chapter on Romania.
50. Gavril Flora and Georgina Szilagyi, 'Church, Identity, Politics: Ecclesiastical Functions and Expectations towards Churches in Post-1989 Romania' in Victor Roudometof, Alexander Agadjanian and Jerry Pankhurst (eds), *Eastern Orthodoxy in a Global Age: Tradition Faces the Twenty-First Century*, Walnut Creek: Altamira Press, 2005, pp. 109–43.
51. Lavinia Stan and Lucian Turcescu, *Religion and Politics in Post-Communist Romania*, Oxford: Oxford University Press, 2007.
52. Liviu Andreescu, 'The Construction of Orthodox Churches in Post-communist Romania', *Europe-Asia Studies*, 2007, 59 (3), pp. 451–80.

2 The Political Control of Orthodoxy and Romanian Nationalism, 1859–1944

1. *L'Empereur Napoléon III et Les Principautés Roumaines*, Paris: E. Dentu, 1858; *La Question d'Orient et la Nation Roumaine*, Paris: Librairie du Luxembourg, 1867.
2. T. W. Riker, *The Making of Roumania: A Study of an International Problem, 1856–1866*, London: Oxford University Press, 1931, pp. 348–9.
3. *Le Protectorat du Czar ou La Roumanie et la Roussie: Nouveaux documents sur la situation européenne*, Paris: Au Comptoir des Imprimeurs-Unis, 1850; M. T. Cidharold,

La Turquie et les Principautés Danubiennes, Paris: E. Dentu, 1857; *L'Autriche et le Prince Roumain*, Paris: E. Dentu, 1859.

4. Dan Berindei, Elisabeta Oprescu, Valeriu Stan, *Documente privind domnia lui Alexandru Ioan Cuza: 1859–1861* [Documents Regarding the Reign of Alexandru Ioan Cuza: 1859–1861], vol. 1, Bucharest: Editura Academiei, 1989.

5. Keith Hitchins, *Romanians. 1774–1866*, Oxford: Clarendon Press, 1996, p. 297.

6. Commission Princière de la Roumanie a l'Exposition universelle de Paris en 1867, *Notice sur la Roumanie principalement au point de vue de son èconomie rurale industrielle et commerciale avec une carte de la Principautè de Roumanie*, Paris: Librairie A. Franck, 1867, p. 17.

7. Ibid., p. 19.

8. Ibid., p. 18.

9. R. W. Seton-Watson, *A History of the Roumanians: From Roman Times to the Completion of Unity*, Cambridge: The University Press, 1934, pp. 306–7.

10. The Orthodox Church had around 70 dedicated monasteries. *Quelques mots sur la sècularisation des biens conventuels en Roumanie*, Paris: E. Dentu, 1864, p. 29.

11. *La France: Le Prince Couza et la Liberté en Orient*, Paris: Chez les Principaux Librairies, 1864; *Quelques mots sur la sècularisation des biens conventuels en Roumanie*, Paris: E. Dentu, 1864.

12. Ministerul Cultelor şi Instrucţiunii, *Casa Bisericii 1902–1919* [The House of the Church 1902–1919], Bucharest: Tipografia Cărţilor Bisericeşti, 1920, p. 6.

13. His decision was approved by the national assembly with a vote of 93 for and 3 against on 25 December 1863. Hitchins, *The Romanians, 1774–1866*, Oxford: Clarendon Press, 1996, p. 313.

14. Ministerul Cultelor şi Instrucţiunii, p. 7.

15. Stelian Izvoranu, 'Sinoadele de sub Regimul lui Cuza Vodă: Importanţa lor pentru viaţa bisericească' [The Synods during the Reign of Cuza: Their Importance for Church Life], *BOR*, 1960, 7–8, pp. 658–82.

16. Ibid, p. 675.

17. Paul E. Michelson, *Conflict and Crisis: Romanian Political Development, 1861–1871*, New York: Garland Pub., 1987; Gerald J. Bobango, *The Emergence of the Romanian National State*, Boulder: Columbia University Press, 1979.

18. *Plebiscitulŭ Poporului Românu* [The Census of the Romanian People], Bucharest: Typografia Cesar Boliac, 1864, pp. 12–3.

19. Ibid., p. 25.

20. He started a national educational programme aimed at reducing the number of illiterates and at offering more land to the peasants. However, the impact of these laws was limited as at the end of the nineteenth century around 78 per cent of Romanians were still illiterate.

21. *1866–1896: Trei-deci de ani de domnie ai Regelui Carol I. Cuvântări şi acte* [1866–1896: Thirty Years of the Reign of King Carol I. Speeches and Documents], vol. I, Bucharest: Institutul de Arte Grafice Carol Göbl, 1897, p. 1.

22. Ibid., pp. 88–9.

23. 'Constituţia din 1866' [The 1866 Constitution] in *Monitorul – Jurnal Oficial al României*, no 142 from 1 June 1866.

24. J. C. Bratiano, *La question religieuse en Roumanie: Lettre a Monsieur le Directeur de l'Opinion Nationale*, Paris: Librairie du Luxembourg, 1866.

25. Nicolae Iorga, *Istoria Bisericii Româneşti şi a vieţii religioase a Românilor* [The History of the Romanian Church and of Religious Life of Romanians], vol. 1, Vălenii-de-Munte: Tipografia 'Neamul Românesc', 1908, p. 313.

26. *Monitorul Oficial* no. 280 from 19 December 1872.
27. Nifon (1850–75); Calinic Miclescu (1875–86); Iosif Gheorghian (1886–93); Ghenadie Petrescu (1893–96); Iosif Gheorghian (1896–1909); Athanasie Mironescu (1909–11); Conon Arămescu-Donici (1912–9). During the time when there was no primate metropolitan, the church was ruled by the following metropolitan deputies: Inochentie Moisiu Ploieşteanul (August–November 1886); Gherasim Timuş Piteşteanul (April–May 1893); Iosif Naniescu, Gherasim Timuş and Partenie Clinceni (May–December 1896); Teodosie Atanasiu Ploieşteanul (July 1911–February 1912); Platon Ciosu Ploieşteanul (January–December 1919).
28. Ploieşteanul for the Metropolitanate of Wallachia; Craioveanul for the Bishopric of Râmnic; Râmniceanul for the Bishopric of Buzău; Piteşteanul for the Bishopric of Argeş; Botoşăneanul for the Metropolitanate of Moldavia; Bacăoanul for the Bishopric of Roman; Bârlădeanul for the Bishopric of Huşi; Gălăţianul for the Bishopric of Dunărea de Jos. Chiru C. Costescu, *Colecţiune de legi, regulamente, acte, deciziuni, circulări, instrucţiuni, formulare şi programe privitoare la Biserică, culte, cler, învăţământ religios, bunuri bisericeşti, etc.* [Collection of Laws, Regulations, Documents, Decisions, Circular Letters, Instructions, Forms and Programmes], Bucharest: Institutul de Arte Grafice C. Sfetea, 1916, 42.
29. Administraţia Casei Bisericii, *Biserica Ortodoxă şi Cultele Străine din Regatul României* [The Orthodox Church and Foreign Confessions in the Romanian Kingdom], Bucharest: Institutul de Arte Grafice Carol Gőbl, 1904, p. 148.
30. Frederick Kellogg, *The Road to Romanian Independence*, West Lafayette: Purdue University Press, 1995; Barbara Jelavich, *Russia and the Formation of the Romanian National State, 1821–1878*, Cambridge: Cambridge University Press, 1984; Charles and Barbara Jelavich, *The Establishment of the Balkan National States, 1804–1920*, Seattle: University of Washington Press, 1977.
31. *1866–1896*, p. 415.
32. Keith Hitchins, *Romanians. 1866–1947*, Oxford: Clarendon Press, 1994; Dennis P. Hupchick, *The Balkans from Constantinople to Communism*, New York: Palgrave Macmillan, 2002, p. 265.
33. Timothy Ware, *The Orthodox Church*, London: Penguin Books, 1997, pp. 278–9.
34. Ministerul Cultelor şi al Instucţiunii Publice, *Acte privitoare la autocefalia Bisericei Ortodoxe a României* [Documents on the Autocephaly of the Orthodox Church of Romania], Bucharest: Tipografia Cărţilor Bisericeşti, 1885, pp. 3–4.
35. Episcopul de Roman Melchisedek, *Raportu despre relaţiunile bisericesci ale clerului orthodoxu românu cu creştinii eterodoxi seu de alt ritu şi cu necredincioşii carii trăiesc în Regatul Românu* [Report on the Clerical Relations of the Romanian Orthodox Clergy with Heterodox Christians or Other Confessions and with Non-believers Who Live in the Romanian Kingdom], Bucharest: Tipografia Cărţilor Bisericeşti, 1882.
36. *Recensământul Populaţiunei din Decembrie 1899* [Census of the Population in December 1899], Bucharest: Eminescu, 1905.
37. Ministère de l'Agriculture, de l'Industrie, du Commerce et des Domaines, *La Roumanie. 1866–1906*, Bucharest: Imprimerie Socec, 1907, p. 111.
38. Melchisedec Episcopul Romanului, *Memoriu despre starea preoţilor din România şi despre posiţiunea lor morală şi materială* [Report on the Clergy's Situation in Romania and on their Moral and Material Situation], Bucharest: Tipografia Cărţilor Bisericeşti, 1888, p. 9.
39. Ibid., p. 13.
40. Ibid., p. 14.

41. Mihai Constandache, 'Măsuri noi de organizare în Biserica Ortodoxă Română la începutul veacului al XX-lea' [New Measures in the Organisation of the Romanian Orthodox Church at the beginning of the 20th Century], *BOR*, 1965, 40, p. 765.

42. The revised law was published in *Regulament pentru punerea în aplicare a Legei asupra clerului mirean și seminariilor sancționată prin înaltul decret regal no 869 din 25 februarie 1906* [Regulations for the Implementation of the Law of the Clergy and Seminaries Approved by the High Royal Decree No. 869 of 25 February 1906], Bucharest: Imprimeria Statului, 1909; Administrația Cassei Bisericii, *Regulamentul Legii Clerului Mirean și Seminariilor* [Regulations of the Law of the Clergy and Seminaries], Bucharest: Tipografia Cărților Bisericești, 1914.

43. Mircea Păcurariu, *Istoria Bisericii Ortodoxe Române* [The History of the Romanian Orthodox Church], Bucharest: IBMBOR, 1981, p. 142.

44. M. Theodorian-Carada, *Politica Religioasă a României: Conferință ținute la Cercul de studii al Partidului Conservator, în ziua de 3 decembrie 1916* [The Religious Politics of Romania: Conference Held at the Centre for Studies of the Conservative Party on 3 December 1916], Bucharest: Tipografia 'Cooperativa', 1916, p. 7.

45. Ibid., p. 8.

46. Ibid., p. 17.

47. He gave the example of some Armenians who left Romania to enlist in the Russian war and suggested that this happened because they did not feel that Romania was their country. Ibid., p. 21.

48. Glenn E. Torrey, *Romania and World War I: A Collection of Studies* (Iași, Portland: Center for Romanian Studies, 1998); Costică Prodan, Dumitru Preda, *The Romanian Army during the First World War*, Bucharest: Univers Enciclopedic, 1998.

49. R. W. Seton-Watson, *A History of the Roumanians* pp. 521–54.

50. Academia Română, *Istoria Românilor, România întregită (1918–1940)* [The History of Romanians, United Romania (1918–1940)], vol. 8, Bucharest: Editura Enciclopedică, 2003, p. 31.

51. ANIC, Fond Miron Cristea, Dossier 2.

52. Ilie Sandru and Valentin Borda, *Un nume pentru istorie – Patriarhul Elie Miron Cristea* [A Name for History – Patriarch Elie Miron Cristea], Târgu-Mureș: Cartea de Editură 'Petru Maior', 1998, pp. 127–8.

53. Ioan Scurtu, Constantin Mocanu, Doina Smârcea, *Documente privind istoria României între anii 1918–1944* [Documents of Romanian History from 1918 to 1944], Bucharest: Editura Didactică și Pedagogică, 1995, p. 38–9.

54. Sandru and Borda, p. 130.

55. Ibid., p. 133.

56. Antonie Plămădeală, *Contribuții istorice privind perioada 1918–1939: Elie Miron Cristea, documente, însemnări și corespondențe* [Historical Contributions on the 1918–1939 Period: Elie Miron Cristea, Documents, Notes and Correspondence], Sibiu: Tiparul Tipografiei Eparhiale, 1987.

57. ANIC, Fond Miron Cristea. Dossier 3.

58. Miron Cristea, *Principii Fundamentale pentru Organizarea Unitară a Bisericii Ortodoxe Române* [Fundamental Principles for the United Organisation of the Romanian Orthodox Church], Bucharest: Tipografia Cărților Bisericești, 1920, p. 3.

59. Ibid., p. 4.

60. Ibid., p. 6.

61. Ibid., p. 11.

62. Ibid., p. 20.

63. Ibid., p. 14.

64. Philip Martineau, *Roumania and her Rulers*, London: Stanley Paul & Co., 1927, p. 107–8.

65. American Committee on the Rights of Religious Minorities, *Roumania: Ten Years After*, Boston: The Beacon Press, 1928, p. 96.

66. Article 5 of *Regulamentul general al Societății Ortodoxe Naționale a Femeilor Române* [General Regulations of the National Orthodox Society of Romanian Women], Bucharest: Tipografia Albina, 1943.

67. Pimen, Mitropolitul Moldovei, *Mărășești, locul biruinței cu biserica neamului* [Mărășești, the Place of Victory with the Church of the People], Neamț: Tipografia Monastirei Neamțu, 1924.

68. Scurtu, p. 53–4.

69. ANIC, Fond Miron Cristea, Dossier 3.

70. Ibid.

71. The new calendar led to a split in the church and the creation of a small fraction, the Orthodox Believers of Old Rite, which continued to preserve the former calendar. The state supported the Orthodox Church and persecuted the leaders of the Old Rite who were seen as not representing the state's interests.

72. 'Cronica bisericească' [Ecclesiastical Chronicle], *BOR*, 1924, 1, p. 33.

73. Academia Română, *Istoria Românilor*, vol. 8, p. 210.

74. 'Cronica Bisericească: Investitura Patriarhului României' [Ecclesiastical Chronicle: The Enthronement of the Romanian Patriarch], *BOR*, 1925, 11, pp. 703–4.

75. Ibid., pp. 704–5.

76. 'România cufundată în jale' [Romania Deepened in Sorrow], *BOR*, 1927, 7, p. 385.

77. *Report of the Conference at Bucharest from June 1st to June 8th, 1935 between the Rumanian Commission on Relations with the Anglican Communion and the Church of England Delegation Appointed by the Archbishop of Canterbury*, London: Church House, 1936.

78. Dr Nicolas Zernov, secretary of the Fellowship of Saint Alban and Saint Sergius, proposed the establishment of the branch in Romania. The main aim of this fellowship was to establish exchanges of student and publication and to broaden knowledge of the other's faith. The fellowship had three main representatives in Romania, but there was no contact between them: Father Nicodim Ioniță at Cernăuți, Father Florian Gâldău in Bucharest and Basil Prisăcaru in Chișinău. The branch had Father Nicodim Ioniță as its chairman, Father Antim Nica as secretary and Metropolitan Visarion Puiu of Bukovina as honorary president. See Letter from Nicolas Zernov to Father Nicodim Ioniță, 2 May 1935; Chairman Father Nicodim Ioniță, General Report, 20 May 1937 Cernăuți, AFSASS. In 1937 the fellowship had only 30 priests in Romania and four British students studied in the country. The Romanian branch received the journal *Sobornost* and published articles translated from English in the periodicals *Raze de Lumina* and *Misionarul*. Letters from Nicolas Zernov to Father Nicodim Ioniță, 6 November 1936; 27 November 1936; Letter from Nicolas Zernov to Walton Hannah, 11 March 1937, AFSASS.

79. Victor Shearburn, Confidential Memorandum to the Executive Committee of the Fellowship, 30 September 1937, AFSASS.

80. P. Stănescu, 'Politicianismul și Biserica' [Politics and the Church], *BOR*, 1926, 8, pp. 448–9; D. Mitrany, *The Land and the Peasant in Rumania*, Cambridge: Cambridge University Press, 1924.

81. The Metropolitanate of Ungrovlahiei had five bishoprics: The Archbishopric of Bucharest, the Bishoprics of Râmnicului-Noului Severin, Buzău, Argeș and Tomis;

The Metropolitanate of Moldavia and Suceava had four bishoprics: The Archbishopric of Iaşi, the Bishoprics of Roman, Huşi and Lower Danube (Galaţi); The Metropolitanate of Ardeal had five bishoprics: The Archbishopric of Alba Iulia and Sibiu, Bishoprics of Arad, Caransebeş, Oradea and Cluj; The Metropolitanate of Bukovina had two bishoprics: The Archbishopric of Cernăuţi and the Bishopric of Hotin (Bălţi); The Metropolitanate of Bessarabia (established in 1927) had two bishoprics: The Archbishopric of Chişinău and the Bishopric of Cetatea Albă- Ismail.

82. The election of bishops remained a complicated procedure. They were elected by a large body, the Electoral Collegium, which was composed of the Clerical National Congress, the members of the vacant Eparchial Assembly and the following named members if they were Orthodox: the president of the Council of Ministers, the minister of Religious Confessions, the president of the Senate, the president of the House of Deputies, the president of the House of Cassation, the president of the Romanian Academy, rectors of the universities and the deans of the faculties of theology. The patriarch was elected by the Clerical National Congress, the members of the Eparchial Assembly of the Archbishopric of Bucharest and the Orthodox members of the House of Deputies and of the Senate. The law stated that after canonical examination by the Holy Synod, the chosen candidate was confirmed and invested by the king and religiously enthroned.

83. The parish had as legislative body the Parish Assembly (*Adunarea Parohială*) and as executive body the Parish Council (*Consiliul Parohial*) composed of 10 to 30 members. The church's local patrimony was under the supervision of the Parish Guardianship (*Epitropia Parohială*) composed of three to five members. The Deanery (*Protopopiat*) had 20 to 50 parishes under its control and followed the same principle in its organisation: the legislative body, the Deanery Assembly (*Adunarea Protopopească*) (15 to 24 members with 1/3 clergy and 2/3 lay members), the executive body and the Deanery Council composed of six members (*Consiliul Protopopesc*). It was financially controlled by four guardians (*epitropi*). The Bishopric had the same composition: the legislative body was the Eparchial Assembly (*Adunarea Eparhială*) (45 to 60 members with 1/3 clergy and 2/3 lay members) and the executive body; the Eparchial Council was composed of six members (*Consiliul Eparhial*) and each bishop had a vicar. This extent of lay participation in the organisation of the church had been in place in the Orthodox Church in Transylvania before unification. Adapting the entire organisation to this formula showed the importance of this region in the life of the church. Keith Hitchins, *Orthodoxy and Nationality: Andreiu Şaguna and the Rumanians of Transylvania, 1846–1873*, Cambridge, MA: Harvard University Press, 1977.

84. Maria Bucur, *Eugenics and Modernisation in Interwar Romania*, Pittsburgh: University of Pittsburgh Press, 2001, p. 73–5.

85. In 1935 the church had ten seminaries included in state budget in the following cities: Bucharest, Cernica, Curtea de Argeş, Râmnicu-Vâlcea, Buzău, Ismail, Iaşi, Galaţi, Roman, Chişinău; in Dorohoi the seminary was supported from students' taxes and local bishoprics. In addition a private seminary, Nifon, functioned in Bucharest from private funding. In these seminaries 213 teachers were teaching 2847 students.

86. In 1935 the church had theological academies in Cluj, Sibiu, Arad, Oradea and Caransebeş and faculties of theology in Bucharest, Cernăuţi and Iaşi, the latter having its headquarters in Chişinău.

87. Mande Parkinson, *Twenty Years in Roumania*, London: George Allen & Unwin, 1921, p. 44.

88. Ioan Crăciun, 'Situația clerului din Ardeal: O comparație' [The Clergy's Situation in Transylvania: A Comparison], *BOR*, 1923, 15, pp. 1114–5.

89. 'Rapoarte generale pe anii 1932–1935 către Cogresul Național Bisericesc din 14 octombrie 1935' [General Reports on 1932–5 to the National Clerical Congress from 14 October 1935], *BOR*, 1936, 1–2, pp. 23–5.

90. Ministère de l'Agriculture, de l'Industrie, du Commerce et des Domaines, *La Roumanie. 1866–1906*, Bucharest: Imprimerie Socec, 1907.

91. Parkinson, p. 48.

92. Miron Cristea, *Memoriu cu privire la trebuințele Bisericii Ortodoxe Române din țară* [Document on the Internal Problems of the Romanian Orthodox Church], 1922 in ANIC, Fond Miron Cristea, Dossier 3.

93. Nicodim, *Biruința nu se poate dobândi numai prin destoinicia clerului, ci prin vitejia întregii oștiri creștine* [Victory Cannot be Obtained only through Clerical Ability, but through the Courage of the Whole Christian Army], Bucharest: Tipografia Cărților Bisericești, 1941, p. 24.

94. *Recensământul Populațiunei din Decembrie 1899* [Census of the Population of December 1899], Bucharest: Eminescu, 1905; Institutul Central de Statistică, *Recensământul General al Populației României din 29 decembrie 1930, publicat de Dr Sabin Manuilă*, [General Census of the Population of Romania of 29 December 1930, published by Dr Sabin Manuilă], vol. 2–4, Bucharest: Monitorul Oficial, Imprimeria Națională.

95. Un om al bisericii [A man of the Church], *Papism și Ortodoxism în Ardeal sau Porfiră și cunună de spini: Studiu statistic bisericesc*, [Papism and Orthodoxism in Transylvania or Porphyry and Thorn Wreath: A Study of Clerical Statistics], Arad: Tiparul tipografiei diecesane, 95 pages in I. Mihalcescu, 'Cronica internă' [Internal Chronicle], *BOR*, 1923, 7, pp. 535–9.

96. Arhiereu Grig. L. Botoșăneanu, 'Biserica Ortodoxă Română și celelalte confesiuni' [The Romanian Orthodox Church and Other Confessions], *BOR*, 1928, 6, p. 487.

97. Nicolae Bălan, *Biserica neamului și drepturile ei: Discurs rostit la discuția generală asupra proiectului de lege a cultelor, în ședinta dela 27 martie 1928, a senatului român* [The Church of the People and its Rights: Speech presented in the General Discussion on the Proposed Law of Religious Confessions, in the Meeting on 27 March 1928 at the Romanian Senate], Sibiu: Tiparul Tipografiei Arhidiecezane, 1928.

98. 'Spicuiri în cuvântările dela Senat despre Legea Cultelor' [Quotations from Senate Speeches on the Law of Religious Confessions], *BOR*, 1928, 5, p. 459–61.

99. Nae Ionescu, 'A fi bun român' [Being a Good Romanian], *Cuvântul*, 6, 2 November 1930.

100. Dumitru Stăniloae, 'Idealul național permanent' [The Permanent National Ideal], *Telegraful român*, 1940, 5, p. 1.

101. D. Staniloae, 'Ortodoxie și latinitate' [Orthodoxy and Latinity], *Gândirea*, 1939, 18 (4), p. 197.

102. Dumitru Stăniloae, 'Prin ce se promovează conștiința națională' [How Might National Conscience be Promoted?], *Telegraful român*, 1940, 6, p. 1.

103. Vasile Gh. Ispir, *Misiunea actuală a Bisericii Ortodoxe Răsăritene (Misiunea Externă a Bisericii Noastre)* [The Present Mission of the Eastern Orthodox Church (The External Mission of Our Church)], Bucharest: Tipografia Cărților Bisericești, 1938.

104. Keith Hitchins, 'Orthodoxism: Polemics over Ethnicity and Religion in Interwar Romania' in Ivo Banac and Katherine Verdery (eds), *National Character and*

National Ideology in Interwar Eastern Europe, New Haven: Yale University Press, 1995, pp. 135–56.

105. Nichifor Crainic, *Ortodoxie și Etnocrație cu o Anexă: Programul Statului Etnocratic* [Orthodoxy and Ethnocracy with an Appendix: The Programme of the Ethnocratic State], Bucharest: Editura Cugetarea, 1937, p. 188.

106. Ibid.

107. His organisation would be renamed in 1930 as the 'Iron Guard', in 1935 as 'Everything for the Country' and in 1940 as the 'Legionnaire Movement'.

108. Irina Livezeanu, *Cultural Politics in Greater Romania: Regionalism, Nation Building, and Ethnic Struggle, 1918–1930*, Ithaca, NY, London: Cornell University Press, 1995, p. 305; Radu Ioanid, *The Sword of the Archangel: Fascist Ideology in Romania*, New York, Boulder: Columbia University Press, 1990; Ioan Scurtu, *Minoritățile naționale din România, 1931–1938* [National Minorities in Romania, 1931–8], Bucharest: Editura Arhivele Statului, 1999; Armin Heinen, *Die Legion 'Erzengel Michael' in Rumänien Soziale Bewegung unde Politische Organisation*, Munich: R. Oldenbourg Verlag, 1986.

109. H. Bolitho, *Romania under King Carol*, London: Eyre & Spottiswoode, 1939.

110. Zigu Ornea, *The Romanian Extreme Right: The Nineteen Thirties*, trans. by Eugenia Maria Popescu, Boulder: East European Monographs, 1999, pp. 288–9.

111. Ibid., p. 4.

112. Francisco Veiga, *La Mistica del Ultranacionalismo: Historia de la Guardia de Hierron. Rumania, 1919–1941*, Bellaterra: Universitat Autonoma de Barcelona, 1989. In Romanian, *Istoria Gărzii de Fier, 1919–1941, Mistica ultranaționalismului*, Bucharest: Humanitas, 1995, p. 231.

113. Alan Scarfe, 'The Romanian Orthodox Church' in Pedro Ramet (ed.), *Eastern Christianity and Politics in the Twentieth Century*, Durham, NC.: Duke University Press, 1988, pp. 215–6.

114. Costantin Iordachi, 'Charisma, religion, and Ideology: Romania's Interwar Legion of the Archangel Michael' in John R. Lampe, Mark Mazower (eds), *Ideologies and National Identities: The Case of Twentieth-Century Southeastern Europe*, Budapest, New York: Central European University Press, 2004, pp. 35–6.

115. 'Cronica bisericească' [Ecclesiastical Chronicle], *BOR*, 1937, 3–4, pp. 244–7.

116. 'Cronica internă' [Internal Chronicle], *BOR*, 1937, 5–6, p. 373.

117. 'Rapoarte generale pe anii 1932–1935 către Cogresul Național Bisericesc din 14 octombrie 1935'.

118. Ministerul Finanțelor, *Expunere de motive la Bugetul general al statului pe exercițiul 1939/1940* [Reasons for the State Budget on 1939/1940], Bucharest: Monitorul Oficial și Imprimeriile Statului, 1939, p. 59; p. 237.

119. In the state budget for 1939/1940 the Ministry of Religious Confessions and Arts received 878.9 million lei; the Ministry of Health and Social Assistance, 927.6 million lei; the Ministry of Agriculture, 770.5 million lei; the Ministry of National Economy, 167.4 million lei; the Ministry of Foreign Affairs, 303.5 million lei.

120. 'Cronica internă' [Internal Chronicle], *BOR*, 1937, 11–12, p. 749.

121. Gh. Velehorschi, *Colaborarea Bisericii Ortodoxe Române cu Straja Țării: Raport pentru Congresul Național Bisericesc* [Collaboration of the Orthodox Church with the Guardians of the Country: Report for the National Clerical Congress], Bucharest: Tipografia Cărților Bisericești, 1939, p. 6.

122. Rebecca Haynes, *Romanian Policy towards Germany, 1936–40*, London: Macmillan, 2000.

123. *Chronicle* by Dr Nicolas Zernov from Father Nicodim Ioniță's letters, [undated], AFSASS.

124. Letter from Walton Hannah to Dr Nicolas Zernov, 29 January 1937; 1 April 1937, AFSASS.
125. The church later established a foundation under the coordination of the Romanian Patriarchate aimed to build a monastery and hospitals in the region. 'Cronica internă' [Internal Chronicle], *BOR*, 1938, 9–10, p. 603.
126. 'Cronica internă' [Internal Chronicle], *BOR*, 1939, 7–8, pp. 469–83.
127. Iuliu Scriban, 'Un dar al lui Dumnezeu' [A Gift from God], *BOR*, 1945, 11–2, pp. 688–698.
128. Ibid.
129. Ibid.
130. 'Însemnări' [Notes], *BOR*, 1940, 1–2, p. 96.
131. 'Pastorala pentru ajutorarea refugiaţilor din Basarabia şi Bucovina' [Pastoral Letter for Helping the Refugees from Bessarabia and Bukovina], *BOR*, 1940, 7–8, pp. 518–20.
132. Gheorghe Vasilescu, 'Nicodim, al doilea patriarh al României' [Nicodim, the Second Patriarch of Romania], *Magazin Istoric*, 1998, 5, pp. 42–5.
133. Marcel-Dumitru Ciucă, Aurelian Teodorescu, Bogdan Florin Popovici, *Stenogramele şedinţelor Consiliului de Miniştri: Guvernarea Ion Antonescu* [The Minutes of the Meetings of the Council of Ministers: Ion Antonescu's Regime], Bucharest: Arhivele Naţionale ale României, 2001, p. 421.
134. Nicodim, *Pastorală* [Pastoral Letter], July 1941; Nicodim, *Sfat duhovnicesc către toată suflarea românească* [Spiritual Advise to all Romanian People], Bucureşti: Tipografia Cărţilor Bisericeşti, 1941.
135. Georgescu, pp. 216–7.
136. Dennis Deletant, *Hitler's Forgotten Ally: Ion Antonescu and His Regime, Romania, 1940–1944*, Basingstoke: Palgrave Macmillan, 2006, p. 168. The religious mission to Transnistria was headed by Archimandrite Iuliu Scriban, from 15 August 1941 to 16 November 1942, followed by Metropolitan Visarion Puiu until 14 December 1943 and Archimandrite Antim Nica.
137. Earl A. Pope, 'Protestantism in Romania' in Sabrina Petra Ramet (ed.), *Protestantism and Politics in Eastern Europe and Russia: The Communist and Postcommunist Eras*, Durham, London: Duke University Press, 1992.
138. Dinu C. Giurescu, *Romania in the Second World War (1939–1945)*, trans. by Eugenia Elena Popescu, Boulder: East European Monographs, 2000.
139. Maurice Pearton, Dennis Deletant, 'The Soviet Takeover in Romania, 1944–48' in Gill Bennett (ed.), *The End of the War in Europe 1945*, London: HMSO, 1996.

3   Orthodoxy and the Installation of Communism, 1944–7

1. Dennis Deletant, *Romania under Communist Rule*, Iaşi, Oxford, Portland: The Center for Romanian Studies, 1999, p. 33; *6 Martie 1945: Începuturile Comunizării României* [6 March 1945: The Beginning of Communism in Romania], Bucharest: Editura Enciclopedică, 1995; Gheorghe Buzatu, Mircea Chiriţoiu, *Agresiunea comunismului în România. Documente din arhivele secrete: 1944–1989* [The Aggression of Communism in Romania: Documents from Secret Archives. 1944–89], Bucharest: Paideia, 1998.
2. In 1925 the Romanian Communist Party had 1661 members. It had 461 members in 1929 and 1635 members in 1937. The drop in number was due to the party being declared illegal by the government in 1924.

3. Vlad Georgescu, *The Romanians: A History*, trans. by Alexandra Bley-Vroman, London, New York: Tauris, 1991, p. 228.

4. As Churchill indicates, the Soviet Union had 90 per cent influence in Romania, 75 per cent in Bulgaria, 50 per cent in Hungary and Yugoslavia and 10 per cent in Greece. Winston Churchill, *The Second World War: Triumph and Tragedy*, vol. 6, London: Cassell, 1954.

5. Nicodim, *Telegrama către Majestatea Sa Regele Mihai* [Telegram to His Majesty King Michael], Bucharest, no publisher, 24 August 1944.

6. Nicodim, *Pastorala 9 Octombrie 1944* [Pastoral Letter 9 October 1944], Bucharest: no publisher, p. 3.

7. Ibid.

8. Ibid., p. 4.

9. Ibid., p. 6.

10. Ibid., p. 8.

11. Ibid., p. 11.

12. Paul Mojzes, *Religion Liberty in Eastern Europe and the USSR Before and After the Great Transformation*, Boulder: East European Monographs, 1992, p. 317.

13. Nicodim, *Pastorala 9 Octombrie 1944*, p. 6.

14. Miron Cristea, *Rânduiala Doxologiei din prilejul încoronării majestăților lor Regelui Ferdinand I și a Reginei Maria* [The Doxology Ceremony during the Enthronement of their Majesties King Ferdinand I and Queen Maria], Bucharest: Tipografia Cărților Bisericești, 1922.

15. Paul D. Quinlan, *Clash over Romania: British and American Policies towards Romania: 1938–1947*, Los Angeles: American Romanian Academy, 1977, p. 128; Dinu Giurescu, *Romania's Communist Takeover: The Rădescu Government*, Boulder: East European Monographs, 1994.

16. Mojzes, p. 318.

17. Ghita Ionescu, *Communism in Rumania, 1944–1962*, Westport, CT: Greenwood Press, 1964, p. 110–11.

18. In the Orthodox Church priests are normally married and in order to be elected a bishop, the candidate should be a monk. Before being elected bishop, Justinian had been a priest under the lay name of Ioan Marina. After the death of his wife a few years earlier, he was eligible to be nominated as a member of the superior church hierarchy. George Stan, *Părintele Patriarch Justinian Marina* [Father Patriarch Justinian Marina], Bucharest: EIBMBOR, 2005.

19. Reuben H. Markham, *Rumania under the Soviet Yoke*, Boston: Meador Publishing, 1949, p. 475–6.

20. Justinian Marina, 'Cooperația și creștinismul' [Cooperation and Christianity], *Renașterea*, 1931, 10, pp. 49–57; 108–16; 419–22.

21. During communism the government had the following ministers of religious confessions: Constantin Burducea, 6 March 1945–30 November 1946; Radu Roșculet, 1 December 1946–29 December 1947; Stanciu Stoian, 30 December 1947–23 April 1951; Vasile Pogăceanu, 23 April 1951–2 June 1952; Petre Constantinescu-Iași, 28 January 1953–19 March 1957; Dumitru Dogaru, general secretary of the Department of Religious Confessions from 1970, president of the Department of Religious Confessions, 1957–1975; Gheorghe Nenciu, 1975–4 February 1977; Ion Roșianu, 4 February 1977–7 May 1984; Ion Cumpănașu, 7 May 1984–18 January 1990.

22. Markham, p. 223.

23. Burducea escaped from Romania in 1948 immigrating to Chile and then to Venezuela where he became a successful businessman. Ex-Priest Turned

Communist Arrives in Rome in Mercedes, Item 9367/56, HU OSA 300–60–1, 530/2808, Religion: Other Religions, 1956–1959.

24. C. Burducea, 'Cuvântul la deschiderea Congresului general al preoţilor ortodocşi şi al slujitorilor tuturor cultelor din România' [The Speech at the Opening of the General Congress of the Orthodox Priests and of all Religious Confessions in Romania] in *Statutul Uniunii Preoţilor Democraţi din România* [The Statute of the Union of the Democratic Priests in Romania], Bucharest: no publisher, 1945, p. 14.

25. Ibid., p. 15.

26. Ibid., p. 16.

27. ASRI, Fond D, Dossier 2327, f. 77, in Păiuşan and Ciuceanu, p. 49.

28. ASRI, Fond D, Dossier 2488, vol. 1, f. 77–84, in Păiuşan and Ciuceanu, pp. 50–5.

29. Ibid.

30. Justinian, *Apostolat social: Pilde şi îndemnuri pentru cler* [Social Apostolate: Examples and Appeals towards Clergy], vol. 2, Bucharest: Tipografia Cărţilor Bisericeşti, 1948, p. 70.

31. Ioan Vască, *O călătorie istorică: Vizita Înalt Prea Sfinţitului Patriarh Nicodim al României la Moscova. Reluarea legăturilor dintre Biserica Ortodoxă Română şi Biserica Ortodoxă Rusă* [A Historic Journey: The Visit of His Holiness Patriarch Nicodim of Romania to Moscow. The Re-establishment of Relations between the Romanian Orthodox Church and the Russian Orthodox Church], Bucharest: Tiparul Cărţilor Bisericeşti, 1947.

32. Ibid., p. 188.

33. Ibid.

34. S. Cândea, *Îndatoriri actuale ale Bisericii Ortodoxe Române* [Current Obligations of the Romanian Orthodox Church], Sibiu: Tipografia Arhidiecezană, 1946.

35. The real figures show that the National Peasant Party won with about 70 per cent of the vote. Deletant, *Romania under Communist Rule*, p. 48; Tismăneanu, *Stalinism for All Seasons: A Political History of Romanian Communism*, p. 288.

36. Ionescu, p. 219.

37. See Ionescu 1964, p. 131.

38. ASRI, Fond D, Dossier 9471, f. 102, in Păiuşan and Ciuceanu, pp. 73–4.

39. ASRI, Fond D, Dossier 9471, f. 82–87, in Păiuşan and Ciuceanu, pp. 71–3.

40. ASRI, Fond D, Dossier 9471, f. 89, in Păiuşan and Ciuceanu, p. 71.

41. ASRI, Fond D, Dossier 9471, f. 102, in Păiuşan and Ciuceanu, pp. 73–4.

42. ASRI, Fond D, Dossier 9471, f. 82–87, in Păiuşan and Ciuceanu, pp. 72–3.

43. Justinian, *Apostolat social*, p. 154.

44. Markham, p. 476.

45. Justinian, *Apostolat social*, p. 147.

46. Nicodim, *Pastorala pentru Republica Populară Română* [Pastoral Letter for the Romanian People's Republic], 1948; for more see Alan Scarfe, 'Patriarch Justinian of Romania: His Early Social Thought', *Religion in Communist Lands*, 5 (3), 1977, pp. 164–9; Ion Raţiu, 'The Communist Attacks on the Catholic and Orthodox Churches in Rumania', *The Eastern Churches Quarterly*, London, 1949, 8, (3), pp. 163–97; A. C. Rădulescu, 'Libertatea religioasă în Republica Populară Română' [Religious Liberty in the Romanian People's Republic], *Ortodoxia*, 1949, 2–3, pp. 3–56; Alan Scarfe, 'The Romanian Orthodox Church' in Pedro Ramet (ed.), *Eastern Christianity and Politics in the Twentieth Century*, Durham, London: Duke University Press, 1988; Tudor R. Popescu, *Salvarea bisericilor de orice rit din România sub ocupaţia militară sovietică* [The Salvation of Church under Any Confession in Romania under the Soviet Military Occupation], Bucharest: Asociaţia Română pentru Educaţie Democratică, 1999.

47. Nicoleta Ionescu Gură, *Stalinizarea României. Republica Populară Română: 1948–1950. Transformări instituţionale* [The Stalinisation of Romania. Romanian People's Republic: 1948–1950. Institutional Transformations], Bucharest: Ed. All, 2005; Anca Şincan, 'Mechanisms of State Control over Religious Denominations in Romania in the late 1940s and early 1950s' in B. Apor, P. Apor and A. Rees (eds), *The Sovietization of Eastern Europe*, Washington: New Academia Publishing, 2008.

4 'The Light Rises from the East': Orthodoxy, Propaganda and Communist Terror, 1947–52

1. The Presidium was composed of Constantin I. Parhon, president of the Romanian Association for Friendship Relations with the Soviet Union, who was named as president of the Presidium, and the following members: Mihail Sadoveanu, president of the House of Deputies, Ştefan Voitec, Minister of Education, Gheorghe Stere, president of the Appeal Court in Bucharest and Ion Niculi, vice-president of the House of Deputies.
2. Thus on 22 January the government decreed the establishment of five national youth building sites in which around 170,000 young people would work voluntarily. Most of these 'volunteers' wanted to avoid accusations of coming from 'unhealthy social origins' and to identify themselves with the working class. Dinu C. Giurescu, *Istoria României în date* [The History of Romania in Dates], Bucharest: Editura Enciclopedică, 2003, pp. 503–4.
3. The census data revealed that Romania covered 237,502 km^2 and had 15,872,624 inhabitants; 23.4 per cent of its population was urban and 76.6 rural; 23.1 per cent of the country was illiterate of which 10.7 urban and 27 per cent rural. The census did not indicate the ethnic configuration of the country but only the spoken language of the population. 85.7 per cent spoke Romanian, 9.4 per cent Hungarian, 2.2 per cent German, 0.9 per cent Yiddish, 0.3 per cent Roma, 0.3 per cent Serb, Croat, Slovenian, 0.2 per cent Russian, 0.2 per cent Turkish, 0.2 per cent Ukrainian, 0.2 per cent Czechoslovak, 0.1 per cent Bulgarian, 0.1 per cent Greek, under 0.1 per cent Polish, under 0.1 per cent Albanian, under 0.1 per cent Armenian, 0.1 per cent other languages and under 0.1 per cent did not declare their language. Anton Golopenţia, Petre Onică, *Recensământul agricol din Republica Populară Română, 25 ianuarie 1948: Rezultate provizorii* [The Agricultural Census from the Romanian People's Republic, 25 January 1948: Provisional Results], Bucharest: Direcţiunea Centrală de Statistică, 1948.
4. Justinian, 'Numărătoarea' [The Counting], *Universul*, 25 January 1948.
5. The Popular Democratic Front was composed of the RWP, the Ploughmen's Front (*Frontul Plugarilor*), the Popular National Party (*Partidul Naţional Popular*), the Liberal National Party – Bejan (*Partidul Naţional Liberal – Bejan*) and the Hungarian Popular Union (*Uniunea Populară Maghiară*).
6. There are suggestions that Patriarch Nicodim was killed by the regime while ill and staying at Neamţ monastery. Cicerone Ionitoiu, *Persecuţia Bisericii din România sub dictatura comunistă* [Church Persecution during the Communist Dictatorship], Freiburg: Coresi, 1983. Dudu Velicu, one of Patriarch Nicodim's advisors, suggests in his diaries that the patriarch died of pneumonia. Dudu Velicu, *Biserica Ortodoxă în perioada sovietizării României: Însemnări zilnice. 1945–1947* [The Orthodox Church during Romania's Sovietization: Dairies. 1945–47], Bucharest: Arhivele Naţionale ale României, 2004; Dudu Velicu,

Biserica Ortodoxă în perioada sovietizării României: Însemnări zilnice II. 1948–1959 [The Orthodox Church during Romania's Sovietization: Diaries. 1948–59], Bucharest: Arhivele Naţionale ale României, 2005.

7. Seven seats for the National Liberal Party, Bejan and two seats for the Democratic Peasant Party, Dr Nicolae Lupu.

8. 98.84 per cent in 1952; 98.88 per cent in 1957; 99.85 per cent in 1965; 99.96 per cent in 1969; 98.80 per cent in 1975; 99.99 per cent in 1980; 97.70 per cent in 1985.

9. Ioan Murariu and Gheorghe Iancu, *Constituţiile Române. Texte. Note. Prezentare comparativă* [Romanian Constitutions. Texts. Notes. Comparative Presentation], Bucharest: Regia autonomă 'Monitorul Oficial', 1995.

10. ASRI, Fond D, Dossier 2488, vol. 1, f. 166–7 in Păiuşan and Ciuceanu, pp. 75–6.

11. ASRI, Fond D, Dossier 2488, vol. 1, f. 171, in Păiuşan and Ciuceanu, pp. 76–7.

12. ASRI, Fond D, Dossier 2488, vol. 1, f. 172, in Păiuşan and Ciuceanu, p. 77.

13. Justinian, IPS Locotenent de Patriarh al României, *Pastorala cu prilejul alegerilor pentru Marea Adunare Naţională-Constituantă* [Pastoral Letter on the Elections to the Great National Assembly], Bucharest: no publisher, 14 March 1948.

14. ASRI, Fond D, Dossier 2488, vol. 1, f. 172, in Păiuşan and Ciuceanu, p. 78.

15. ASRI, Fond D, Dossier 9471, vol. 1, f. 372–4, in Păiuşan and Ciuceanu, pp. 79–80.

16. ASRI, Fond D, Dossier 2488, vol. 1, f. 174, in Păiuşan and Ciuceanu, p. 78.

17. Paul Caravia, Virgiliu Constantinescu, Flori Stănescu, *The Imprisoned Church: Romania, 1944–1989*, Bucharest: The Romanian Academy, 1999, p. 28.

18. Keith Hitchins, 'The Romanian Orthodox Church and the State' in *Religion and Atheism in the USSR and Eastern Europe*, Bohdan R. Bociurkiw and John W. Strong (eds), Toronto: University of Toronto Press, 1975, p. 316; Rumanian National Committee, *Persecution of Religion in Rumania*, Washington: no publisher, 1949.

19. Robert Tobias, *Communist-Christian Encounter in East Europe*, Indianapolis: School of Religion Press, 1956, p. 325.

20. Gh. Vintilescu, 'Raportul sectorului administrativ bisericesc din administraţia patriarhală' [Report of the Clerical Administrative Sector from the Patriarchal Administration], *BOR*, 1949, 7–10, p. 48.

21. Ibid., p. 46.

22. Henry Roberts, *Rumania: Political Problems of an Agrarian State*, New Haven: Yale University Press, 1951.

23. Emil Ciurea, 'Religious Life' in Alexandre Cretzianu (ed.), *Captive Rumania: A Decade of Soviet Rule*, London: Atlantic Press, 1956, pp. 174–5.

24. Marius Bucur and Lavinia Stan, *Persecuţia Bisericii Catolice în România. Documente din arhiva Europei Libere, 1948–1960* [The Persecution of the Catholic Church in Romania. Documents from the Archive of Radio Free Europe, 1948–60], Târgu-Lăpuş: Galaxia Gutenberg, 2004.

25. Ioan Lăcustă, *1948–1952. Republica Populară şi România* [People's Republic and Romania], Bucharest: Curtea Veche, 2005, p. 51.

26. Ibid., p. 53.

27. Ibid., pp. 57–8.

28. Ibid., p. 59.

29. Ovidiu Marina, *30 de zile în URSS* [30 days in the USSR], Bucharest: Editura Institutului Biblic şi de Misiune Ortodoxă, 1949.

30. Vintilescu, p. 65.

31. The Romanian delegation presented the following reports: 'The Attitude of the Vatican towards Orthodoxy in the Last Thirty Years'; 'The Possibility of Recognising

the Validity of Priesthood in the Anglican Church'; 'The Orthodox Church and the Ecumenical Movement'; and a short note on the problem of the religious calendar. 'Referatele prezentate de delegația Bisericii Ortodoxe Române la Conferința Ortodoxă de la Moscova' [Reports Presented by the Delegation of the Romanian Orthodox Church at the Orthodox Conference in Moscow], *Ortodoxia*, 1949, 1, p. 29.

32. *Actes de la Conférence des Chef et des Représentants des Eglises Orthodoxes Autocéphales Réunis a Moscou a L'Occasion de la Célébration Solennelle des Fêtes du 500eme Anniversaire de L'Autocéphalie de L'Eglise Orthodoxe Russe, 8–18 Juillet 1948*, Moscow: Editions du Patriarcat de Moscou, 1950.

33. *Monitorul Oficial* no. 176 from 2 August 1948.

34. *Monitorul Oficial* no. 177 from 3 August 1948.

35. 'Legea pentru Regimul general al Cultelor religioase, Monitorul Oficial no. 177 from 4 August 1948' [Law on the General Regime of Religious Confessions] in *Legea și Statutele Cultelor Religioase din Republica Populară Română* [The Law and Statutes of Religious Confessions in the Romanian People's Republic], Bucharest: Editura Ministerului Cultelor, 1949.

36. The Theological Institutes in Bucharest had 13 professors, in Sibiu 10 professors and in Cluj 10 professors.

37. The first director was Bishop Partenie Ciopron. Neamț seminary had 56 students, Agapia 51 and Plumbuita 23. Vintilescu, p. 55.

38. Ciurea, pp. 169–70.

39. Cristian Vasile, 'Atitudini ale clericilor ortodocși și catolici față de URSS și față de regimul de tip sovietic (1944–1948)' [Attitudes of Orthodox and Catholic Clergy towards the USSR and the Soviet Regime (1944–48)] in Ovidiu Bozgan (ed.), *Biserică, Putere, Societate: Studii și Documente* [Church, Power, Society: Studies and Documents], Bucharest: Editura Universității din București, 2001, pp. 155–81.

40. ASRI, Fond D, Dossier 2488, vol. 1, f. 157–60 in Păiușan and Ciuceanu, p. 81.

41. Ibid.

42. The Office 1 from the Direction 1, Service 3 of the General Direction of the People's Security reported on the Orthodox confession with the indicative 131; Office 2 reported on the Roman Catholic confession with the indicative 132 and on Protestant confessions with the indicative 133. Adrian Nicolae Petcu, 'Securitatea și cultele în 1949' [The *Securitate* and Religious Confessions in 1949] in Adrian Nicolae Petcu (ed.), *Partidul, Securitatea și Cultele, 1945–1989* [Party, *Securitate* and Religious Confessions], Bucharest: Nemira, 2005, pp. 124–222.

43. Dennis Deletant, *Ceaușescu and the Securitate: Coercion and Dissent in Romania, 1965–1989*, London: Hurst, 1995, p. 392.

44. Cristian Vasile, *Între Vatican și Kremlin: Biserica Greco-Catolică în timpul regimului comunist* [Between the Vatican and the Kremlin: The Greek Catholic Church during the Communist Regime], Bucharest: Curtea Veche, 2003; Cristian Vasile, *Istoria Bisericii Greco-Catolice sub regimul comunist, 1945–1989: Documente și mărturii* [The History of the Greek Catholic Church under the Communist Regime, 1945–89: Documents and Testimonies], Iași: Polirom, 2003.

45. André Kom, 'Unificarea Bisericii Unite cu Biserica Ortodoxă Română în 1948' [The Unification of the Uniate Church with the Orthodox Church in 1948] in Ovidiu Bozgan (ed.), *Studii de Istoria Bisericii* [Church History Studies], Bucharest: Editura Universității din București, 2000, pp. 88–124.

46. Ovidiu Bozgan, 'Nunțiatura Apostolică din România în anii 1948–1950' [The Apostolic Nunciature in Romania between 1948–50] in Ovidiu Bozgan (ed.),

Biserică, Putere, Societate. Studii și Documente [Church, Power, Society. Studies and Documents], Bucharest: Editura Universității din București, 2001, p. 132.
47. *Stenogramele ședințelor Biroului Politic al Comitetului Central al Partidului Muncitoresc Român, vol. 1, 1948* [The Minutes of the Meetings of the Political Office of the Central Committee of the Romanian Workers' Party. vol. 1, 1948], Bucharest: Arhivele Naționale ale României, 2002, pp. 267–9.
48. *'Cuvântarea IPSS Justinian Patriarhul României rostită cu ocazia unirii'* [The Speech of His Holiness Justinian, the Patriarch of Romania Given on the Occasion of Unity], Sighet: Tipografia Gheorghe Cziple, 1949.
49. British Legation, Bucharest, Item 261/2/49, TNA PRO FO 371/78623.
50. ASRI, Fond D, Dossier 2488, vol. 1, f 186–92, in Păiușan and Ciuceanu, p. 85.
51. ASRI, Fond D, Dossier 2488, f 103, in Păiușan and Ciuceanu, p. 94.
52. The decree was abolished on 31 December 1989.
53. *Stenogramele ședințelor Biroului Politic al Comitetului Central al Partidului Muncitoresc Român vol. 1, 1948*, pp. 380–6.
54. ASRI, Fond D, Dossier 7755, vol. 3, f 55, in Păiușan and Ciuceanu, p. 94.
55. Benedict Ghiuș, CNSAS, I2723.
56. ASRI, Fond D, Dossier 7755, f. 137 in Păiușan and Ciuceanu, p. 119.
57. ASRI, Fond D, Dossier 7755, vol. 3, f. 159, in Păiușan and Ciuceanu, p. 98.
58. ASRI, Fond D, Dossier 9471, f. 121, in Păiușan and Ciuceanu, p. 105.
59. ASRI, Fond D, Dossier 7755, f. 127, in Păiușan and Ciuceanu, p. 115.
60. ASRI, Fond D, Dossier 7755, vol. 3, f. 123 in Păiușan and Ciuceanu, p. 115.
61. ASRI, Fond D, Dossier 7755, vol. 3, f. 133, in Păiușan and Ciuceanu, p. 117.
62. ASRI, Fond D, Dossier 7755, vol. 3, f. 131–2, in Păiușan and Ciuceanu, p. 118.
63. *Monitorul Oficial* no. 22, 27 January 1949.
64. Vintilescu, p. 57.
65. Vintilescu, p. 58.
66. 'Deschiderea cursurilor institutului teologic universitar din București' [The Opening of the Courses at the Theological Institute in Bucharest], *BOR*, 1949, 1–2, p. 9.
67. Ibid., p. 17.
68. Ibid., p. 18.
69. Ibid., p. 20–1.
70. ASRI, Fond D, Dossier 9471, f. 163, in Păiușan and Ciuceanu, p. 114.
71. 'Statutul pentru funcționarea și organizarea Bisericii Ortodoxe Române' [The Statute for the Functioning and Organisation of the ROC] in *Monitorul Oficial* no 47 from 25 February 1949, republished in *BOR*, 1949, 1–2.
72. The Metropolitanate of Wallachia with three eparchial sees: the Archbishopric of Bucharest and the Bishoprics of Buzău and Constanța; the Metropolitanate of Moldavia and Suceava with four eparchial sees: the Archbishopric of Iași; the Bishoprics of Suceava; Roman and Huși; Galați; the Metropolitanate of Transylvania with three eparchial sees: the Archbishopric of Alba-Iulia and Sibiu; the Bishopric of Vad, Feleac and Cluj; the Bishopric of Oradea; the Metropolitanate of Oltenia with two eparchial sees: the Archbishopric of Craiova and the Bishopric of Râmnic and Argeș; the Metropolitanate of Banat with two eparchial sees: the Archbishopric of Timișoara and Caransebeș and the Bishopric of Arad, Ienopol and Halmagiu.
73. Ciurea, p. 170.
74. The Holy Synod was composed of all members of the church hierarchy and presided over by the patriarch. The patriarch was helped in his activities by the Patriarchal Administration which was composed of: two vicar-bishops, six administrative advisors, the Patriarchate Office and the Inspection and Control Offices.

The vicar-bishops were elected by the Holy Synod and the advisors by the National Clerical Assembly on the patriarch's approval. Each advisor was responsible for one of the followings divisions: ecclesiastical, cultural, missionary and foreign affairs, theological education, economic and monasteries (art. 34). The Patriarchate Office, ruled by a director, was the body that put into practice all legislative and executive decisions (art. 35), while the Inspection and Control Offices were composed of three inspectors.

75. The main tasks of the National Clerical Assembly were in the economic and administrative fields as it was the body that could make any changes regarding the eparchial sees (art. 20 e). The assembly was composed of three members of each eparchy (two lay members and one priest) and the members of the Holy Synod with the patriarch as its president (art 21).

76. The National Clerical Council was 'the supreme administrative organ for all church affairs and the executive organ of the Holy Synod and the National Clerical Assembly' (art. 25). It was composed of six lay members and three clerics elected by the National Clerical Assembly and six administrative councillors with the patriarch as its president.

77. *Stenogramele şedinţelor Biroului Politic şi ale Secretariatului Comitetului Central al PMR vol. II, 1949* [The Minutes of the Meetings of the Political Office of the Central Committee of the Romanian Workers' Party, vol. II, 1949], Bucharest: Arhivele Naţionale ale României, 2003, p. 116.

78. Ibid., p. 117.

79. On 2 March Nicolae Ceauşescu was appointed Minister of Agriculture, being responsible for the implementation of the new law. However, with the evolution of peasant resistance to collectivisation and Soviet pressure after the death of Stalin to reform the political class in Romania, Gheorghiu-Dej would claim that sole responsibility for the peasant purges lay not with Ceauşescu but with Luca, Pauker and Teohari Georgescu.

80. They were renamed 'farms of agricultural production' (*gospodării agricole de producţie*) in 1965.

81. Vintilescu, p. 68.

82. Justinian, *Cuvântul Bisericii Ortodoxe Române în chestiunea libertăţii religioase* [The Position of the Romanian Orthodox Church on Religious Freedom], Bucharest: Institutul Biblic şi de Misiune Ortodoxă, 1949, pp. 5–6.

83. Ibid., p. 7.

84. Ibid., p. 8.

85. Ibid, p. 9.

86. Vintilescu, p. 64.

87. Tobias, p. 333.

88. 'Omagiul adus IPS Patriarh Justinian de Adunarea Naţională Bisericească' [Homage to His Holiness Justinian by the National Clerical Assembly], *BOR*, 1949, 7–10, p. 1.

89. 'Colegiul Electoral pentru alegerea episcopului Eparhiei Romanului şi Huşilor' [The Electoral Collegium for the Election of the Bishop of the Bishopric of Roman and Huşi], *BOR*, 1949, 7–10, p. 158.

90. 'Cuvântarea PS Episcop Teofil Herineanu' [The Speech of His Holiness Bishop Teofil Herineanu], *BOR*, 1949, 7–10, p. 165.

91. Tobias, p. 334.

92. The following religious leaders participated in the meeting: Dr Friederich Müller, bishop of the Evangelic Church; Gheorghe Argay, bishop of the Lutheran-Evangelic

Church; Dr Moses Rosen, chief rabbi of the Jewish community; Dr Simen Dănilă, bishop of the Unitarian Church; Archimandrite Vasken Balgian, chief of the Armenian Church; Metropolitan Tihon of the Old Rite Christians (Lipovenian) Church; and Mehmet, the Mufti of the Muslim community. 'Consfătuirea de la Sfânta Patriarhie a Bisericii Ortodoxe Române' [The Meeting at the Holy Patriarchate of the Romanian Orthodox Church], *Ortodoxia*, 1949, 2–3, p. 8.

93. 'Cuvântarea Domnului Dr Petru Groza' [The Speech of Dr Petru Groza], *Ortodoxia*, 1949, 2–3, p. 46.

94. 'Cuvântarea Domnului Vasile Luca, Vice Preşedinte al Consiliului de Miniştri' [The Speech of Vasile Luca, Vice-President of the Council of Ministers], *Ortodoxia*, 1949, 2–3, p. 47–8.

95. '"La 21 octombrie 1949": Cuvântul IPS Patriarh Justinian' [21 October 1949: The Speech of His Holiness Patriarch Justinian], *Ortodoxia*, 1949, 4, p. 12.

96. 'Deciziunea nr 107', *BOR*, 1949, 11–2, pp. 102–11.

97. ASRI, Fond D, Dossier 2488, vol. 1, f. 415–33 in Păiuşan and Ciuceanu, p. 165; Justinian Moisescu, CNSAS, I185025.

98. ASRI, Fond D, Dossier 2488, vol. 1, f. 415–33 in Păiuşan and Ciuceanu., p. 166.

99. Ibid., p. 167.

100. Ibid., p. 168.

101. Ibid., p. 169.

102. Ibid., p. 174.

103. Ibid., p. 175.

104. George Rosu, Mircea Vasiliu and George Crisan, 'Church and State in Romania' in Vladimir Gsovski (ed.), *Church and State Behind the Iron Curtain: Czechoslovakia, Hungary, Poland, Romania with an Introduction on the Soviet Union*, New York: Frederick A. Pragaer, 1956, pp. 292–3.

105. 'Sărbătorirea a 25 de ani de la înfiinţarea Patriarhiei Române' [The Celebration of 25 Years from the Establishment of the Romanian Patriarchate], *BOR*, 1950, 3–6, p. 168.

106. The patriarch mentioned the following names which were considered saints by the people: Nicodim of Tismana, Leonte of Rădăuţi, Ioan from Prislop, Daniil Sihastru, Stephen the Great, Iosif the New from Partoş, Ioan from Râşca, Calinic the Saint, Antipa from Calapodeşti; and other saints whose relics were in Romania: Saint Ioan the New from Suceava, Saint Dumitru the New, Saint Antonie from Vâlcea and Paisie from Neamţ.

107. ASRI, Fond D, Dossier 2488, vol. 1, f. 223–8 in Păiuşan and Ciuceanu, p. 180–3.

108. 'Sărbătorirea a 25 de ani de la înfiinţarea Patriarhiei Române', *BOR*, 1950, 3–6, p. 185.

109. 'Biserica Ortodoxă Română şi Apărarea Păcii: Pastorala Sfântului Sinod către clerul şi creştinii ortodocşi şi către toţi creştinii din Republica Populară Română' [The Romanian Orthodox Church and the Defence of Peace: The Pastoral Letter of the Holy Synod to the Clergy and Orthodox Christians and to all Christians in the Romanian People's Republic], *BOR*, 1950, 3–6, p. 164.

110. Ibid., p. 166.

111. Ibid.

112. Chesarie Păunescu, CNSAS, I185031, 3 vols.

113. A few months later he would be the first hierarch from the Holy Synod who would have to retire. The new positions of these hierarchs were: Metropolitan Efrem Enăchescu, became abbot at Cernica monastery; Bishop Partenie Ciopron, abbot at Suceava monstery; Bishop Veniamin Nistor, abbot at Holy Trinity – Alba

Iulia monastery; Bishop Policarp Moruşcă, monk at Holy Trinity – Alba Iulia monastery; Bishop Emilian Antal, abbot at Cozia-Vâlcea monastery; Bishop Eugeniu Laiu, abbot at Cocoş-Tulcea monastery; Bishop Atanasie Dincă, abbot at Saint Apostles – Huşi monastery; Bishop Teodor Scobăreţ, monk at Holy Trinity – Sibiu monastery; Bishop Galaction Cordun, monk at Căldăruşani monastery; Bishop Veniamin Pocitan, monk at Cernica monastery; Metropolitan Tit Simedrea, monk at Dărvari monastery; and Metropolitan Nifon Criveanu, monk at Domniţa Bălaşa monastery.

114. On 16 December 1949, Patriarch Alexius of Moscow sent a letter to the Holy Synod in Bucharest asking for the names of those saints who were celebrated in Romania. Deciding which saints should be venerated was suggestive in the propagandistic clashes between the Catholic and Orthodox churches, between West and East. The Orthodox feared that the Catholics were claiming that their churches were not following the true Christian doctrine because their people did not rise to the level of sainthood.

115. ASRI, Fond D, Dossier 7755, vol. 3, f. 282 in Păiuşan and Ciuceanu, p. 184.

116. Ibid.

117. ASRI, Fond D, Dossier 7755, vol. 3, f. 284 in Păiuşan and Ciuceanu, pp. 184–5.

118. 'În întreaga ţară preoţimea semnează alături de întregul popor apelul pentru interzicerea armei atomice' [All over the Country the Clergy Sign with All of the People the Appeal for the Abolition of the Atomic Bomb], *Ortodoxia*, 1950, 1, p. 30.

119. The BBC Romanian Service was considered the most important radio station and one of the most truthful and reliable sources of information. Information broadcast on this station had an impact and was influential. For example, from 22 December 1947 to 4 January 1948 the BBC conducted a survey of its Romanian broadcasts and received 728 letters, of which 699 were from Romania and 29 from abroad. Report on Listener Research Fortnight, July 1948, E3/298/1, Romanian Service, BBC WA; Hugh Carleton Green, Head of East Europe Service, Religious Broadcasting to Eastern Europe, 16 September 1949, Policy Religion European Service; D. M. Graham, Assistant Head of East European Broadcasts, Religious Element in Broadcasts to Eastern Europe, [undated], Policy Religion European Service File 2, 1947–1954; BBC WA. The church hierarchy was especially affected by the broadcasts as after 15 September 1949 the BBC began presenting religious programmes for Romania, Bulgaria and the Soviet Union. These lasted 15 minutes on Sundays and religious festivals, and included prayers, hymns and an address from a dissident. The address was usually given by Father Florian Gâldău or Father Vasile Leu both of whom had escaped from Romania in 1948 and who accused the hierarchy of transforming the church into a branch of the Soviet Union. Father Gâldău and Father Leu were recommended by Canon Herbert Waddams from the Church of England Council on Foreign Relations in a letter to Reverend Francis House. In the following years the Romanian Church hierarchy believed that Cecil Douglas Horsley, the bishop of Gibraltar, was the one who facilitated Gâldău's talks at the BBC, and for this reason the regime refused his entry visa to Romania after 1948. Letter from Canon Herbert Waddams, CECFR to Reverend Francis House, 23 September 1949, Policy Religion European Service, File 2, 1947–1954, BBC WA.

120. After the Second World War, many Soviet and Romanian companies merged, with the word Sovrom added to their name.

121. ASRI, Fond D, Dossier 1775, f. 290 in Păiuşan and Ciuceanu, p. 185.

122. Deletant, *Communist Terror in Romania*, pp. 94–8.
123. George N. Shuster, *Religion behind the Iron Curtain*, New York: The Macmillan Company, 1954, p. 241.
124. Raoul Bossy, 'Religious Persecutions in Captive Romania', *Journal of Central European Affairs*, 1955, 15 (2), p. 174.
125. Robert Tobias, *Communist – Christian Encounter in East Europe*, Indianapolis: School of Religion Press, 1956, p. 335.
126. 'Alianţa încheiată de Bisericile ortodoxe rusă şi română, în vederea desfăşurării unei acţiuni comune pentru apărarea păcii' [The Alliance established between the Russian and Romanian Orthodox churches regarding Common Action for the Defence of Peace], *Ortodoxia*, 1950, 2, p. 176–7.
127. Ibid., p. 195–6.
128. British Embassy, Moscow, Item 178.1/10/50, TNA PRO FO 371/88127.
129. ASRI, Fond D, Dossier 7755, vol. 3, f. 250–4 in Păiuşan and Ciuceanu, pp. 186–90.
130. Ibid.
131. ASRI, Fond D, Dossier 2488, vol. 1, f. 609–18 in Păiuşan and Ciuceanu, pp. 190–5.
132. Robert Levy, *Ana Pauker: The Rise and Fall of a Jewish Communist*, Berkeley: University of California Press, 2001; Dennis Deletant, 'New Light on Gheorghiu-Dej's Struggle for Dominance in the Romanian Communist Party', *Slavonic and East European Review*, 1995, 73 (4), pp. 659–90.
133. Tismăneanu, *Stalinism for All Seasons: A Political History of Romanian Communism*, p. 127.
134. The law reinforced the decision to expel on 6 July the leading members of the Roman Catholic Church in Romania, the Papal Nuncio, Archbishop Gerald Patrick O'Hara and two of his assistants, Monsignor John C. Kirk and Monsignor Del Mesti de Schonberg, while the other members of the Nuncio's administration were imprisoned. Shuster, p. 235. Ovidiu Bozgan, 'Nunţiatura Apostolică din România în anii 1948–1950', p. 130–54. The full text of accusations was published in newspaper *Universul*, 6 July 1950.
135. The letter was two-thirds written by the patriarch and one-third by Liviu Stan. See ASRI, Fond D, Dossier 2488, vol. 1, f. 104–14 in Păiuşan and Ciuceanu, pp. 198–204.
136. ASRI, Fond D, Dossier 2488, vol. 1, f. 104–14 in Păiuşan and Ciuceanu, pp. 198–204.
137. Ibid.
138. Adrian Nicolae Petcu, 'Cazul Episcopului Nicolae Popovici' [The Case of Bishop Nicolae Popovici] in Miruna Tătaru-Cazaban (ed.), *Teologie şi politică: De la sfinţii părinţi la Europa unită* [Theology and Politics: From the Holy Fathers to United Europe], Bucharest: Anastasia, 2004, pp. 219–82.
139. Paul Caravia, Virgiliu Constantinescu and Flori Stănescu, *The Imprisoned Church: Romania, 1944–1989*, Bucharest: The Romanian Academy, 1999, p. 32.
140. ASRI, Fond D, Dossier 2488, vol. 1, f. 116–25 in Păiuşan and Ciuceanu, pp. 220–5.
141. Ibid.
142. An example in this sense was the case of the former prime minister General Rădescu who was in Paris in 1949 attempting to seek greater unity of the Romanian community. Part of the community was for maintaining close relations with the Catholic Church, which was seen as possibly having an influence on the political situation in Romania, while another group was against the

Catholics. Doreen Berry, Interviews with Romanians in Paris, 30 September 1949, E1/1172, Countries Romania: Romanian Service, File 1–2, 1940–51, BBC WA.

143. Gerald J. Bobango, *The Romanian Orthodox Episcopate of America: The First Half Century, 1929–1979*, Jackson, Michigan: The Romanian-American Heritage Center, 1979, p. 185.

144. Visarion Puiu, CNSAS, SIE142, microfilm.

145. Bishop Moruşcă Policarp Watched by Communist Police, Item 00406, HU OSA 300–60–1, 530/2808, Religion: Other Religions, 1948–55.

146. Antim Nica, CNSAS, I701, 5 vols.

147. Andrei Moldovan, I185029, CNSAS, p. 8.

148. Ibid., p. 1.

149. Ibid., p. 27.

150. Ibid., p. 66.

151. Ibid., p. 92.

152. British Legation, Bucharest, Item 218 (1783/3/50), TNA PRO, FO 371/88127.

153. Moldovan file, p. 113.

154. *The Canonical Status of the Romanian Orthodox Episcopate of America*, no place: Episcopal Council of the Romanian Orthodox Episcopate of America, 1954; Gerald J. Bobango, *Religion and Politics: Bishop Valerian Trifa and His Times*, Boulder: East European Monographs, 1981.

155. *Universul*, 29 November 1950.

156. Tobias, p. 336.

157. Tobias, p. 337.

158. Shuster, p. 241.

159. British Legation, Bucharest, Item 37 (1782/6/52), TNA PRO FO 371/100788.

160. Antim Târgovişteanul, 'Instalarea Prea Fericitului Macarie în scaunul de Mitropolit şi Întâistătător al Bisericii Ortodoxe Autocefale din Polonia' [The Enthronement of His Beatitude Makary as Metropolitan and Head of the Autocephalous Orthodox Church in Poland], *BOR*, 1951, 7–9, p. 327.

161. 'Consfătuirea de la Moscova a Întâi-Stătătorilor Bisericilor Ortodoxe din Antiohia, Rusia, Georgia, România şi Bulgaria' [The Conference in Moscow of the Heads of the Orthodox Churches from Antioch, Russia, Georgia, Romania and Bulgaria], *BOR*, 1951, 7–9, pp. 286–7.

162. 'Vizita IPS Patriarch Alexandru III al Antiohiei în ţara noastră' [The Visit of His Eminence Patriarch Alexander III of Antioch to Our Country], *BOR*, 1951, 7–9, p. 317.

163. 'Vizita Întâistătătorului Bisericii Ortodoxe din Albania în ţara noastră' [The Visit of the Head of the Orthodox Church of Albania to Our Country], *BOR*, 1952, 1–2, p. 45.

164. At the age of 81, Bishop Pacha was sentenced to 18 years in prison but was released on 31 July 1954 and he died shortly after.

165. Ciurea, 'Religious Life', p. 189; Bucur and Stan, pp. 56–8.

166. *Digest* 2, 1 December 1951, AKI.

167. Girl Pupils Suspended for Making the Sign of the Cross, Item 11563/52, HU OSA 300–60–1, 522/2803, Religion: Obstacles to Worship God, 1951–65.

168. Romania, Item 4164, HU OSA 300–60–1, 522/2804, Religion: Persecution, 1951–5.

169. The journal *Ortodoxia* dedicated a special number for this event.

170. Enache, pp. 130–1.

171. Pierre Gherman, *L'âme roumaine écartelée: Fait et documents*, Paris: Les Editions du Cèdre, 1955, pp. 9–14.

172. British Legation, Holy See, Item 34 (49/10/52), TNA PRO FO 371/100788.
173. Ghita Ionescu, pp. 203–4.
174. On hearing these accusations Luca fainted, aware of their impact on his personal and public life. Tismăneanu, p. 130.
175. Justinian, 'Pastorala pentru muncile de primăvară' [Pastoral Letter on Work in the Spring], *BOR*, 1952, 1–2, pp. 5–10.
176. British Legation, Bucharest, Item 118, TNA PRO FO 371/100744.
177. The Political Office was reduced from 13 to 9 members, the Organisational Office from 17 to 11 and the Secretariat from 7 to 5.
178. British Legation, Bucharest, Item 1782/10/52, TNA PRO FO 371/100788.
179. British Legation, Bucharest, Item 1782/11/52, TNA PRO FO 371/100788.
180. 'Pastorala Sfântului Sinod al Bisericii Ortodoxe Române către clerul şi credincioşii ortodocşi din RPR' [Pastoral Letter of the Holy Synod of the Romanian Orthodox Church to the Clergy and Orthodox Christians in the PRR], *BOR*, 1952, 6–8, p. 272.
181. 'Clerul şi credincioşii ortodocşi trebuie să desfăşoare o activitate mai vie pentru apărarea păcii: Hotărârile Sfântului Sinod şi ale Adunării Naţionale Bisericeşti' [The Clergy and Orthodox Christians Have to Develop Livelier Activity for the Defence of Peace: The Decisions of the Holy Synod and the National Clerical Assembly], *BOR*, 1952, 6–8, p. 304.
182. Decision no. 10937/1952 of the Ministry of Religious Confessions.
183. 'Regulament de organizarea şi funcţionarea instituţiilor de învăţământ pentru pregătirea personalului bisericesc şi rectrutarea corpului didactic din Patriarhia Română' [Regulations for the Organisation and the Functioning of the Educational Institutions for the Training of Clerical Personnel and the Recruitment of the Didactic Body in the Romanian Patriarchate], *BOR*, 1952, 9–10, p. 641–94.
184. 'Importante hotărâri luate de Sfântul Sinod în sesiunea din iunie 1952' [Important Decision taken by the Holy Synod at its Meeting in June 1952], *BOR*, 1952, 6–8, p. 612.
185. Proiectul de Constituţie a Republicii Populare Române: Atitudinea Bisericii Ortodoxe Române faţă de proiectul noii Constituţii a Statului nostru democratic-popular [The Draft Constitution of the Romanian People's Republic: The Attitude of the Romanian Orthodox Church towards the Draft Constitution of our Democratic-Popular State], *BOR*, 1952, 6–8, pp. 390–9.
186. Articles 1–3 of the 'Constitution of the Romanian People's Republic' in *Buletinul Oficial al Marii Adunări Naţionale a Republicii Populare Române* [Official Bulletin of the Great National Assembly of the Romanian People's Republic] no 1 from 27 September 1952.
187. George Schöpflin, 'The Ideology of Rumanian Nationalism', *Survey*, 1974, 20, 2–3, pp. 77–104.
188. Rosu et al., pp. 282.
189. James Hemminges, David Donaldson, Hugh Baillie, Robert Bonner and Donald Sutherland. 'Vizita unui grup de muncitori englezi la Patriarhia Română' [The Visit of a Group of British Workers to the Romanian Patriarchate], *BOR*, 1952, 9–10, pp. 517–21.
190. 180,000 people were interned in labour camps and 40,000 people were forced to build the Danube-Black Sea Canal. Deletant, *Romania under Communist Rule*, p. 75.
191. A suggestive depiction of the dimensions of state terror in Romania after the installation of the communist regime can be seen in the statement of Teohari Georgescu on 15 September, written in prison at the communists' request in

which he presented the 'achievements' of his Ministry of Internal Affairs. He stated that during the seven years in which he ruled the ministry, more than 100,000 people were arrested including virtually all political leaders who had connections with the previous regime.

192. 'Chemare către clerul şi credincioşii tuturor cultelor din Republica Populară Română' [The Calling to the Clergy and Faithful of All Confessions in the Romanian People's Republic], *BOR*, 1952, 11–12, pp. 701–5.
193. *Scânteia*, 26 November 1952.
194. Gheorghe Gheorghiu-Dej, *Articole şi Cuvântări* [Articles and Speeches], Bucharest: Editura pentru literatura politică, 1952.
195. 'Congresul naţional pentru apărarea păcii' [The National Congress for the Defence of Peace], *BOR*, 1952, 11–12, p. 753.
196. Ibid, p. 758–61.
197. *Digest* 8, 7 June 1952, AKI.
198. *Digest* 7, 3 May 1952, AKI.
199. Anti Religious Propaganda among the Youth, Item 11920/52, HU OSA 300–60–1, 522/2801.
200. *Digest* 13, 1 November 1952, AKI.
201. Christmas Just Another Working Day in Romania, Item 00876/53, HU OSA 300–60–1, 522/2803.
202. Regime's Device to Nip Christmas Celebrations, Item 1492/53, HU OSA 300–60–1, 522/2803.
203. Orthodox Priest Imprisoned, Item 11175/52, HU OSA 300–60–1, 522/2804, Religion: Persecution, 1951–5.
204. Arrest of Orthodox Priest, Item 15079/52, HU OSA 300–60–1, 522/2804, Religion: Persecution, 1951–5.
205. Soviet Intrigues in the Orthodox Church, Item 1035, HU OSA 300–60–1, 518/2800, Religion: General, 1948–58.
206. British Legation, Bucharest, Item 1781/6, TNA PRO FO 371/1/128953.
207. Other representatives of religious confessions were allowed to study in Bucharest. In 1951–2, two Baptist students, Alexa Popovici and Nicolae Covaci and a reformed priest, Dohi Arpad, studied in Bucharest, while an Orthodox priest studied in the Theological Protestant Institute in Cluj. 'Din activitatea institutelor teologice ale Bisericii Ortodoxe Române' [From the Activity of the Theological Institutes of the ROC], *BOR*, 1952, 9–10, p. 569–71.
208. Väinö Kalev Ihatsu stayed in Romania from 14 December 1950 until 15 July 1954; Väinö Kalev Ihatsu, CNSAS, I185023, 2 vols.
209. Alexandru Ciurea, CNSAS, I185022.
210. Iuhani Homanen stayed in Romania from 17 December 1951 to 17 April 1952.
211. Hieromonk Apostolos Panaiotis Dimelis came in Romania on 24 November 1956; Apostolos Panaiotis Dimelis, CNSAS, I185028.
212. Al. I. Ciurea, 'Studenţi străini la Institutul Teologic din Bucureşti' [Foreign Students in the Theological Institute in Bucharest], *BOR*, 1957, 3–4, pp. 211–7.
213. The following quotations are from Justinian, the 'Sovrom Patriarch', Item 1944/54, HU OSA 300–60–1, 522/2801, Religion: Communist Influenced Religious Movements, 1951–68.
214. Other reports from the Radio Free Europe suggest that people heard rumours that Justinian had a mistress living with him at the patriarchal palace. Patriarch Marina Justinian, Item 3227, HU OSA 300–60–1, 522/2804, Religion: Persecution, 1951–5.

5 Orthodoxy and the Romanian Road to Communism, 1953–5

1. 'Vizita studenților Institutului Teologic Protestant din Cluj la Patriarhia Română' [The Visit of Students from the Protestant Theological Institute in Cluj to the Romanian Patriarchate], *BOR*, 1953, 2–3, p. 210.
2. 'Cultele religioase din lumea întreagă și-au exprimat adânca lor durere cu prilejul încetării din viață a lui Iosif Vissarionovici Stalin' [Religious Confessions from the Whole World Expressed their Deep Sorrow at the Death of Iosif Vissarionovici Stalin'], *BOR*, 1953, 2–3, pp. 136–7.
3. Ibid., 138.
4. British Legation, Bucharest, Item 1784/2/53, TNA PRO FO 371/106489.
5. 'Cultele religioase din lumea întreagă și-au exprimat adânca lor durere cu prilejul încetării din viață a lui Iosif Vissarionovici Stalin', p. 139.
6. Ibid.
7. How Bucharest Learned of Stalin's Fatal Stroke, Item 06614/53, HU OSA, 300–60–1, 530/2808 Religion: Other Religions 1948–55.
8. Ibid.
9. Ibid.
10. Ibid.
11. King Carol married Elena Lupescu in Brazil in summer 1947. Elena Lupescu died on 7 July 1977 and was buried near King Carol in Portugal.
12. 'Un grup de oaspeți din țările Americii Latine au vizitat Patriarhia Română' [Guests from Latin America Visited the Romanian Patriarchate], *BOR*, 1953, 4, p. 338.
13. 'Regulamentul pentru organizarea și funcționarea administrativă și disciplinară a mânăstirilor' [Regulations for the Organisation and the Administrative and Disciplinary Working of the Monasteries], *BOR*, 1953, 4, pp. 394–420.
14. Article 7 of these regulations stated that monasteries had the following duties: monks and nuns should conduct courses for illiterate people of all ages; in order to comply with the new religious and political environment, monasteries should organise missionary courses on social and religious teachings; those monasteries which had monastic schools had to combine teachings with manual work, dedicating two hours for learning and six hours for working in crafts. Regulations allowed the old tradition of the church, in that any person could stay three days in the monastery without paying for food or accommodation.
15. The Orthodox monasteries became real trading craft centres supporting the state in developing local manufacture. The most important monasteries were: Plumbuita specialised in painting and sculpture, Curtea de Argeş monastery in ceramic, Hurezu monastery in painting and ceramic and Agapia monastery in embroidery and carpets.
16. 'Cuvantul Întâistătătorului Bisericii Ortodoxe Romane pentru împlinirea la timp a muncilor agricole de primăvară' [The Speech of the Holiness of the Romanian Orthodox Church for the Accomplishment in Time of the Spring Agricultural Works], *BOR*, 1953, 1, pp. 3–9.
17. Justinian offered the following gifts: an encolpion with the apostles' faces, a patriarchal staff, an icon of the Virgin Mary and religious instruments. Justinian, *Apostolat social*, 1955, p. 72.
18. Sebastian Ruşan, metropolitan of Moldavia and Suceava, and Nicolae Bălan, metropolitan of Transylvania were decorated with the 'Star of the Romanian People's

Republic', second class, while Valerian Zaharia, bishop of Oradea, was awarded the same medal but third class. Alexandru Ionescu, the priest from the Romanian Patriarchate involved in the official meetings regarding the battle for peace, was awarded 'Order of the Defence of the Fatherland', second class, while three low-profile priests, Alexandru Vlaiculescu, Ioan Gagiu and Al. Cerna-Rădulescu, were awarded the 'Star of the Romanian Peoples' Republic', fourth class. British Legation, Bucharest, Item 1782/4/53, TNA PRO FO 371/106489. 'Înalt Prea Sfinţitul Patriarh Justinian şi alţi reprezentanţi ai clerului ortodox au fost decoraţi cu ordine ale Republicii Populare Române' [His Holiness Patriarch Justinian and Other Representatives of the Orthodox Clergy Were Decorated with Orders of the Romanian People's Republic], *BOR*, 1953, 5–6, p. 423.

19. On 4 July 1951 the congress of the Romanian community in the US in Chicago, Illinois, rejected all claims of Moldovan to be their religious leader, and voted that the Romanian Orthodox Church in America should be completely autonomous in administrative and canonical affairs from any control from Romania. In addition, the congress nominated two people as their own candidates for the position of bishop: Archimandrite Ştefan Lucaciu, a refugee who was living in Paris and Valerian D. Trifa, the managing editor of the Romanian language weekly *Solia* in Cleveland, Ohio. Lucaciu received 26 votes, Trifa 34, one ballot was blank and one was annulled. Trifa was elected the first bishop of the Romanian community outside the jurisdiction of the Holy Synod. He was consecrated bishop in Philadelphia in 1952 by three hierarchs of the Ukrainian Orthodox Church in exile: John Theodorovich of Philadelphia, metropolitan of the Autocephalous Ukrainian Orthodox Church in America; Archbishop Mstyslav of Pereyaslav of New York and Archbishop Hennady of Sicheslaw of Chicago. Bishop Moldovan objected to the congress's actions and brought the case to the civil courts, even reaching the Supreme Court of the US; however, all decisions went against him. The headquarters of the Romanian Episcopate were in Jackson, Michigan, rather than in Detroit where Moldovan remained bishop under the jurisdiction of the Holy Synod in Romania.

20. Valerian D. Trifa was the nephew of Iosif Trifa who started 'The Lord's Army' in Sibiu, a movement within the Orthodox Church in the interwar period. During the Second World War he was arrested by the Germans and sent to Buchenwald and Dachau, from where he was freed. He stayed five years in Italy and went to the US in 1950.

21. 'Lucrările Sfântului Sinod al Bisericii Ortodoxe Române în sesiunea ordinară pe anul 1953' [The Meetings of the Holy Synod of the ROC in its Ordinary Session in 1953], *BOR*, 1953, 7–8, p. 711.

22. Father Florian Gâldău was a former advisor of the patriarch who, anticipating his arrest in 1948 by the communists, escaped to Yugoslavia where he was imprisoned for a few months. He then managed to reach Austria and was issued a temporary residence permit by the British Foreign Office. Gâldău was already known in the West, especially in London, having been a member of the committee for dialogue between the Anglican and Romanian Orthodox churches in 1935 and of the Romanian delegation led by Patriarch Miron Cristea which visited London in 1936. Gâldău became the priest of the Romanian community which included some former members of the Iron Guard and supporters of King Michael. In 1955 he moved to New York. Lambeth Palace Archives (LPA), Correspondence and Papers of John Albert Douglas, 52, ff. 309–10, 311, 321–3. Florian Gâldău, CNSAS, SIE75, 3 vols, microfilm; I3714, 2 vols.

23. British Legation, Bucharest, Item 1782/5/53, TNA PRO FO 371/106489.
24. Orthodox Priests Applaud RFE's Campaign Against Corrupt Patriarch, Item 3933/56, HU OSA Reel 67.
25. British Legation, Bucharest, Item 1784/4/53, TNA PRO FO 371/106489. Archbishop Cisar died soon after, on 7 January 1954, in his residence at the age of 73. The attitude of the regime could be seen through that the press release on his death was published only after his burial on 9 January. British Legation, Bucharest, Item 1783/1/54, TNA PRO FO 371/111657.
26. M. J. Rura, *Reinterpretation of History as a Method of Furthering Communism in Rumania*, Washington: Georgetown University Press, 1961; Pavel Ţugui, *Istoria şi limba română în vremea lui Gheorghiu-Dej: Memoriile unui fost şef de secţie a CC al PMR* [Romanian History and Language in Gheorghiu-Dej's Time: The Memoirs of a Former Head of Section of the Central Committee of the RWP], Bucharest: Editura Ion Cristoiu, 1999.
27. Lucreţiu Pătrăşcanu, *Sub trei dictaturi* [Under Three Dictatorships], Bucharest: Editura Politică, 1970. Pătrăşcanu was rehabilitated by Nicolae Ceauşescu at the Plenum of the Central Committee from 22 to 25 April 1968.
28. Ghita Ionescu, p. 232.
29. *BOR*, 1954, 5, pp. 529–47.
30. *Romînia Liberă*, 4 September 1954. The same attitude could be seen on 22 September 1954 when at his 70th birthday celebrations in Iaşi, in a public ceremony led by Justinian, Metropolitan Ruşan was awarded the highest distinction, which until then had been awarded only to the patriarch: the 'Star of the Romanian People's Republic', first class. *Scînteia*, 24 September 1954.
31. *Scînteia*, 11 December 1955; 13 December 1955.
32. British Legation, Bucharest, Item 1781/6/54, TNA PRO FO 371/111657; *Scînteia* 4, 5, 14 September 1954.
33. Foreign Office, London, Item 1781/6, TNA PRO FO 371/111657.
34. *The Churches of Europe under Communist Governments: A Survey Presented with the Fifteenth Report of the Church of England Council on Foreign Relations and Communicated to the February Session of the Church Assembly, 1954, by Order of the Archbishop of Canterbury*, Westminster: Church Information Board, 1954, p. 13.
35. Ibid., 16.
36. Ibid., 17.
37. Stanley Evans and John Bliss, *The Church in Rumania*, London: British Rumanian Friendship Association, 1954.
38. Ibid., 5.
39. Ibid., 30.
40. Ibid., 32.
41. Ibid., 33.
42. Church of England Council on Foreign Relations, Item 1781/24, TNA PRO FO 371/111657.
43. *Romînia Liberă*, 12 October 1954.
44. *Romînia Liberă*, 26 October 1954.
45. On 26 October both patriarchs consecrated the newly renovated 'Saint Elijah–Colţea Inn' church which was offered to the Bulgarian community living in Bucharest.
46. Romania: Vital Statistics, Item 4032/55, HU OSA Reel 53.
47. Romania: Religion, Item 146/55, HU OSA 300–60–1, 533/2904 Resistance and Criticism of the Regime, Religions, 1952–69.

48. The Warsaw Pact was signed by Bulgaria, Czechoslovakia, the German Democratic Republic, Hungary, Poland, Romania and the Soviet Union. Albania joined in 1962 but withdrew in 1968.
49. 'Reclădirea mânăstirilor Dealul, Ghighiu şi Schitul Maicilor' [The Rebuilding of Dealu, Ghighiu and Schitul Maicilor Monasteries], *BOR*, 1955, 7, pp. 590–1.
50. *Scînteia*, 17 July 1955.
51. *Romînia Liberă*, 19 July 1955.
52. The communiqué was signed by Patriarch Christopher of Alexandria; Patriarch Alexius of Moscow; Patriarch Justinian of Romania; Melkisedeck, Catholicos of Georgia; Metropolitan Niphon of Iliopol; Bishop Vasilii of Sergiopol; Metropolitan Makary of Warsaw; Metropolitan Elevteri of Prague and Archimandrite Maxim from Bulgaria, *Romînia Liberă*, 29 July 1955.
53. 'Marile festivităţi religioase ale Bisericii Ortodoxe Romîne din octombrie 1955' [The Great Religious Festivities of the ROC in October 1955], *BOR*, 1955, 11–2, p. 994.
54. Teodor M. Popescu, 'Însemnătatea canonizării sfinţilor romîni' [The Importance of the Canonisation of the Romanian Saints], *BOR*, 1953, pp. 493–502.
55. Justinian sent letters of invitation to Patriarch Athenagoras of Constantinople, Patriarch Christopher of Alexandria, Patriarch Alexius of Moscow, Patriarch Vikentije of Serbia, Patriarch Kiril of Bulgaria and Archbishop Spyridon of Greece.
56. Metropolitan Gregory of Leningrad, Bishop Paladios of Volinie and Roven in Ukraine and Monk Constantine Nachaev, professor at the Theological Academy of Zagorsk in Moscow.
57. Archbishop Athenagoras Kabadas of Thyatira from London; Bishop Jacob of Malta from Geneva.
58. Patriarch Kiril, Metropolitan Sofronius of Tirnovo, Archimandrite Josef Dicov, Archdeacon Dr Basil Valeanov Stoeanov, Professor Stefan Zankov and Archimandrite Atanasi Boncev.
59. Metropolitan Chrysostomos of Philippi, Metropolitan Yacovos of Attika and Archimandrite Panteleimon Karanikola.
60. *BOR*, 1955, 11–2, p. 1005.
61. Saint Parascheva from Iaşi, Saint Filofteia from Curtea de Argeş, Saint Dimitrie the New from Bucharest, Saint Grigorie Decapolitul from Bistriţa-Vâlcii monastery.
62. Visarion Sarai came to Transylvania in 1744 and preached the Orthodox to preserve their faith and not to convert to Catholicism. Sofronie from Cioara was against the Catholics and started an armed resistance in mountains in Transylvania. Oprea Miclăuş had the same attitude and went to Vienna defending the religious rights of Romanians in Transylvania. *The Rumanian Orthodox Church*, Bucharest: The Bible and Orthodox Missionary Institute, 1962, p. 60–2.
63. *BOR*, 1955, 11–2, p. 1126.
64. Ibid., p. 1028–9.
65. Ibid., p. 1036.
66. On 14 October, Justinian attended the festivities at Curtea de Argeş monastery for the cult of Saint Mucenita Filofteia, on 16 October in Râmnicu Vâlcii for cult of Saint Grigorie Decapolitul, on 21 October at Cernica monastery for the cult of Saint Hierarch Calinic and on 27 October in Bucharest for the cult of Saint Dimitrie Basarabov. Ibid., p. 1235–6.
67. Romanian Church Jubilee, Item 11019/55, October 1955, HU OSA 300–60–1, 522/2801, Religion: Communist Influenced Religious Movements, 1951–68.

68. Official Report of the Ecumenical Delegation to the Festivities in Romania on the Occasion of the Seventieth Anniversary of the Autocephalous Romanian Church, Item 1185/56, October 1955, HU OSA 300–60–1, 522/2801.
69. Ibid.
70. Ibid.
71. Ibid.
72. Confidential Report, 17 November 1955, LPA, Church of England Council on Foreign Relations, File OC 158.
73. On 31 May 1955, the Ecumenical Patriarchate decided the canonisation of Saint Nicodim. On 19 July 1962 the Russian Orthodox Church canonised Saint John the Russian. On 26 June 1962 the Bulgarian Orthodox Church canonised Paisi from Hilandar and on 31 December 1964, Saint Sofroni from Vrata. The decision to have a common system for future canonisations was made at the first Pan-Orthodox Conference in Rhodes in 1961. Liviu Stan, 'Soborul panortodox din Rodos' [The Pan-Orthodox Synod in Rhodes], *Mitropolia Olteniei*, 1961, 11–2, p. 728; Liviu Stan, 'Despre rânduiala canonizării solemne a sfinţilor în Ortodoxie' [About the System of Solemn Canonisation of Saints in Orthodoxy], *Mitropolia Olteniei*, 1955, 7–9.
74. A list of members of foreign clerical delegations was published in *Scînteia* on 8 and 9 October 1955. Without giving details, the article only stated that they arrived on the invitation of the Romanian Orthodox Church which celebrated 70 years of autocephaly. In addition, *Scînteia* of 20 October 1955 mentioned the arrival in Romania of the Greek delegation, while *Romînia Liberă* of 25 October 1955 stated that the Ministry of Religious Confessions organised a reception for foreign guests emphasising general good relations between church and state. The religious procession from the Patriarchal Palace to Saint Spiridon Cathedral was described in *Scînteia* on 11 October 1955, mentioning the presence of the foreign delegation but without reference to the canonisation of the Romanian saints.
75. Vasile Leu, CNSAS, I185024, 4 vols; Son of a Former Bishop Arrested in Austria by Russians, Item 9174/55, HU OSA Reel 60.
76. Kidnapped Bishop Confined in Piteşti Prison, Item 2612/56, HU OSA 300–60–1, 523/2804, Religion: Persecution, 1956–76.
77. Partidul Muncitoresc Romîn, *Congresul al II-lea al PMR, 23–28 Dec. 1955* [The Second Congress of the RWP, 23–28 December 1955], Bucharest: Editura Politică, 1956.
78. The party numbered 593,398 members, a reduced number due to political purges in 1952.

6   Religious Diplomacy and Socialism, 1956–9

1. Gheorghiu-Dej was leading the Romanian delegation at the Moscow Congress, which was composed of Iosif Chişinevschi, Miron Constantinescu and Petre Borilă.
2. Report by Sir Wavell Wakefield, MP, 18 July 1956, LPA CECFR File 158.
3. Danish Church Delegation Invited to Visit Romania, Item 2916/56, 8 March 1956, HU OSA 300–60–1, 522/2801, Religion: Communist Influences Religious Movements, 1951–1968.
4. Anti-Communist Leaflets Tell Constanţa Citizens Days of Shaken Communist Regime Are Numbered, Item 7354/56, HU OSA 300–60–1, 532/2900, Resistance and Criticism of the Regime: General, 1951–69.

5. The results revealed that Romania numbered 17,489,450 people of which 85.7 per cent were Romanians, 9.1 per cent Hungarians, 2.2 per cent Germans, 0.6 Roma and 3.0 per cent other minorities; 10.1 per cent of the population was illiterate; 31.3 per cent of the population lived in cities and the remainder of 68.7 per cent was rural. Ioan Scurtu, Ion Alexandru, Ion Bulei, Ion Mamina, *Enciclopedia de istorie a României* [The Encyclopaedia of Romanian History], Bucharest: Meronia, 2002, pp. 389–99.

6. The Problem of the Students and the 1956 Movement, Item 2431/62, HU OSA 300–60–1, 532/2903, Resistance and Criticism of the Regime: Political, 1956–71.

7. S. Verona, *Military Occupation and Diplomacy: Soviet Troops in Romania. 1944–1958*, Durham, London: Duke University Press, 1992; Dennis Deletant, 'Impactul revoltei maghiare în România' [The Impact of the Hungarian Revolt in Romania] in Analele Sighet 8, *Anii 1954–1960: Fluxurile şi refluxurile stalinismului* [The Years 1954–60: Ebb and Flow of Stalinism], Bucharest: Fundaţia Academia Civică, 2000, pp. 598–602.

8. *Mitropolia Banatului*, 1956, 11–12.

9. Letter from the Bishop of Gibraltar to Canon Waddams, 21 January 1957, LPA CECFR File 158.

10. In reality Bishop Horsley did not have any connection with Father Gâldău's appointment to the BBC. Letter from the Bishop of Gibraltar to Canon Waddams, 4 February 1957, LPA CECFR File 158.

11. Report: The Romanian Orthodox Church, Bishop of Gibraltar, January 1957, LPA CECFR File 158.

12. Note for file, 27 February 1957, LPA CECFR File 158.

13. British Legation, Bucharest, Item 1782/1/57, TNA PRO FO 371/128953.

14. *Ortodoxia*, 1956, 4.

15. Letter from Patriarch Justinian to Archbishop Fischer, 26 January 1957, LPA CECFR File 158.

16. Ibid.

17. *Romînia Liberă*, 28 February 1956.

18. *Romînia Liberă*, 17 March 1956.

19. ASRI, Fond D, Dossier 2488, vol. 1, f. 415–33 in Păiuşan and Ciuceanu, p. 166.

20. *Romînia Liberă*, 19 September 1956.

21. Note, 15 November 1956, HU OSA 300–60–1, 530/2808, Religion: Other Religions, 1956–9.

22. After the death of Justinian in 1977 Metropolitan Justin would become patriarch.

23. 'Alegerea, învestiturea şi înscăunarea IPS Mitropolit Iustin al Moldovei şi Sucevei' [The Election and Enthronement of His Holiness Justin, Metropolitan of Moldavia and Suceava], *BOR*, 1957, 1–2, pp. 39–40.

24. Ibid., p. 62–3.

25. British Legation, Bucharest, Item 1782/6/57, TNA PRO FO 371/1/128953.

26. Monasteries Tolerated as Source of Cheap Labour, Item 319/58, HU OSA 300–60–1, 518/2800, Religion: General, 1948–58.

27. Report, 5 October 1956, LPA CECFR File 158.

28. *Communist Exploitation of Religion*, p. 15.

29. Foreign Office, London, Item 1781/1, TNA PRO FO 371/1/128953.

30. British Legation, Bucharest, Item 1782/5/57, TNA PRO FO 371/1/128953.

31. Radio Bucharest Redefection, 22 April 1957, HU OSA 300–60–1, 532/2812, Religion: Religious Holidays, 1955–1979.

32. *Scînteia*, 24 May 1957; *Romînia Liberă*, 26 May 1957; 28 May 1957.

33. British Legation, Bucharest, Item 1781/4, TNA PRO FO 371/1/128953.
34. *Romînia Liberă*, 20 December 1957.
35. *Scînteia*, 2 June 1957.
36. *Scînteia*, 12 June 1957.
37. *Scînteia*, 11 June 1957.
38. *Romînia Liberă*, 11 September 1957.
39. An Impression of Petru Groza, Item 2776/54, HU OSA 300–60–1, 530/2808 Religion: Other Religions, 1948–55.
40. The funeral was attended by the following religious leaders: Patriarch Justinian; Nicolae Colan, metropolitan of Transylvania; Justin Moisescu, metropolitan of Moldavia and Suceava; Dr Fr Mueller, bishop of the Evangelical Church; Aladar Arday, bishop of the Reformed Church; Dr Elek Kiss, Unitarian bishop; Traian Jovanelli, vicar of the Roman Catholic Archbishopric in Bucharest; Gheorghe Argay, Lutheran bishop; Dr Moses Rozen, chief rabbi; and Iacub Mehmet, the mufti.
41. *Romînia Liberă*, 10 January 1958.
42. British Legation, Bucharest, Item 1015/11/58, LPA CECFR File 158.
43. N. Cazacu, 'Vizite primite de Prea Fericitul Patriarh Justinian' [Visits Received by His Holiness Patriarch Justinian], *BOR*, 1958, 7–8, pp. 721–4.
44. Interview of the author with Bishop John R. Satterthwaite, 2 February 2006.
45. Report of the Church of England Council on Foreign Relations, 29 March 1957, LPA Lambeth Conference Papers 196, f. 97.
46. Letter to Archbishop Fisher, LPA Correspondence and Papers of Geoffrey Francis Fisher, Archbishop of Canterbury, 1945–1961, 184, f. 158, 4 March 1957; Notes of discussion at Council of Lambeth Conference and Proposals for Anglican/Orthodox Discussions, the Bishop of Chichester 15 November 1957, LPA CECFR File 158.
47. I. Rămureanu, 'Direct Talks for Union between the Orthodox and Anglican churches from 1920 till Today', *Romanian Orthodox Church and the Church of England*, Bucharest: Biblical and Orthodox Missionary Institute, 1976, pp. 45–66.
48. Herbert Waddams, Interview with the Patriarch Justinian, 8 April 1958, LPA CECFR File 158.
49. Letter from Canon Waddams to Patriarch Justinian, 22 September 1958, LPA CECFR File 158.
50. *Ortodoxia*, 1958, 2.
51. *Romînia Liberă*, 10 May 1958.
52. Romanian delegation was composed of Patriarch Justinian, Metropolitan Justin of Moldavia and Suceava, Metropolitan Vasile of Banat and Father Simeon Neaga as translator. Simeon Neaga, 'Vizita delegaţiei Bisericii Ortodoxe Romîne în Uniunea Sovietică' [The Visit of the ROC to the Soviet Union], *BOR*, 1958, 7–8, p. 587.
53. *Romînia Liberă*, 31 May 1958; 6 June 1958.
54. Easter Messages by the Patriarchs of Russia and Romania, Item 6191/56, HU OSA 300–60–1, 511/2607 Propaganda Speeches of VIPs Justinian [Patriarch 1955–6].
55. 'Inaugurarea unor aşezăminte bisericeşti construite sau restaurate la centrele patriarhal şi mitropolitan din Bucureşti' [The Inauguration of Some Clerical Buildings Constructed or Renovated at the Patriarchate and Metropolitan Centres in Bucharest], *BOR*, 1958, 7–8, p. 684.
56. 'Vizita în ţara noastră a Prea Fericitului Patriarh Hristofor al Alexandriei' [The Visit to Our Country of His Holiness Patriarch Christopher of Alexandria], *BOR*, 1958, 7–8, p. 632.

57. 'Conferinţa Mişcării creştine pentru pace de la Debreţin' [The Conference of the Christian Movement for Peace in Debrecen], *BOR*, 1958, 10–1, pp. 1069–72.
58. Alexandru Ionescu, 'Conferinţa de la Lambeth 1958' [The Lambeth Conference, 1958], *BOR*, 1959, 1, pp. 73–98.
59. Reception of Official Visitors, 5 July 1958, LPA Lambeth Conference Papers, 196, f. 244.
60. Enache, pp. 381–400; Vasile, *Biserica Ortodoxă Română*, pp. 257–60.
61. Father Dumitru Stăniloae was arrested in 1958 and sentenced to five years imprisonment. He was released in 1963. Paul Caravia et al., p. 367.
62. Professor Theodor M. Popescu was arrested in 1959 and sentenced to 15 years imprisonment. He was released in 1963. Adrian Nicolae Petcu, 'Profesorul Theodor M. Popescu şi regimul comunist' [Professor Theodor M. Popescu and the Communist Regime] in Consiliul Naţional pentru Studierea Arhivelor Securităţii, *Arhivele Securităţii* [Securitate Archives], vol. 1, Bucharest: Pro Historia, 2002, pp. 80–96.
63. *Digest* 113, 4 March 1961, AKI.
64. Cristian Vasile, 'Autorităţile comuniste şi problema mânăstirilor ortodoxe în anii '50' [The Communist Authorities and the Problem of the Monasteries in the 1950s], in Analele Sighet 8, *Anii 1954–1960: Fluxurile şi refluxurile stalinismului* [The Years 1954–60: Ebb and Flow of Stalinism], Bucharest: Fundaţia Academia Civică, 2000, pp. 179–89.
65. *Mitropolia Moldovei şi Sucevei*, 1958, September-October, p. 694.
66. British Legation, Bucharest, Item 1783/1/59, TNA PRO FO 371/143380.
67. Ibid.
68. Canon Waddams, Romanian attacks on the Church [1958?]; Canon Waddams, Note, 30 December 1958, LPA CECFR File 158.
69. World Council of Churches, Confidential Memorandum from WA Visser't Hooft to Canon Waddams, 13 January 1959, LPA CECFR File 158.
70. World Council of Churches, Current Developments in the Eastern European Churches, no. 1, January 1959, AKI.
71. British Legation, Bucharest, Item 1784/11/59, TNA PRO FO 371/143380.
72. Church Has Lost Christian Meaning, Item 640/60, HU OSA 300–60–1, 522/2803, Religion: Obstacles to Worship God, 1951–65.
73. Ioan Dură, *Monahismul românesc în anii 1948–1989* [Romanian Monastic Life, 1948–89], Bucharest: Harisma, 1994; Vasile Manea, *Preoţi ortodocşi în închisorile comuniste* [Orthodox Priest in Communist Jails], Bucharest: Editura Patmos, 2000; A. Raţiu, W. Virtue, *Stolen Church: Martyrdom in Communist Romania*, Hungtinton: Our Sunday Visitor, 1978.
74. Anti-religious measures, Item 388/61, HU OSA 300–60–1 523/2804 Religious Persecution 1956–76'.
75. Translation from 'The Rumanian Nation', Paris, February 1961, LPA CECFR File 159.
76. State and Religion, Item 4433/59, HU OSA 300–60–1, 522/2803 Religion: Obstacles to Worship God, 1951–65.
77. The Orthodox Church in Romania, Item 3345/59, HU OSA 300–60–1, 522/2801 Religion: Communist Influenced Religious Movements, 1951–68.
78. *Scînteia Tineretului*, 19 September 1959.
79. *Digest* 97, 7 November 1959, AKI.
80. *Scînteia Tineretului*, 2 August 1959.
81. Memorandum, Interview between Patriarch Justinian and Reverend John R. Satterthwaite, 7 March 1959, LPA CECFR File 163/1–2; British Legation, Bucharest, Item 1781/3, TNA PRO FO 371/143380.

82. Ibid.
83. Ibid.
84. Ibid.
85. Ibid.
86. British Legation, Bucharest, Item 1783/4/59, TNA PRO FO 371/143380.
87. Extract from a visit of Reverend Satterthwaite to Professor and Mrs Popescu, 8 March 1959, LPA CECFR File 163/1–2.
88. Letter from Patriarch Justinian to Archbishop Fischer, 9 March 1959, LPA CECFR File 163/1–2.
89. It is still unclear if Justinian ever received all the books or if they were confiscated by the Romanian customs. Letter from Reverend Satterthwaite to Patriarch Justinian, 28 July 1959, LPA CECFR File 163/1–2.
90. Letter from Canon Waddams to Visser't Hooft, 16 March 1959, LPA CECFR File 163/1–2.
91. Father Seşan attended another session of the Prague Conference between 16 and 19 April 1959 while the Holy Synod issued a message to the World Council of Peace in Stockholm on 8 May 1959. 'Conferinţa Mişcării creştine pentru pace de la Praga' [The Conference of the Christian Movement for Peace from Prague], *BOR*, 1959, 5–6, pp. 456–64.
92. *Romînia Liberă*, 5 August 1959.
93. British Legation, Bucharest, Item 1784/11/59, TNA PRO FO 371/143380.
94. Doreen Berry, Record of Conversation with Mr Rotunda, Head of the Romanian Desk of the Voice of America, 14 November 1959, BBC WA, E1/2308/2, Romania A-Z.

7 Between Moscow and London: Romanian Orthodoxy and National Communism, 1960–5

1. Francis H. House, 'Summary of Comparisons of the Situation and Life of the Orthodox Churches in Bulgaria, Romania and the USSR, from WCC, Strictly Confidential, Not To Be Quoted or Published', 21 April 1961, LPA CECFR File 159.
2. Constantin Noica and Dinu Pillat were sentenced to 25 years in prison; Alexandru Paleologu to 14 years and Nicolae Steinhardt to 7 years. *Prigoana: Documente ale Procesului C. Noica, C. Pillat, N. Steinhardt, Al. Paleologu, A. Acterian, S. Al-George, Al. O. Teodoreanu* [The Persecution: Documents of the Trial of C. Noica, C. Pillat, N. Steinhardt, Al. Paleologu, A. Acterian, S. Al-George, Al. O. Teodoreanu], Bucharest: Editura Vremea, 1996.
3. 'Pastorala de Sfintele Paşti a Prea Fericitului Patriarch Justinian' [Pastoral Letter for Easter of His Holiness Patriarch Justinian], *BOR*, 1960, 3–4, pp. 187–91; 'Pastorala de Sfintele Paşti a Prea Fericitului Patriarh Justinian' [Pastoral Letter for Easter of His Holiness Patriarch Justinian], *BOR*, 1961, 3–4, pp. 217–20.
4. Partidul Muncitoresc Romîn, *Congresul al III-lea al PMR, 20–25 iunie 1960* [The Third Congress of the RWP, 20–25 June 1960], Bucharest: Editura Politică, 1960.
5. 'Directives of the Third Congress of the Rumanian Workers' Party for the Plan of Development of the National Economy in the Years 1960–5 and for the Long-Term Economic Programme', *Scînteia*, 19 May 1960.
6. *Scînteia*, 26 June 1960.
7. *Romînia Liberă*; *Scînteia*, 9 October 1960.
8. 'Pastorala de Crăciun a Prea Fericitului Patriarh Justinian' [Pastoral Letter for Christmas of His Holiness Patriarch Justinian], *BOR*, 1960, 11–2, p. 973.

9. 'Scrisoarea irenică trimisă de Prea Fericitul Patriarh Justinian cu prilejul Sfintelor Sărbători ale Naşterii Domnului 1960' [The Irenic Letter Sent by His Holiness Patriarch Justinian on the Occasion of the Lord's Birth 1960], *BOR*, 11–2, pp. 975–6.

10. Hintikka, p. 38.

11. H. M. Waddams, *Anglo-Russian Theological Conference, Moscow 1956: A Report of a Theological Conference held between Members of a Delegation from the Russian Orthodox Church and a Delegation from the Church of England: With a Preface by A. M. Ramsey, Archbishop of York*, London: The Faith Press, 1957.

12. Letter from Visser's Hooft to Patriarch Justinian, 1 May 1956; Letters from Patriarch Justinian to Visser's Hooft, 29 June 1956 and 13 July 1956, AWCC GS. 42.4. 065.

13. Letter from the Foreign Office to Canon Satterthwaite, 25 April 1961, LPA CECFR File 159.

14. Notes of a Short Informal Visit to the Rumanian Orthodox Church, 29 March to 4 April 1961, by Francis H. House. AWCC GS. 42.4. 065.

15. Hintikka, p. 39.

16. British Legation, Bucharest, Item 1782, TNA PRO FO 371/159527.

17. Francis H. House, 21 April 1961, LPA CECFR File 159.

18. Notes of a Short Informal Visit to the Rumanian Orthodox Church.

19. Notes of a Short Informal Visit to the Rumanian Orthodox Church.

20. Letter from Bishop Antim to Visser't Hooft, 1 April 1961, AWCC GS. 42.4. 065.

21. Letter from Patriarch Justinian to Visser't Hooft, 10 June 1961, AWCC GS. 42.4. 065.

22. Appendix to the Minutes of Executive Committee, Visit to Geneva of Bishop Nikodim, February 6–10, 1961, Strictly confidential. AWCC Divisions of Ecumenical Actions. 423.029.

23. Notes of Conversations with Bishop Nikodim, 13 March 1961, AWCC Divisions of Ecumenical Actions. 423.029.

24. Hintikka, p. 42.

25. For more see Adrian Hastings, *A History of English Christianity, 1920–1985*, London: Collins, 1986.

26. Church of England Council on Inter-Church Relations, Item 1781/1, TNA PRO FO 371/159527.

27. Telegram, 12 June 1961, LPA CECFR File 163/1–2.

28. 'Noul Primat al Bisericii Anglicane şi Arhiepiscop de Canterbury, Dr Arthur Michael Ramsey' [Dr Arthur Michael Ramsey, The New Primate of the Anglican Church and the Archbishop of Canterbury], *BOR*, 1962, 1–2, pp. 67–70.

29. Report on the Official Visits to Geneva of Representatives of the Ecumenical Patriarchate and the Romanian and Bulgarian Orthodox Churches, by Francis H. House, 17 July 1961, AWCC GS. 42.41.03.

30. Hintikka, pp. 43–7.

31. Letter from the Holy Synod of the ROC to the WCC, 15 September 1961, AWCC GS. 42.4. 065.

32. The Romanian delegation to the Third Assembly of WCC was composed of Metropolitan Justin and Father Alexandru Ionescu of the Orthodox Church; Professor Herman Binder of the Evangelical-Lutheran Church, Sibiu; Professor Tiberiu Kosma of the Hungarian-Lutheran Church, Cluj; Bishop Al. Butti of the Reformed Church, Oradea and Iosef Chivu as translator. Metropolitan Justin was elected in the Central Committee of the WCC. Alexandru Ionescu, 'Lucrările celei de a III-a Adunări Generale a Consiliului Ecumenic al Bisericilor' [The Meetings of

the Third General Assembly of the World Council of Churches], *BOR*, 1962, 3–4, pp. 270–310. Three other churches in Romania had already been members of this organisation since 1948: the Hungarian Lutheran Church in Romania; the Protestant Evangelical Church of the Augsburgian Confession; the Transylvanian Reformed Church. Current developments in the Eastern European Churches, August 1961, no. 3, Romania, LPA CECFR File 159.

33. The ROC sent the first students in 1963: Ion Bria (33 years old), Ilie Georgescu (32 years old) and Constantin Drăgușin (31 years old). Bria was the only one to speak English, the other two spoke only French. AWCC GS. 42.4.065.

34. *The Rumanian Orthodox Church*, Bucharest: The Bible and Orthodox Missionary Institute, 1962.

35. Interview with General Secretary Emilio Castro: 'Romania: The WCC and the Churches during the Silent Years', 12 January 1990, AWCC GS. 42.3.060.

36. Letter from Father Florian Gîldău to Visser't Hooft, 4 April 1979, AWCC Visser't Hooft, General Correspondence. 994.1.60/1. It remains unclear if he wrote this book; it has never been published.

37. Letter from Visser't Hooft to Father Florian Gîldău, 23 April 1979, AWCC Visser't Hooft, General Correspondence. 994.1.60/1.

38. Moscow and Satellite Churches Refuse 1951 All-Orthodox Invitation of Istanbul Patriarchate, Item 03361/53, 1951, HU OSA 300–60–1, 522/2801, Religion: Communist Influenced Religious Movements, 1951–1968.

39. The Romanian delegation left Bucharest on 19 September and returned on 7 October 1961. It was composed of Metropolitan Justin, Bishop Nicolae Corneanu, Bishop Andrei Moldovan, Nicolae Chițescu, Professor Liviu Stan, Father Grigore Cernăianu and Hieromonk Lucian Florea.

40. *Digest* 120, 7 October 1961, AKI.

41. N. Chițescu, 'Note și impresii de la Conferința Panortodoxă de la Rhodos' [Notes and Impressions from the Pan-Orthodox Conference in Rhodes], *BOR*, 1961, 9–10, pp. 854–89.

42. Anca Șincan, 'Inventing Ecumenism? Inter-Confessional Dialogue in Transylvania, Romania in the 1960s', *Religion in Eastern Europe*, 2006, 26 (3), pp. 1–16.

43. British Legation, Bucharest, Item 1781/4, TNA PRO FO 371/159527.

44. Gheorghe Gheorghiu-Dej, 'Report of the Delegation of the Rumanian Workers' Party Which Attended the 22nd Congress of the CPSU: Submitted to the Plenum of the CC, RWP Held Between November 30 and December 5, 1961', in Gheorghe Gheorghiu-Dej, *Articles and Speeches: June 1960–December 1962*, Bucharest: Meridiane Publishing House, 1963, p. 286; Gheorghe Gheorghiu-Dej, *Articole și Cuvântări* [Articles and Speeches], Bucharest: Editura Politică, 1963.

45. Ibid., p. 287–8.

46. He died at Cernica monastery in 1969: Mircea Păcurariu, *Dicționarul teologilor români* [The Dictionary of Romanian Theologians], Bucharest: Editura Enciclopedică, 2002.

47. Metropolitan Lăzărescu Charged with Embezzlement, Item 1117/66, HU OSA 300–60–1, 528/2808, Religion: Orthodox Church, 1964–6.

48. 'Lucrările Sfîntului Sinod al Bisericii Ortodoxe Romîne în sesiunea anului 1961' [The Meetings of the Holy Synod of the ROC in its 1961 Session], *BOR*, 1962, 1–2, p. 178.

49. Caravia et al., p. 27.

50. *Glasul Bisericii*, March–April, 1962.

51. *Scînteia*, 3 June 1962.

52. 'Sanctitatea Sa Alexei, Patriarhul Moscovei şi a toată Rusia, oaspete al Bisericii Ortodoxe Romîne' [His Sanctity Alexius, Patriarch of Moscow of all Russia, Guest of the ROC], *BOR*, 1962, 5–6, pp. 418–50.
53. George Enache, Adrian Nicolae Petcu, 'Biserica Ortodoxă Română şi Securitatea: Note de lectură' [The Romanian Orthodox Church and the *Securitate*: Notes], in Gheorghe Onişoru (ed.), *Totalitarism şi rezistenţă, teroare şi represiune în România comunistă* [Totalitarism and Resistance, Terror and Repression in Communist Romania], Bucharest: Consiliul Naţional pentru Studierea Arhivelor Securităţii, 2001.
54. Memorandum by Reverend M. A. Halliwell, 31 August 1962, LPA CECFR File 159.
55. Bishop Nica became archbishop of Lower Danube two years before Justinian died.
56. *Romînia Liberă*, 24 October 1962.
57. 'Vizita făcută Bisericii Ortodoxe Romîne între 23 octombrie şi 5 noiembrie 1962 de o delegaţie a Bisericii Ortodoxe Sîrbe, în frunte cu Sanctitatea Sa Gherman, Arhiepiscop de Ipec şi Patriarh al Serbiei şi Macedoniei' [The Visit to the ROC between 23 October and 5 November 1962 by a Delegation of the Serbian Orthodox Church, led by His Holiness German, Archbishop of Ipec and Patriarch of Serbia and Macedonia], *BOR*, 1962, 11–2, pp. 986.
58. The delegation was composed of Archbishop Iacobos of North and South Americas, president of the World Council of Churches; Dr Franklin Clark Fry, president of the World Lutheran Federation and president of the Executive Committee of the World Council of Churches; Dr W. A. Visser't Hooft, general secretary of the World Council of Churches and Bishop Emilianos, the representative of the Ecumenical Patriarchate to the World Council of Churches. Vintilă Popescu, 'Vizita unor reprezentanţi ai Consiliului Ecumenic al Bisericilor în ţara noastră ca invitaţi ai Bisericii Ortodoxe Romîne' [The Visit of Some Representatives of the WCC in Our Country as Guests of the ROC], *BOR*, 1962, 11–2, pp. 1072–100; *Romînia Liberă*, 10 November 1962; 17 November 1962.
59. Ioan G. Coman, 'Concilile ecumenice şi importanţa lor pentru viaţa bisericii' [The Ecumenical Councils and their Importance to Church Life], *Ortodoxia*, July–September 1962.
60. Ovidiu Bozgan, *Cronica unui eşec previzibil: România şi Sfântul Scaun în epoca pontificatului lui Paul al VI-lea (1963–1978)* [The Chronicle of a Predicted Failure: Romania and the Holy See during the Pontificate of Paul VI (1963–78)], Bucharest: Curtea Veche, 2004.
61. 'Pastorala de Sfintele Paşti a Prea Fericitului Patriarh Justinian' [Pastoral Letter for Easter of His Holiness Patriarch Justinian], *BOR*, 1963, 3–4, pp. 185–9.
62. British Legation, Bucharest, Item 1781/2, TNA PRO FO 371/171918.
63. *Religious Digest* 148, February 1964, AKI.
64. *Romînia Liberă*, 18 December 1962.
65. Erich Ditmak Tappe, CNSAS, I138685, 6 vols.
66. Eric Tappe, My Audience with the Patriarch Justinian, Note 22 April 1963, LPA CECFR File 163/1–2.
67. *Romînia Liberă*, 19 March 1963.
68. 'Adormirea întru Domnul a Prea Sfinţitului Andrei Moldovanu, Episcopul romînilor ortodocşi din America' [The Death of His Holiness Andrei Moldovanu, Bishop of Orthodox Romanians in the USA], *BOR*, 1963, 3–4, pp. 215–22.
69. British Embassy, Washington, Item 1781/3, TNA PRO FO 371/171918.
70. Ibid.
71. British Legation, Bucharest, Item 1781/3, TNA PRO FO 371/171918.

72. 'Lucrările Sfîntului Sinod al Bisericii Ortodoxe Romîne în sesiunea ordinară a anu-lui 1963' [The Meeting of the Holy Synod of the ROC in its 1963 Ordinary Session], *BOR*, 11–2, p. 1241.

73. Foreign Office, Memorandum, Item 1781/3, TNA PRO FO 371/171918.

74. Confidential. Note for File by Canon Satterthwaite, 26 June 1963, LPA CECFR File 159.

75. *Romînia Liberă*, 21 June 1963; 6 July 1963.

76. *Romînia Liberă*, 13 July 1963; 23 July 1963.

77. J. F. Brown, 'Rumania Steps Out of Line', *Survey*, 1963, 49, pp. 19–35.

78. As a listener wrote in a letter to the BBC in 1963, the Romanian broadcasts were extremely important for ordinary people who were thus able to hear a different side to communist propaganda. The listener even proclaimed that 'the BBC Romanian Service is considered almost a national institution in Romania'. BBC External Broadcasting, Audience Research Report, 1963, BBC WA, E3/158/1, Romanian Service Reports, 1955–1974.

79. British Legation, Bucharest, Item 1781/7, TNA PRO FO 371/171918.

80. Ibid.

81. Church of England Council on Foreign Relations, Confidential, Memorandum on visit to Romania 17th – 24th October 1963, by John R. Sattertwaite, TNA PRO FO 371/171918.

82. British Legation, Bucharest, Item 1781/7, TNA PRO FO 371/171918.

83. Church of England Council on Foreign Relations, Confidential, Memorandum on visit to Romania 17th – 24th October 1963, by John R. Sattertwaite, TNA PRO FO 371/171918.

84. Letter from Patriarch Justinian to Archbishop Ramsey, 23 October 1963, LPA CECFR File 163/1–2.

85. Letter from Archbishop Ramsey to Patriarch Justinian, 7 November 1963, LPA CECFR File 180/1–2.

86. Church of England Council on Foreign Relations, 2 November 1963, TNA PRO FO 371/171918.

87. The Foreign Office agreed with the visit even after the death of Gheorghiu-Dej. Confidential Memorandum, 14 April 1965, LPA CECFR File 160/1–2.

88. British Legation, Bucharest, Item 1781, TNA PRO FO 371/171918.

89. 'Lucrările Sfîntului Sinod al Bisericii Ortodoxe Romîne în sesiunea ordinară a anu-lui 1963' [The Meeting of the Holy Synod of the ROC in its 1963 Ordinary Session], *BOR*, 11–12, p. 1236.

90. Letter from Patriarch Justinian to Archbishop Ramsey, 15 February 1964, LPA CECFR File 179.

91. Canon Satterthwaite, Note for File, 6 March 1964, LPA CECFR File 179.

92. Letter from Archbishop Ramsey to Patriarch Justinian, 13 March 1964, LPA CECFR File 179.

93. Letter from Patriarch Justinian to Archbishop Ramsey, 9 April 1964, LPA CECFR File 179.

94. *Scînteia*, 23 April 1964.

95. Signs of changes in foreign trade between Romania and the Soviet Union were evi-dent from the end of the 1960s. Foreign trade decreased from 51.5 per cent in 1958 to 47.3 per cent in 1959 and 40.1 per cent in 1960. Michael Shafir, pp. 364–77; R. L. Braham, 'Rumania: Onto the Separate Path', *Problems of Communism*, 1964, 13, pp. 65–89; George Cross, 'Rumania: The Fruits of Autonomy', *Problems of Communism*, 1966, 15, pp. 16–28; R. V. Burks, 'The Rumanian National

Deviation: An Accounting', in Kurt London (ed.), *Eastern Europe in Transition*, Baltimore: John Hopkins Press, 1966; Stephen Fischer-Galaţi, 'Rumania and the Sino-Soviet Split' in Kurt London (ed.), *Eastern Europe in Transition*; John Michael Montias, 'Backgrounds and Origins of the Rumanian Dispute with Comecon', *Soviet Studies*, 1964, pp. 125–52; John Michael Montias, *Economic Development in Communist Rumania*, Cambridge, MA: The MIT Press, 1967; Maurice Pearton, *Oil and the Romanian State*, London: Oxford University Press, 1971.

96. Doreen Marguerite Berry, CNSAS, I151122, 3 vols.

97. Letter from Doreen Berry to Canon Satterthwaite, 30 April 1964, LPA CECFR File 159.

98. Confidential Report of Doreen Berry to Canon Satterthwaite on her visit to Romania, 15 June 1964, LPA CECFR File 159.

99. Confidential Report of Doreen Berry to Canon Satterthwaite on her visit to Romania, 15 June 1964, LPA CECFR File 159.

100. Letter from Patriarch Justinian to Archbishop Ramsey, 4 December 1964, LPA CECFR File 163/1–2.

101. Constantin Brâncoveanu was born in 1875 and died in 1967. He received a state pension and an additional one from the patriarch for his position at Bălaşa Church. His son was imprisoned from 1948 to 1953 and again from 1959 to 1964, the latter time for listening to the BBC. After a general amnesty in 1964, he managed to leave Romania and settled in England with his family. I am grateful to Professor Constatin Brancovan for this information. Email to the author, 22 March 2006.

102. This word is unclear in the original document apart from the first two letters 'be'.

103. Confidential Report of Doreen Berry to Canon Satterthwaite on her visit to Romania, 15 June 1964, LPA CECFR File 159.

104. Ibid.

105. Letter from Patriarch Justinian to Archbishop Ramsey, 22 June 1964, LPA CECFR File 163/1–2.

106. Gheorghe Apostol, first vice-president of the Council of Ministers; Grigore Geamănu, secretary of the State Council; Corneliu Mănescu, Minister of Foreign Affairs; Dumitru Dogaru, general secretary of the Department of Religious Confessions.

107. 'Vizita Majestăţii Sale Haile Selassie I, Împăratul Etiopiei, la Prea Fericitul Patriarch Justinian' [The Visit of His Majesty Haile Selassie I, the Emperor of Ethiopia to His Holiness Patriarch Justinian], *BOR*, 1964, 9–10, pp. 799–800.

108. Stelian Tănase, *Elite şi societate: Guvernarea Gheorghiu-Dej, 1948–1965* [Elites and Society: The Gheorghiu-Dej's Regime, 1948–65], Bucharest: Humanitas, 1998; Vladimir Tismăneanu, *Fantoma lui Gheorghiu-Dej* [The Phantom of Gheorghiu-Dej], Bucharest: Editura Univers, 1995.

109. 'Amintirea Preşedintelui Gheorghe Gheorghiu-Dej va trăi veşnic în inima poporului romîn' [The Memory of President Gheorghe Gheorghiu-Dej will Always Live in the Heart of the Romanian People], *BOR*, 1965, 3–4, pp. 173–5.

110. Churches Position in Romania, Item F-94, HU OSA 300–60–1, 518/2800, Religion: General, 1959–1969.

111. Confidential Memorandum. 14 April 1965, LPA CECFR File 160/1–2.

112. Church of England Council on Foreign Relations, Item 1781/3, TNA PRO FO 371/182733.

113. Letter from John Satterthwaite to Foreign Office, 14 May 1965, TNA PRO FO 371/182733.

114. Canon Satterthwaite, Note for file, 31 May 1965, LPA CECFR File 179. The British-Romanian Association in London was founded in 1956 by Ecaterina Iliescu, Ion Raţiu, Horia Georgescu, Marie-Jeanne Macdonald and Gladys Wilson. I am grateful to Nicolae Raţiu for providing this information.
115. Archbishop Ramsey came together with Stanley Eley, bishop of Gibraltar; Canon John Satterthwaite, general secretary of the Church of England Council on Foreign Relations; Canon John Findlow, associate secretary of the Church of England Council on Foreign Relations and Reverend John G. B. Andrew, the archbishop's Domestic Chaplain.
116. British Embassy, Bucharest, Item 1784, TNA PRO FO 371/182733.
117. On 3 June the archbishop visited the de Argeş monastery and the Argeş hydro-electric station, and on 4 June 1965 visited a simple parish, Leordeni and Târgovişte, Viforâta, Dealu and Ghighiu monasteries.
118. *Scînteia*, 8 June 1965.
119. 'Vizita Arhiepiscopului de Canterbury, Dr Arthur Michael Ramsey, Primat a toată Anglia şi Mitropolit' [The Visit of Dr Arthur Michael Ramsey, the Archbishop of Canterbury, Primate of all England and Metropolitan], *BOR*, 1965, 5–6, pp. 382–451.
120. Canon Satterthwaite, Note for file, 8 June 1965, LPA CECFR File 163/1–2.
121. Ibid.
122. Similar positions are presented in Gheorghe Gaston Marin, *În serviciul României lui Gheorghiu-Dej: Însemnări din viaţă* [Serving Gheorghiu-Dej's Romania: Notes from My Life], Bucharest: Editura Evenimentul Românesc, 2000; Paul Sfetcu, *13 ani în anticamera lui Dej* [Thirteen Years in Dej's Antichamber], Bucharest: Editura Fundaţiei Culturale Române, 2000.
123. Foreign Office, Item 1781//11, Confidential. Mr Glass's Despatch no 34 (8) of the 11th of June 1963, TNA PRO FO 371/182733.
124. Ibid.
125. Ibid.
126. *Romînia Liberă*, 9 June 1965.
127. *Lumea* 24, 10 June 1965.
128. British Embassy, Bucharest, Item 1781/13, TNA PRO FO 371/182733.
129. Justinian came to London with Bishop Antim Târgovişteanul, vicar of the Patriarchate and secretary of the Holy Synod; Father Ioan Gagiu, director of the Patriarchate Administration; Father Niţişor Cazacu, secretary of the National Ecclesiastical Council and of Patriarchate Office; Deacon Constantin Dumitrescu from the Patriarchal Cathedral.
130. Letter from Patriarch Justinian to Archbishop Ramsey, 107, f. 171, 25 November 1965; Letter from Windsor Castle, 107, f. 176, 15 June 1966, LPA Correspondence and Papers of Michael Ramsey, Archbishop of Canterbury, 1961–74; LPA CECFR Files 165 and 166.
131. On 22–23 May 1966, Nicolae Ceauşescu visited Putna monastery; on 10–11 June 1966, Cozia monastery; on 12–13 June 1966, Tismana monastery and on 8–9 September 1966, Neamţ monastery. ANIC, CC of CRP, Dossier 203/1965.

Conclusion

1. The exact figure of those imprisoned is still unknown. The National Institute for the Study of Totalitarianism identified 1725 Orthodox clergymen arrested during the Cold War period. Caravia et al., 1998.

2. Edward Behr, *Kiss the Hand You Cannot Bite: The Rise and Fall of the Ceauşescus*, New York: Willard Books, 1991; Deletant, *Ceauşescu and the Securitate*; Mary Ellen Fischer, *Nicolae Ceauşescu: A Study in Political Leadership*, Boulder: Lynne Rienner, 1989.

3. Vladimir Tismăneanu, Dorin Dobrincu, Cristian Vasile, *Comisia prezidenţială pentru analiza dictaturii comuniste din România: Raport final* [Presidential Commission for the Analysis of the Communist Dictatorship in Romania: Final Report], Bucharest: Humanitas, 2007.

4. John Richard Satterthwaite, ASRI, I185027.

5. The official Romanian publication in English, *The Romanian Orthodox Church*, stated that in 1972 the church was composed of 8185 parishes; 11,722 churches; 8564 priests and 78 deacons. *ROC*, 1972, 2 (1), p. 19.

Bibliography

Archival sources

Consiliul Naţional pentru Studierea Arhivelor fostei Securităţi, National Council for the Study of the Archives of the former *Securitate* (CNSAS), *Arhiva Serviciului Român de Informaţii*, The Archive of the Romanian Information Service, Bucharest (ASRI)
Doreen Marguerite Berry, I151122, 3 vols.
Alexandru Ciurea, I185022.
Florian Gâldău, SIE75, 3 vols, microfilm; I3714, 2 vols.
Benedict Ghiuş, I2723.
Ihatsu Vania Kalev, I 185023, 2 vols.
Vasile Leu, I185024, 4 vols.
Justinian Moisescu, I185025.
Andrei Moldovan, I185029.
Antim Nica, I701, 5 vols.
Dimelis Panaiotis Apostolos, I185028.
Chesarie Păunescu, I185031, 3 vols.
Visarion Puiu, SIE142, microfilm.
Sebastian Ruşan, I2193, 3 vols.
John Richard Satterthwaite, I185027.
Erich Ditmak Tappe, I138685, 6 vols.

Arhivele Naţionale Istorice Centrale, The Central National Historical Archives, Bucharest (ANIC)
Miron Cristea
Central Committee of the Romanian Workers' Party.

Lambeth Palace Archives, London (LPA)
Correspondence and Papers of Canon John Albert Douglas, Honorary General Secretary of the Church of England Council on Foreign Relations, 1933–45: 52.
Correspondence and Papers of Geoffrey Francis Fisher, Archbishop of Canterbury, 1945–61.
Correspondence and Papers of Michael Ramsey, Archbishop of Canterbury, 1961–74.
Lambeth Conference Papers: 196.
Church of England Council on Foreign Relations, Files:
158: Romania. General Situation, 1944–59.
159: Romania. General Situation, 1960–4.
160/1–2: Romania. General Situation, 1965–9.
163/1–2: Romania. Contact Patriarch Justinian.
165: Romania. Contact Patriarch Justinian. Visit to England, 1966. Programme, Speeches, Reports.
166: Romania. Contact Patriarch Justinian. Visit to England, 1966. Correspondence, Arrangements.
179: Romania. Visit by Archbishop Ramsey, June 1965.
180/1–2: Orthodox Churches. Romania. Anglican Visits 1959–79.

The National Archives of the United Kingdom, Public Record Office, Foreign Office, London (TNUK PRO FO)
371/100744; 371/94472; 371/78623; 371/88127; 371/100788; 371/106489; 371/111657; 371/128953; 371/1/128953; 371/143380; 371/159527; 371/171918; 371/182733; 371/159527.

Archives of Keston Institute, Oxford (AKI)
Digest; Religious Digest.
Richard Wurmbrand.

Archives of the Fellowship of Saint Alban and Saint Sergius, Oxford (AFSASS)
Romanian Orthodox Church.

BBC Written Archives, Reading (BBC WA)
E3/298/1, Romanian Service.
Policy Religion European Service File 2, 1947–54.
E1/2308/2, Romania A-Z.
E3/158/1, Romanian Service Reports, 1955–74.

Radio Free Europe Archives, Budapest (HU OSA)
300–60–1, 511/2607, Propaganda Speeches of VIPs Justinian [Patriarch 1955–6].
300–60–1, 518/2800, Religion: General, 1948–58.
300–60–1, 518/2800, Religion: General, 1959–69.
300–60–1, 522/2801, Religion: Communist Influenced Religious Movements, 1951–68.
300–60–1, 522/2802, Religion: Influence of Communist Ideology, 1948–87.
300–60–1, 522/2803, Religion: Obstacles to Worship God, 1951–65.
300–60–1, 522/2804, Religion: Persecution, 1951–5.
300–60–1, 523/2804, Religion: Persecution, 1956–76.
300–60–1, 528/2808, Religion: Orthodox Church, 1964–6.
300–60–1, 530/2808, Religion: Other Religions 1948–55.
300–60–1, 530/2808, Religion: Other Religions, 1956–9.
300–60–1, 532/2812, Religion: Religious Holidays, 1955–79.
300–60–1, 532/2900, Resistance and Criticism of the Regime: General, 1951–69.
300–60–1, 532/2903, Resistance and Criticism of the Regime: Political, 1956–71.
300–60–1, 533/2904, Resistance and Criticism of the Regime, Religions, 1952–69.
Reels 53; 60; 67.

The Archives of the World Council of Churches, Geneva (AWCC)
WCC General Secretariat. Correspondence Member Churches, 1938–93. 42.4.065.
WCC General Secretariat. Correspondence Member Churches, 1938–93. 42.4.066.
WCC General Secretariat. Country Files and Correspondence (1938) 1946–95: Europe. 42.3.060.
WCC General Secretariat. Orthodox Churches: Special Section in WCC/General Secretariat Archives. 42.41.03.
WCC General Secretariat. Orthodox Churches: Special Section in WCC/General Secretariat Archives. 42.41.02.
Visser't Hooft, General Correspondence. 994.1.60/1.
Divisions of Ecumenical Actions. 423.029.

Periodicals and newspapers

Biserica Orthodoxă Română: from 1954 *Biserica Orthodoxă Romînă* [The Romanian Orthodox Church].

Cuvântul [The Word].
Glasul Bisericii [The Voice of the Church].
Glasul Patriei [The Voice of the Fatherland].
Lumea [The World].
Lumea noastră [Our World].
Magazin Istoric [Historical Journal].
Mitropolia Banatului [The Metropolitanate of Banat].
Mitropolia Moldovei şi Sucevei [The Metropolitanate of Moldavia and Suceava].
Mitropolia Olteniei [The Metropolitanate of Oltenia].
Monitorul Oficial [Official Monitor].
Ortodoxia [Orthodoxy].
Renaşterea [Rebirth].
România Liberă: from 1954 *Romînia Liberă* [Free Romania].
Scânteia: from 1954 *Scînteia* [The Spark].
Scînteia Tineretului [The Spark of the Youth].
Studii Teologice [Theological Studies].
Telegraful Român [Romanian Telegraph].
The Guardian.
Universul [The Universe].

Published collections of documents

1866–1896. Trei-deci de ani de domnie ai Regelui Carol I: Cuvântări şi acte [1866–96. Thirty Years of the Reign of King Carol I: Speeches and Documents], vol. I, Bucharest: Institutul de Arte Grafice Carol Göbl, 1897.

Administraţia Casei Bisericii, *Biserica Ortodoxă şi Cultele Străine din Regatul României* [The Orthodox Church and Foreign Confessions in the Romanian Kingdom], Bucharest: Institutul de Arte Grafice Carol Göbl, 1904.

Administraţia Cassei Bisericii, *Regulamentul Legii Clerului Mirean şi Seminariilor* [Regulations of the Law of the Clergy and Seminaries], Bucharest: Tipografia Cărţilor Bisericeşti, 1914.

Berindei, Dan; Oprescu, Elisabeta; Stan, Valeriu, *Documente privind domnia lui Alexandru Ioan Cuza: 1859–1861* [Documents Regarding the Reign of Alexandru Ioan Cuza: 1859–61], vol. 1, Bucharest: Editura Academiei, 1989.

Bucur, Marius; Stan, Lavinia, *Persecuţia Bisericii Catolice în România: Documente din arhiva Europei Libere 1948–1960* [The Persecution of the Catholic Church in Romania: Documents from the Archive of Radio Free Europe, 1948–60], Târgu-Lăpuş: Galaxia Gutenberg, 2004.

Buzatu, Gheorghe; Chiriţoiu, Mircea, *Agresiunea comunismului în România. Documente din arhivele secrete: 1944–1989* [The Aggression of Communism in Romania. Documents from Secret Archives: 1944–89], Bucharest: Paideia, 1998.

Costescu, Chiru C., *Colecţiune de legi, regulamente, acte, deciziuni, circulări, instrucţiuni, formulare şi programe* [Collection of Laws, Regulations, Documents, Decisions, Circular Letters, Instructions, Forms and Programmes], Bucharest: Institutul de Arte Grafice C. Sfetea, 1916.

Legea şi statutele cultelor religioase din Republica Populară Română [Law and Statutes of Religious Confessions in the Romanian People's Republic], Bucharest: Editura Ministerului Cultelor, 1949.

Ministerul Cultelor şi Instrucţiunii, *Casa Bisericii 1902–1919* [The House of the Church 1902–19], Bucharest: Tipografia Cărţilor Bisericeşti, 1920.

Ministerul Cultelor şi al Instucţiunii Publice, *Acte privitoare la autocefalia Bisericei Ortodoxe a României* [Documents Concerning the Autocephaly of the Orthodox Church of Romania], Bucharest: Tipografia Cărţilor Bisericeşti, 1885.

Ministère de l'Agriculture, de l'Industrie, du Commerce et des Domaines, *La Roumanie. 1866–1906*, Bucharest: Imprimerie Socec, 1907.

Murariu, Ioan; Iancu, Gheorghe, *Constituţiile Române. Texte. Note. Prezentare comparativă* [Romanian Constitutions. Texts. Notes. Comparative Presentation], Bucharest: Regia autonomă 'Monitorul Oficial', 1995.

Păiuşan, Cristina; Ciuceanu, Radu, *Biserica Ortodoxă Română sub regimul communist 1945–1958*, [The Romanian Orthodox Church under the Communist Regime 1945–58], vol. I, Bucharest: Institutul Naţional Pentru Studiul Totalitarismului. Colecţia Documente, 2001.

Plămădeală, Antonie, *Contribuţii istorice privind perioada 1918–1939: Elie Miron Cristea, documente, însemnări şi corespondenţe* [Historical Contributions on the 1918–39 Period: Elie Miron Cristea, Documents, Notes and Correspondence], Sibiu: Tiparul Tipografiei Eparhiale, 1987.

Prigoana: Documente ale Procesului C. Noica, C. Pillat, N. Steinhardt, Al. Paleologu, A. Acterian, S. Al-George, Al. O. Teodoreanu [Persecution: Documents of the Trial of C. Noica, C. Pillat, N. Steinhardt, Al. Paleologu, A. Acterian, S. Al-George, Al. O. Teodoreanu], Bucharest: Editura Vremea, 1996.

Regulament pentru punerea în aplicare a Legei asupra clerului mirean şi seminariilor sancţionată prin înaltul decret regal no 869 din 25 februarie 1906 [Regulations for the Implementation of the Law of the Clergy and Seminaries Approved by the High Royal Decree No. 869 of 25 February 1906], Bucharest: Imprimeria Statului, 1909.

Scurtu, Ioan; Mocanu, Constantin; Smârcea, Doina, *Documente privind istoria României între anii 1918–1944* [Documents of Romanian History from 1918 to 1944], Bucharest: Editura Didactică şi Pedagogică, 1995.

Stenogramele şedinţelor Biroului Politic al Comitetului Central al Partidului Muncitoresc Român, Vol. I. 1948 [The Minutes of the Meetings of the Political Office of the Central Committee of the Romanian Workers' Party], Bucharest: Arhivele Naţionale ale României, 2002.

Stenogramele şedinţelor Biroului Politic şi ale Secretariatului Comitetului Central al PMR, vol. II. 1949 [The Minutes of the Meetings of the Political Office of the Central Committee of the Romanian Workers' Party], Bucharest: Arhivele Naţionale ale României, 2003.

Speeches

Bălan, Nicolae, *Biserica neamului şi drepturile ei. Discurs rostit la discuţia generală asupra proiectului de lege a cultelor, în şedinta dela 27 martie 1928, a senatului român* [The Church of the People and its Rights. Speech Presented in the General Discussion on the Proposed Law of Religious Confessions, in the Meeting on 27 March 1928 at the Romanian Senate], Sibiu: Tiparul Tipografiei Arhidiecezane, 1928.

Burducea, C., 'Cuvântul la deschiderea Congresului general al preoţilor ortodocşi şi al slujitorilor tuturor cultelor din România', [The Speech at the Opening of the General Congress of the Orthodox Priests and of all Religious Confessions in Romania] in *Statutul Uniunii Preoţilor Democraţi din România, [The Statute of the Union of the Democratic Priests in Romania]*, Bucharest: no publisher, 1945.

'*Cuvântarea IPSS Justinian Patriarhul României rostită cu ocazia unirii*' [The Speech of His Holiness Justinian, the Patriarch of Romania Given on the Occasion of Unity], Sighet: Tip. Gheorghe Cziple, 1949.

Gheorghiu-Dej, Gheorghe, *Articole şi Cuvântări* [Articles and Speeches], Bucharest: Editura pentru literatura politică, 1952.

Gheorghiu-Dej, Gheorghe, 'Report of the Delegation of the Rumanian Workers' Party Which Attended the 22nd Congress of the CPSU. Submitted to the Plenum of the CC, RWP Held Between November 30 and December 5, 1961' in *Articles and Speeches: June 1960–December 1962*, Bucharest: Meridiane Publishing House, 1963.

Gheorghiu-Dej, Gheorghe, *Articole şi Cuvântări* [Articles and Speeches], Bucharest: Editura Politică, 1963.

Justinian, *Apostolat social* [*Social Apostolate*], 12 vols, Bucharest: Tipografia Cărţilor Bisericeşti, 1948–76.

Justinian, IPS Locotenent de Patriarh al României, *Pastorala cu prilejul alegerilor pentru Marea Adunare Naţională-Constituantă* [Pastoral Letter on the Elections to the Great National Assembly], Bucharest, 14 March 1948.

Justinian, *Cuvântul Bisericii Ortodoxe Române în chestiunea libertăţii religioase* [The Word of the Romanian Orthodox Church Regarding Religious Freedom], Bucharest: Institutul Biblic şi de Misiune Ortodoxă, 1949.

Nicodim, *Pastorală* [Pastoral Letter], July 1941.

Nicodim, *Sfat duhovnicesc către toată suflarea românească* [Spiritual Advise to all Romanian People], Bucureşti: Tipografia Cărţilor Bisericeşti, 1941.

Nicodim, *Biruinţa nu se poate dobândi numai prin destoinicia clerului, ci prin vitejia întregii oştiri creştine* [Victory cannot be Obtained only through Clerical Ability, but through the Courage of the Whole Christian Army], Bucharest: Tipografia Cărţilor Bisericeşti, 1941.

Nicodim, *Telegrama către Majestatea Sa Regele Mihai I* [Telegram to His Majesty King Michael I], Bucharest: no publisher, 24 August 1944.

Nicodim, *Pastorala 9 Octombrie 1944* [Pastoral Letter 9 October 1944], Bucharest: no publisher.

Nicodim, *Pastorala pentru Republica Populară Română* [Pastoral Letter for the People's Republic of Romania], Bucharest: no publisher, 1948.

Articles and monographs

6 Martie 1945: Începuturile Comunizării României [6 March 1945. The Beginning of Communism in Romania], Bucharest: Editura Enciclopedică, 1995.

Academia Română, *Istoria Românilor, România întregită (1918–1940)* [The History of Romanians: United Romania (1918–1940)], vol. 8, Bucharest: Editura Enciclopedică, 2003.

Actes de la Conférence des Chef et des Représentants des Eglises Orthodoxes Autocèphales Rèunis a Moscou a L'Occasion de la Cèlèbration Solennelle des Fêtes du 500eme Anniversaire de L'Autocèphalie de L'Eglise Orthodoxe Russe, 8–18 Juillet 1948, Moscow: Editions du Patriarcat de Moscou, 1950.

Alexander, Stella, *Church and State in Yugoslavia since 1945*, Cambridge: Cambridge University Press, 1979.

American Committee on the Rights of Religious Minorities, *Roumania: Ten Years After*, Boston: The Beacon Press, 1928.

Andreescu, Liviu, 'The Construction of Orthodox Churches in Post-communist Romania', *Europe-Asia Studies*, 2007, 59 (3), pp. 451–80.

Behr, Edward, *Kiss the Hand You Cannot Bite: The Rise and Fall of the Ceauşescus*, New York: Willard Books, 1991.

Armstrong, John A., *Nations before Nationalism*, Chapel Hill: The University of North Carolina Press, 1982.

Barker, Ernest, *Social and Political Thought in Byzantium from Justinian I to the last Palaeologus*, Oxford: Clarendon Press, 1957.

Betea, Lavinia, *Alexandru Bârlădeanu despre Dej, Ceauşescu şi Iliescu. Convorbiri* [Alexandru Bârlădeanu on Dej, Ceauşescu and Iliescu. Conversations], Bucharest: Editura Evenimentul Românesc, 1998.

Betea, Lavinia, *Maurer şi lumea de ieri. Mărturii despre stalinizarea României* [Maurer and Yesterday's World. Testimonies on the Stalinisation of Romania], Arad: Fundaţia Ioan Slavici, 1995.

Bobango, Gerald J., *The Romanian Orthodox Episcopate of America: The First Half Century, 1929–1979*, Jackson, Michigan: The Romanian-American Heritage Center, 1979.

Bobango, Gerald J., *The Emergence of the Romanian National State*, Boulder: Columbia University Press, 1979.

Bobango, Gerald J., *Religion and Politics: Bishop Valerian Trifa and His Times*, New York: East European Monographs, 1981.

Boia, Lucian, *History and Myth in Romanian Consciousness*, Budapest: Central European University Press, 2001.

Bolitho, H., *Romania under King Carol*, London: Eyre & Spottiswoode, 1939.

Bossy, Raoul, 'Religious Persecution in Captive Romania', *Journal of Central European Affairs*, 1955, 15 (2), pp. 161–81.

Bozgan, Ovidiu, 'Nunţiatura Apostolică din România în anii 1948–1950' [The Apostolic Nunciature in Romania between 1948–50] in Ovidiu Bozgan (ed.), *Biserică, Putere, Societate. Studii şi Documente* [Church, Power, Society. Studies and Documents], Bucharest: Editura Universităţii din Bucureşti, 2001.

Bozgan, Ovidiu, *Cronica unui eşec previzibil: România şi Sfântul Scaun în epoca pontificatului lui Paul al VI-lea (1963–1978)* [The Chronicle of a Predicted Failure: Romania and the Holy See during the Pontificate of Paul VI (1963–78)], Bucharest: Curtea Veche, 2004.

Braham, R. L., 'Rumania: Onto the Separate Path', *Problems of Communism*, 1964, 13, pp. 65–89.

Bratiano, J. C., *La question religieuse en Roumanie: Lettre a Monsieur le Directeur de l'Opinion Nationale*, Paris: Librairie du Luxembourg, 1866.

Brown, J. F., 'Rumania Steps Out of Line', *Survey*, 1963, 49, pp. 19–35.

Buchenau, Klaus, *Orthodoxie und Katholizismus in Jugoslawien 1945–1991: Ein serbisch-kroatischer Vergleich*, Wiesbaden: Harrassowitz 2004.

Burks, R. V., 'The Rumanian National Deviation: An Accounting', in Kurt London (ed.), *Eastern Europe in Transition*, Baltimore: John Hopkins Press, 1966.

Caravia, Paul; Constantinescu, Virgiliu; Stănescu, Flori, *The Imprisoned Church, Romania, 1944–1989*, Bucharest: The National Institute for the Study of Totalitarianism, 1999.

Cândea, S., *Îndatoriri actuale ale Bisericii Ortodoxe Române* [Current Obligations of the Romanian Orthodox Church], Sibiu: Tipografia Arhidiecezană, 1946.

Chadwick, Owen, *The Christian Church in the Cold War*, London: Penguin, 1993.

Chirot, Daniel, *Social Change in a Peripheral Society: The Creation of a Balkan Colony*, New York, London: Academic Press, 1976.

Churchill, Winston, *The Second World War: Triumph and Tragedy*, vol. 6, London: Cassell, 1954.

Chrypinski, Vincent C., 'The Catholic Church in Poland, 1944–1989' in Pedro Ramet (ed.), *Catholicism and Politics in Communist Societies*, Durham, London: Duke University Press, 1990.

Chumachenko, Tatiana A., *Church and State in Soviet Russia: Russian Orthodoxy from World War II to the Khrushchev Years*, trans. by Edward E. Roslof, Armonk, London: M. E. Sharpe, 2002.

Cidharold, M. T., *La Turquie et les Principautès Danubiennes*, Paris: E. Dentu, 1857.

Ciurea, Emil, 'Religious Life' in Alexandre Cretzianu (ed.), *Captive Rumania: A Decade of Soviet Rule*, London: Atlantic Press, 1956.

Commission Princière de la Roumanie a l'Exposition universelle de Paris en 1867, *Notice sur la Roumanie principalement au point de vue de son èconomie rurale industrielle et commerciale avec une carte de la Principautè de Roumanie*, Paris: Librairie A. Franck, 1867.

Communist Exploitation of Religion: Hearing before the Subcommittee to Investigate the Administration of the Internal Security Act and Other Internal Security Laws of the Committee on the Judiciary United States Senate, Eighty-Ninth Congress, Second Session, Testimony of Rev. Richard Wurmbrand, London: European Christian Mission, 6 May 1966.

Conquest, Robert, *Religion in the USSR*, London: The Bodley Head, 1968.

Crainic, Nichifor, *Ortodoxie și Etnocrație cu o Anexă: Programul Statului Etnocratic* [Orthodoxy and Ethnocracy with an Appendix: The Programme of the Ethnocratic State], Bucharest: Editura Cugetarea, 1937.

Cristea, Miron, *Principii Fundamentale pentru Organizarea Unitară a Bisericii Ortodoxe Române* [Fundamental Principles for the United Organisation of the Romanian Orthodox Church], Bucharest: Tipografia Cărților Bisericești, 1920.

Cristea, Miron, *Rânduiala Doxologiei din prilejul încoronării majestăților lor Regelui Ferdinand I și a Reginei Maria* [The Doxology Ceremony during the Enthronement of their Majesties King Ferdinand I and Queen Maria], București: Tipografia Cărților Bisericești, 1922.

Cross, George, 'Rumania: The Fruits of Autonomy', *Problems of Communism*, 1966, 15, pp. 16–28.

Cross, F. L., *The Oxford Dictionary of the Christian Church*, Oxford, New York: Oxford University Press, 2005.

Curtiss, John Shelton, *The Russian Church and the Soviet State, 1917–1950*, Boston: Litlle, Brown & Company, 1953.

Deletant, Dennis, 'New Light on Gheorghiu-Dej's Struggle for Dominance in the Romanian Communist Party', *Slavonic and East European Review*, 1995, 73 (4), pp. 659–90.

Deletant, Dennis, *Ceaușescu and the Securitate: Coercion and Dissent in Romania, 1965–1989*, London: Hurst, 1995.

Deletant, Dennis, 'România sub regimul comunist (decembrie 1947–decembrie 1989)' [Romania under the Communist Regime (December 1947-December 1989)] in Mihai Bărbulescu, Dennis Deletant, Keith Hitchins, Șerban Papacostea, Pompiliu Teodor (eds), *Istoria României* [The History of Romania], Bucharest: Editura Enciclopedică, 1998.

Deletant, Dennis, *Communist Terror in Romania: Gheorghiu-Dej and the Police State, 1948–1965*, London: Hurst, 1999.

Deletant, Dennis, *Romania under Communist Rule*, Iași, Oxford, Portland: The Center for Romanian Studies, 1999.

Deletant, Dennis, 'Impactul revoltei maghiare în România' [The Impact of the Hungarian Revolt in Romania] in *Analele Sighet 8, Anii 1954–1960: Fluxurile și refluxurile stalinismului* [The Years 1954–60: Ebb and Flow of Stalinism], Bucharest: Fundația Academia Civică, 2000.

Deletant, Dennis, *Hitler's Forgotten Ally: Ion Antonescu and His Regime, Romania, 1940–44*, Basingstoke: Palgrave Macmillan, 2006.

Dură, Ioan, *Monahismul românesc în anii 1948–1989* [Romanian Monastic Life, 1948–89], Bucharest: Harisma, 1994.

Enache, George, *Ortodoxie și putere politică în România contemporană* [Orthodoxy and Political Power in Contemporary Romania], Bucharest: Nemira, 2005.

Enache, George and Petcu, Adrian Nicolae 'Biserica Ortodoxă Română și Securitatea. Note de lectură' [The Romanian Orthodox Church and the Securitate. Notes], in Gheorghe Onișoru (ed.), *Totalitarism și rezistență, teroare și represiune în România comunistă* [Totalitarism and Resistance, Terror and Repression in Communist Romania], Bucharest, 2001.

Enache, George; Petcu, Adrian Nicolae, 'Biserica Ortodoxă Română și Securitatea. Note de lectură' [The Romanian Orthodox Church and the Securitate. Notes], in Gheorghe Onișoru (ed.), *Totalitarism și rezistență, teroare și represiune în România comunistă* [Totalitarism and Resistance, Terror and Repression in Communist Romania], Bucharest: Consiliul Național pentru Studierea Arhivelor Securității, 2001.

Evans, Stanley; Bliss, John, *The Church in Rumania*, London: British Rumanian Friendship Association, 1954.

Fischer, Mary Ellen, *Nicolae Ceaușescu: A Study in Political Leadership*, Boulder: Lynne Rienner, 1989.

Fischer-Galați, Stephen, 'Rumania and the Sino-Soviet Split' in Kurt London (ed.), *Eastern Europe in Transition*, Baltimore: John Hopkins Press, 1966.

Fischer-Galați, Stephen, *The New Rumania: From People's Democracy to Socialist Republic*, Cambridge, MA: MIT Press, 1967.

Fischer-Galați, Stephen, *Twentieth Century Romania*, New York: Columbia University Press, 1991.

Fletcher, William C., *Soviet Believers: The Religious Sector of the Population*, Lawrence: Regents Press of Kansas, 1981.

Flora, Gavril and Szilagyi, Georgina, 'Church, Identity, Politics: Ecclesiastical Functions and Expectations towards Churches in Post-1989 Romania' in Victor Roudometof, Alexander Agadjanian, Jerry Pankhurst (eds), *Eastern Orthodoxy in a Global Age: Tradition Faces the Twenty-First Century*, Walnut Creek, CA: Altamira Press, 2005.

Floyd, David, *Rumania: Russia's Dissident Ally*, London, Dunmow: Pall Mall Press, 1965.

Gabor, Adrian and Petcu, Adrian Nicolae, 'Biserica Ortodoxă Română și puterea comunistă in timpul patriarhului Justinian' [The Romanian Orthodox Church and Communist Power during Patriarch Justinian], în *Anuarul Facultății de Teologie Ortodoxă a Universității București*, Bucharest: Editura Universității din București, 2002.

Gabor, Adrian, 'Note de lectură asupra Raportului Tismăneanu' [Notes on Tismăneanu's Report] in *Anuarul Facultății de Teologie Ortodoxă "Patriarhul Justinian Marina"*, Bucharest: Editura Universității din București, 2006, pp. 185–208.

Georgescu, Vlad, *The Romanians: A History*, trans. by Alexandra Bley-Vroman, London, New York: Tauris, 1991.

Gilberg, Trond, 'Religion and Nationalism in Romania' in Pedro Ramet (ed.), *Religion and Nationalism in Soviet and East European Politics*, Durham: Duke University Press, 1984.

Gillet, Olivier, *Religion et Nationalisme: L'Idèologie de l'Eglise Orthodoxe Roumaine sous le Règime Communiste*, Bruxelles: Universitè de Bruxelles, 1997.

Giurescu, Dinu C., *Romania in the Second World War (1939–1945)*, trans. by Eugenia Elena Popescu, Boulder: East European Monographs, 2000.

Giurescu, Dinu C., *Istoria României în date* [The History of Romania in Dates], Bucharest: Editura Enciclopedică, 2003.

Giurescu, Dinu, *Romania's Communist Takeover: The Rădescu Government*, Boulder: East European Monographs, 1994.

Golopenția, Anton; Onică, Petre, *Recensământul agricol din Republica Populară Română, 25 ianuarie 1948: rezultate provizorii* [The Agricultural Census from the Romanian People's Republic, 25 January 1948: Provisional Results], Bucharest, Direcțiunea Centrală de Statistică, 1948.

Grègoire, Henri, 'The Byzantine Church' in Norman H. Bayner and H. St. L. B. Moss (eds), *Byzantium. An Introduction to East Roman Civilization*, Oxford: Clarendon Press, 1948.

Grossu, Sergiu, *Calvarul României creștine* [The Sufferings of Christian Romania], Chișinău: Convorbiri literare & ABC Dava, 1992.

Grossu, Sergiu, *Biserica persecutată: Cronica a doi români în exil la Paris* [The Persecuted Church: The Chronicle of Two Romanians in Paris], Bucharest: Compania, București, 2004.

Gură, Nicoleta Ionescu, *Stalinizarea României. Republica Populară Română: 1948–1950. Transformări instituționale* [The Stalinisation of Romania. Romanian People's Republic: 1948–50. Institutional Transformations], Bucharest: Editura All, 2005.

Hastings, Adrian, *A History of English Christianity, 1920–1985*, London: Collins, 1986.

Hastings, Adrian, *The Construction of Nationhood: Ethnicity, Religion and Nationalism*, Cambridge, New York: Cambridge University Press, 1997.

Haynes, Rebecca, *Romanian Policy towards Germany, 1936–40*, London: Macmillan, 2000.

Heinen, Armin *Die Legion 'Erzengel Michael' in Rumänien Soziale Bewegung unde Politische Organisation*, Munich: R. Oldenbourg Verlag, 1986.

Hintikka, Kaisamari, *The Romanian Orthodox Church and the World Council of Churches, 1961–1977*, Helsinki: Luther-Agricola-Society, 2000.

Hitchins, Keith, 'The Romanian Orthodox Church and the State' in B. R. Bociurkiw and J. W. Strong (eds), *Religion and Atheism in the USSR and Eastern Europe*, London: Macmillan, 1975, pp. 314–27.

Hitchins, Keith, *Orthodoxy and Nationality: Andreiu Șaguna and the Rumanians of Transylvania, 1846–1873*, Cambridge: Harvard University Press, 1977.

Hitchins, Keith, *Romanians. 1866–1947*, Oxford: Clarendon Press, 1994.

Hitchins, Keith, 'Orthodoxism: Polemics over Ethnicity and Religion in Interwar Romania' in Ivo Banac and Katherine Verdery (eds), *National Character and National Ideology in Interwar Eastern Europe*, New Haven: Yale University Press, 1995.

Hitchins, Keith, *Romanians. 1774–1866*, Oxford: Clarendon Press, 1996.

Hupchick, Dennis P., *The Balkans from Constantinople to Communism*, New York: Palgrave Macmillan, 2002.

Hussey, J. M., *The Orthodox Church in the Byzantine Empire*, Oxford: Oxford University Press, 2004.

Hutchinson, John and Smith, Anthony (eds), *Nationalism*, Oxford, New York: Oxford University Press, 1994.

Ioanid, Radu, *The Sword of the Archangel: Fascist Ideology in Romania*, New York, Boulder: Columbia University Press, 1990.

Ionescu, Ghita, *Communism in Rumania, 1944–1962*, London, New York, Toronto: Oxford University Press, 1964.

Ionițoiu, Cicerone, *Persecuția Bisericii din România sub dictatura comunistă* [Church Persecution during the Communist Dictatorship], Freiburg: Coresi, 1983.

Iorga, Nicolae, *Istoria Bisericii Româneşti şi a vieţii religioase a Românilor* [The History of the Romanian Church and of Religious Life of Romanians], vol. 1, Vălenii-de-Munte: Tipografia 'Neamul Românesc', 1908.

Institutul Central de Statistică, *Recensământul General al Populaţiei României din 29 decembrie 1930, publicat de Dr Sabin Manuilă*, [General Census of the Population of Romania of 29 December 1930, Published by Dr Sabin Manuilă], vols 2–4, Bucharest: Monitorul Oficial, Imprimeria Naţională, 1938.

Ispir, Vasile Gh., *Misiunea actuală a Bisericii Ortodoxe Răsăritene (Misiunea Externă a Bisericii Noastre)* [The Present Mission of the Eastern Orthodox Church (The External Mission of Our Church)], Bucharest: Tipografia Cărţilor Bisericeşti, 1938.

Janz, Denis R., *World Christianity and Marxism*, New York, Oxford: Oxford University Press, 1998.

Jelavich, Barbara, *Russia and the Formation of the Romanian National State, 1821–1878*, Cambridge: Cambridge University Press, 1984.

Jelavich, Charles and Barbara, *The Establishment of the Balkan National States, 1804–1920*, Seattle: University of Washington Press, 1977.

Jowitt, Kenneth, *Revolutionary Breakthroughs and National Development: The Case of Romania, 1944–1965*, Berkeley, Los Angeles, University of California Press, 1971.

Kellogg, Frederick, *The Road to Romanian Independence*, West Lafayette: Purdue University Press, 1995.

King, Robert R., *A History of the Romanian Communist Party*, Stanford: Hoover Institution Press, 1980.

Kitromilides, Paschalis M., 'Imagined Communities and the Origins of the National Question in the Balkans', *European History Quarterly*, 1989, 19 (2), pp. 149–92.

Kitromilides, Paschalis M., *Enlightenment, Nationalism, Orthodoxy: Studies in the Culture and Political Thought of South-East Europe*, Aldershot: Variorum, 1994.

Kolarz, Walter, *Religion in the Soviet Union*, London, New York: Macmillan, 1961.

Kom, Andrè, 'Unificarea Bisericii Unite cu Biserica Ortodoxă Română în 1948' [The Unification of the Uniate Church with the Orthodox Church in 1948] in Ovidiu Bozgan (ed.), *Studii de Istoria Bisericii* [Church History Studies], Bucharest: Editura Universităţii din Bucureşti, 2000, pp. 88–124.

Kubik, Jan, *The Power of Symbols Against the Symbols of Power: The Rise of Solidarity and the Fall of State Socialism in Poland*, University Park: Pennsylvania State University Press, 1994.

L'Autriche et le Prince Roumain, Paris: E. Dentu, 1859.

La France: Le Prince Couza et la Libertè en Orient, Paris: Chez les Principaux Librairies, 1864.

La Question d'Orient et la Nation Roumaine, Paris: Librairie du Luxembourg, 1867.

Lăcustă, Ioan, *1948–1952: Republica Populară şi România* [1948–1952: People's Republic and Romania], Bucharest: Curtea Veche, 2005.

L'Empereur Napolèon III et Les Principautès Roumaines, Paris: E. Dentu, 1858.

Le Protectorat du Czar ou La Roumanie et la Roussie: Nouveaux documents sur la situation europèenne, Paris: Au Comptoir des Imprimeurs-Unis, 1850.

Leustean, Lucian N., 'Ethno-symbolic Nationalism, Orthodoxy and the Installation of Communism in Romania, 23 August 1944 to 31 December 1947', *Nationalities Papers*, 2005, 33 (4), pp. 439–58.

Leustean, Lucian N., 'The Political Control of Orthodoxy in the Construction of the Romanian State, 1859–1918', *European History Quarterly*, 2007, 37 (1), pp. 60–81.

Leustean, Lucian N. '"There is No Longer Spring in Romania, It is All Propaganda": Orthodoxy and Sovietization, 1950–52', *Religion, State and Society*, 35 (1), pp. 43–68.

Leustean, Lucian N., 'Constructing Communism in the Romanian People's Republic. Orthodoxy and State, 1948–49', *Europe-Asia Studies*, 2007, 59 (2), pp. 303–29.

Leustean, Lucian N., 'Between Moscow and London: Romanian Orthodoxy and National Communism, 1960–65', *Slavonic and East European Review*, 2007, 85 (3), pp. 491–522.

Leustean, Lucian N., '"For the Glory of Romanians": Orthodoxy and Nationalism in Greater Romania, 1918–45', *Nationalities Papers*, 35 (3), pp. 717–42.

Leustean, Lucian N., 'Religious Diplomacy and Socialism: The Romanian Orthodox Church and the Church of England, 1956–59', *East European Politics and Society*, 2008, 22 (1), pp. 7–43.

Leustean, Lucian N., 'Orthodoxy and Political Myths in Balkan National Identities', *National Identities*, 2008, 10 (forthcoming).

Leustean, Lucian N. (ed.), *Eastern Christianity and the Cold War, 1945–91*, London: Routledge (forthcoming 2009).

Levy, Robert, *Ana Pauker: The Rise and Fall of a Jewish Communist*, Berkeley: University of California Press, 2001.

Livezeanu, Irina, *Cultural Politics in Greater Romania: Regionalism, Nation Building, and Ethnic Struggle, 1918–1930*, Ithaca, London: Cornel University Press, 1995.

Manea, Vasile, *Preoți ortodocși în închisorile comuniste* [Orthodox Priest in Communist Jails], Bucharest: Editura Patmos, 2000.

Marin, Gheorghe Gaston, *În serviciul României lui Gheorghiu-Dej. Însemnări din viață* [Serving Gheorghiu-Dej's Romania. Notes from My Life], Bucharest: Editura Evenimentul Românesc, 2000.

Marina, Ovidiu, *30 de zile în URSS* [30 days in the USSR], Bucharest: Editura Institutului Biblic și de Misiune Ortodoxă, 1949.

Martin, David, *Does Christianity Cause War?*, London: Clarendon Press, 1997.

Martineau, Philip, *Roumania and her Rulers*, London: Stanley Paul & Co., 1927.

Melchisedek, Episcopul de Roman, *Raportu despre relațiunile bisericesci ale clerului orthodoxu românu cu creștinii eterodoxi seu de alt ritu și cu necredincioșii carii trăiesc în Regatul Românu* [Report on the Clerical Relations of the Romanian Orthodox Clergy with Heterodox Christians or Other Confessions and with Non-believers Who Live in the Romanian Kingdom], Bucharest: Tipografia Cărților Bisericești, 1882.

Melchisedec Episcopul Romanului, *Memoriu despre starea preoților din România și despre posițiunea lor morală și materială* [Report on the Clergy Situation in Romania and on their Moral and Material Situation], Bucharest: Tipografia Cărților Bisericești, 1888.

Meyendorff, John, *Byzantine Theology: Historical Trends and Doctrinal Themes*, New York: Fordham University Press, 1974.

Meyendorff, John, *The Byzantine Legacy in the Orthodox Christianity*, Crestwood: St Vladimir's Seminary Press, 1982.

Michelson, Paul E., *Conflict and Crisis: Romanian Political Development, 1861–1871*, New York: Garland Pub., 1987.

Mitrany, D., *The Land and the Peasant in Rumania*, Cambridge: Cambridge University Press, 1924.

Mojzes, Paul, *Religion Liberty in Eastern Europe and the USSR Before and After the Great Transformation*, Boulder: East European Monographs, 1992.

Montias, John Michael, 'Backgrounds and Origins of the Rumanian Dispute with Comecon', *Soviet Studies*, 1964, pp. 125–52.

Montias, John Michael, *Economic Development in Communist Rumania*, Cambridge, MA: The MIT Press, 1967.

Mylonas, Christos, *Serbian Orthodox Fundamentals: The Quest for an Eternal Identity*, Budapest, New York: Central European University Press, 2003.

Oprea, Marius, *Moştenitorii Securităţii* [The *Securitate*'s Successors], Bucharest: Humanitas, 2004.

Papathomas, G. D., *Le Patriarcat Oecumenique de Constantinople (y compris la Politeia monastique du Mont Athos) dans l'Europe unie*, Katerini, Athens: Editions Epektasis, 1998.

Parkinson, Mande, *Twenty Years in Roumania*, London: George Allen & Unwin, 1921.

Partidul Muncitoresc Romîn, *Congresul al II-lea al PMR, 23–28 Dec. 1955* [The Second Congress of the RWP, 23–28 December 1955], Bucharest: Editura Politică, 1956.

Partidul Muncitoresc Romîn, *Congresul al III-lea al PMR, 20–25 iunie 1960*, [The Third Congress of the RWP, 20–25 June 1960], Bucharest: Editura Politică, 1960.

Păcurariu, Mircea, *Istoria Bisericii Ortodoxe Române* [The History of the Romanian Orthodox Church], Bucharest: IBMBOR, 1981.

Păcurariu, Mircea, *Dicţionarul teologilor români* [The Dictionary of Romanian Theologians], Bucharest: Editura Enciclopedică, 2002.

Pătrăşcanu, Lucreţiu, *Sub trei dictaturi* [Under Three Dictatorships], Bucharest: Editura Politică, 1970.

Pearton, Maurice, *Oil and the Romanian State*, London: Oxford University Press, 1971.

Pearton, Maurice; Deletant, Dennis, 'The Soviet Takeover in Romania, 1944–48' in Gill Bennett (ed.), *The End of the War in Europe 1945*, London: HMSO, 1996.

Pelikan, Jaroslav, *The Christian Tradition. A History of the Development of Doctrine*, vol. 2: *The Spirit of Eastern Christendom (600–1700)*, Chicago, London: The University of Chicago Press, 1974.

Perica, Vjekoslav, *Balkan Idols: Religion and Nationalism in Yugoslav States*, Oxford: Oxford University Press, 2002.

Persak, Krzysztof; Kamiński, Lukasz (eds), *A Handbook of the Communist Security Apparatus in East Central Europe, 1944–1989*, Warsaw: Institute of National Remembrance, 2005.

Petcu, Adrian Nicolae, 'Cazul Episcopului Nicolae Popovici' [The Case of Bishop Nicolae Popovici] in Miruna Tătaru-Cazaban (ed.), *Teologie şi politică: De la sfinţii părinţi la Europa unită* [Theology and Politics: From the Holy Fathers to United Europe], Bucharest: Anastasia, 2004.

Petcu, Adrian Nicolae, 'Securitatea şi cultele în 1949' [The Securitate and Religious Confessions in 1949] in Adrian Nicolae Petcu (ed.), *Partidul, Securitatea şi Cultele, 1945–1989* [Party, Securitate and Religious Confessions, 1945–89], Bucharest: Nemira, 2005.

Petcu, Adrian Nicolae, 'Profesorul Theodor M. Popescu şi regimul comunist' [Professor Theodor M. Popescu and the Communist Regime] in Consiliul Naţional pentru Studierea Arhivelor Securităţii, *Arhivele Securităţii* [The Securitate Archives], vol. 1, Bucharest: Pro Historia, 2002.

Pimen, Mitropolitul Moldovei, *Mărăşeşti, locul biruinţei cu biserica neamului*, [Mărăşeşti, the Place of Victory with the Church of the People], Neamţ: Tipografia Monastirei Neamţu, 1924.

Plebiscitulŭ Poporului Românu [Census of the Romanian People], Bucharest: Typografia Cesar Boliac, 1864.

Pope, Earl A., 'Protestantism in Romania' in Sabrina Petra Ramet (ed.), *Protestantism and Politics in Eastern Europe and Russia: The Communist and Postcommunist Eras*, Durham, London: Duke University Press, 1992.

Popescu, Tudor R., *Salvarea bisericilor de orice rit din România sub ocupaţia militară sovietică* [The Salvation of Church under Any Confession in Romania under the Soviet Military Occupation], Bucharest: Asociaţia Română pentru Educaţie Democratică, 1999.

Pospielovsky, Dimitri V., *A History of Soviet Atheism in Theory and Practice, and the Believer*, London: Macmillan, 1987.

Prodan, Costică; Preda, Dumitru, *The Romanian Army during the First World War*, Bucharest: Univers Enciclopedic, 1998.

Quelques mots sur la sècularisation des biens conventuels en Roumanie, Paris: E. Dentu, 1864.

Quinlan, Paul D., *Clash over Romania. British and American Policies towards Romania: 1938–1947*, Los Angeles: American Romanian Academy, 1977.

Raikin, Spas T., 'The Communists and the Bulgarian Orthodox Church, 1944–1948: The Rise and Fall of Exarch Stefan', *Religion in Communist Lands*, 1984, 12, pp. 281–91.

Raikin, Spas T., 'The Bulgarian Orthodox Church' in Pedro Ramet (ed.), *Eastern Christianity and Politics in the Twentieth Century*, Durham, London: Duke University Press, 1988.

Ramet, Pedro (ed.), *Eastern Christianity and Politics in the Twentieth Century*, Durham, London: Duke University Press, 1988.

Ramet, Pedro (ed.), *Catholicism and Politics in Communist Societies*, Durham, London: Duke University Press, 1990.

Ramet, Sabrina Petra (ed.), *Protestantism and Politics in Eastern Europe and Russia: The Communist and Postcommunist Eras*, Durham, London: Duke University Press, 1992.

Ramet, Sabrina Petra (ed.), *Religious Policy in the Soviet Union*, Cambridge: Cambridge University Press, 1993.

Ramet, Sabrina Petra, 'The Serbian Church and the Serbian Nation' in Sabrina Petra Ramet and Donald W. Treadgold (eds), *Render Unto Caesar. The Religious Sphere in World Politics*, Washington: The American University Press, 1995.

Ramet, Sabrina P., *Nihil Obstat: Religion, Politics, and Social Change in East-Central Europe and Russia*, Durham, London: Duke University Press, 1998.

Rațiu, Ion, 'The Communist Attacks on the Catholic and Orthodox Churches in Rumania', *The Eastern Churches Quarterly*, London, 8 (3), 1949, pp. 163–97.

Rațiu, A.; Virtue, W., *Stolen Church: Martyrdom in Communist Romania*, Hungtinton: Our Sunday Visitor, 1978.

Recensământul Populațiunei din Decembrie 1899 [Census of the Population in December 1899], Bucharest: Eminescu, 1905.

Regulamentul general al Societății Ortodoxe Naționale a Femeilor Române [General Regulations of the National Orthodox Society of Romanian Women], Bucharest: Tipografia Albina, 1943.

Report of the Conference at Bucharest from June 1st to June 8th, 1935 between the Rumanian Commission on Relations with the Anglican Communion and the Church of England Delegation Appointed by the Archbishop of Canterbury, London: Church House, 1936.

Riis, Carsten, *Religion, Politics, and Historiography in Bulgaria*, Boulder, New York: East European Monographs, 2002.

Riker, T. W., *The Making of Roumania: A Study of an International Problem, 1856–1866*, London: Oxford University Press, 1931.

Roberts, Henry, *Rumania: Political Problems of an Agrarian State*, New Haven: Yale University Press, 1951.

Romania. Direcția Generală de Statistică, *Recensămîntul populației din 21 februarie 1956, Rezultate Generale* [21 February 1956 Census, General Results], Bucharest: Direcția Generală de Statistică, 1959.

Romania. Direcția Centrală de Statistică, *Recensămîntul populației și locuințelor din 15 martie 1966: Rezultate generale* [The Census of Population and Buildings on 15 March 1966: General Results], vol. 1, Bucharest: Direcția Centrală de Statistică, 1969.

Rosu, George; Vasiliu, Mircea and Crisan, George, 'Church and State in Romania' in Vladimir Gsovski (ed.), *Church and State Behind the Iron Curtain: Czechoslovakia, Hungary, Poland, Romania with an Introduction on the Soviet Union*, New York: Frederick A. Pragaer, 1956.

Rumanian National Committee, *Persecution of Religion in Rumania*, Washington: no publisher, 1949.

Runciman, Steven, *Byzantine Civilisation*, London: Edward Arnold, 1933.

Runciman, Steven, *The Great Church in Captivity: A Study of the Patriarchate of Constantinople from the Eve of the Turkish Conquest to the Greek War of Independence*, Cambridge: Cambridge University Press, 1968.

Runciman, Steven, *The Orthodox Churches and the Secular State*, Trentham, Auckland: University Press & Oxford University Press, 1971.

Rura, M. J., *Reinterpretation of History as a Method of Furthering Communism in Romania*, Washington: Georgetown University Press, 1961.

Sandru, Ilie; Borda, Valentin, *Un nume pentru istorie – Patriarhul Elie Miron Cristea* [A Name for History – Patriarch Elie Miron Cristea], Târgu-Mureş: Cartea de Editură 'Petru Maior', 1998.

Scarfe, Alan, 'The Romanian Orthodox Church' in Pedro Ramet (ed.), *Eastern Christianity and Politics in the Twentieth Century*, Durham, London: Duke University Press, 1988.

Scarfe, Alan, 'Patriarch Justinian of Romania: His Early Social Thought', *Religion in Communist Lands*, 1977, 5 (3), pp. 164–9.

Schöpflin, George, 'The Ideology of Rumanian Nationalism', *Survey*, 1974, 20 (2–3), pp. 77–104.

Scurtu, Ioan, *Minorităţile naţionale din România, 1931–1938* [National Minorities in Romania, 1931–38], Bucharest: Editura Arhivele Statului, 1999.

Scurtu, Ioan; Alexandru, Ion; Bulei, Ion; Mamina, Ion, *Enciclopedia de istorie a României* [The Encyclopaedia of Romanian History], 3rd edition, Bucharest: Editura Meronia, 2002.

Seton-Watson, R. W., *A History of the Roumanians: From Roman Times to the Completion of Unity*, Cambridge: The University Press, 1934.

Sfetcu, Paul, *13 ani în anticamera lui Dej* [Thirteen Years in Dej's Antichamber], Bucharest: Editura Fundaţiei Culturale Române, 2000.

Shafir, Michael, 'Romanian Foreign Policy under Dej and Ceauşescu' in George Schöpflin (ed.), *The Soviet Union and Eastern Europe: A Handbook*, Oxford, New York: Muller, Blond & White, 1986, pp. 364–77.

Smith, Anthony D., *Chosen People: Sacred Sources of National Identity*, Oxford: Oxford University Press, 2003.

Spinka, Matthew, *The Church in Soviet Russia*, New York: Oxford University Press, 1956.

Stan, George, *Părintele Patriarch Justinian Marina* [Father Patriarch Justinian Marina], Bucharest: EIBMBOR, 2005.

Stan, Lavinia and Turcescu, Lucian, *Religion and Politics in Post-Communist Romania*, Oxford: Oxford University Press, 2007.

Stroyen, William B., *Communist Russia and the Russian Orthodox Church, 1943–1962*, Washington: The Catholic University of America Press, 1967.

Şincan, Anca, 'Inventing Ecumenism? Inter-Confessional Dialogue in Transylvania, Romania in the 1960s', *Religion in Eastern Europe*, 2006, 26 (3), pp. 1–16.

Şincan, Anca, 'Mechanisms of State Control over Religious Denominations in Romania in the late 1940s and early 1950s' in B. Apor, P. Apor, A. Rees (eds), *The Sovietization of Eastern Europe*, Washington: New Academia Publishing, 2008.

Tănase, Stelian, *Elite și societate: Guvernarea Gheorghiu-Dej, 1948–1965* [Elites and Society: The Gheorghiu-Dej's Regime], Bucharest: Humanitas, 1998.

The Canonical Status of the Romanian Orthodox Episcopate of America, no place: Episcopal Council of the Romanian Orthodox Episcopate of America, 1954.

The Churches of Europe under Communist Governments: A Survey Presented with the Fifteenth Report of the Church of England Council on Foreign Relations and Communicated to the February Session of the Church Assembly, 1954, by Order of the Archbishop of Canterbury, Westminster: Church Information Board, 1954.

Theodorian-Carada, M., *Politica Religioasă a României. Conferință ținute la Cercul de studii al Partidului Conservator, în ziua de 3 decembrie 1916* [The Religious Politics of Romania. Conference Held at the Centre for Studies of the Conservative Party on 3 December 1916], Bucharest: Tipografia 'Cooperativa', 1916.

The Rumanian Orthodox Church, Bucharest: The Bible and Orthodox Missionary Institute, 1962.

Tismăneanu, Vladimir, *Fantoma lui Gheorghiu-Dej* [The Phantom of Gheorghiu-Dej], Bucharest: Editura Univers, 1995.

Tismăneanu, Vladimir, *Stalinism for All Seasons: A Political History of Romanian Communism*, Berkeley, Los Angeles, London: University of California Press, 2003.

Tismăneanu, Vladimir; Dobrincu, Dorin; Vasile, Cristian, *Comisia prezidențială pentru analiza dictaturii comuniste din România: Raport final* [Presidential Commission for the Analysis of the Communist Dictatorship in Romania: Final Report], Bucharest: Humanitas, 2007.

Tobias, Robert, *Communist-Christian Encounter in East Europe*, Indianapolis: School of Religion Press, 1956.

Torrey, Glenn E., *Romania and World War I: A Collection of Studies*, Iași, Portland: Center for Romanian Studies, 1998.

Troncotă, Cristian, *Istoria securității regimului comunist din România, vol. 1, 1948–1964* [The History of the *Securitate* of the Communist Regime in Romania. vol. 1, 1948–64], Bucharest: Institutul Național pentru Studiul Totalitarismului, 2003.

Țugui, Pavel, *Istoria și limba română în vremea lui Gheorghiu-Dej: Memoriile unui fost șef de secție a CC al PMR* [Romanian History and Language in Gheorghiu-Dej's Time: The Memoirs of a Former Head of Section of the Central Committee of the RWP], Bucharest: Editura Ion Cristoiu, 1999.

Ure, P. N., *Justinian and His Age*, Wesport: Greenwood Press, 1979.

Vască, Ioan, *O călătorie istorică: Vizita Înalt Prea Sfințitului Patriarh Nicodim al României la Moscova. Reluarea legăturilor dintre Biserica Ortodoxă Română și Biserica Ortodoxă Rusă* [A Historic Journey: The Visit of His Holiness Patriarch Nicodim of Romania to Moscow. The Re-establishment of Relations between the Romanian Orthodox Church and the Russian Orthodox Church], Bucharest: Tiparul Cărților Bisericești, 1947.

Vasile, Cristian, 'Autoritățile comuniste și problema mânăstirilor ortodoxe în anii '50' [The Communist Authorities and the Problem of the Monasteries in the 1950s], in *Analele Sighet 8, Anii 1954–1960: Fluxurile și refluxurile stalinismului* [The Years 1954–60: Ebb and Flow of Stalinism], Bucharest: Fundația Academia Civică, 2000.

Vasile, Cristian, 'Atitudini ale clericilor ortodocși și catolici față de URSS și față de regimul de tip sovietic (1944–1948)' [Attitudes of Orthodox and Catholic Clergy towards the USSR and the Soviet Regime (1944–48)] in Ovidiu Bozgan (ed.), *Biserică, Putere, Societate. Studii și Documente* [Church, Power, Society. Studies and Documents], Bucharest: Editura Universității din București, 2001.

Vasile, Cristian, *Între Vatican și Kremlin: Biserica Greco-Catolică în timpul regimului comunist* [Between the Vatican and the Kremlin: The Greek Catholic Church during the Communist Regime], Bucharest: Curtea Veche, 2003.

Vasile, Cristea, *Istoria Bisericii Greco-Catolice sub regimul comunist 1945–1989: Documente și mărturii* [The History of the Greek Catholic Church under the Communist Regime 1945–1989: Documents and Testimonies], Iași: Polirom, 2003.

Vasile, Cristian, *Biserica Ortodoxă Română în primul deceniu communist* [The Romanian Orthodox Church in the First Communist Decade], Bucharest: Curtea Veche, 2005.

Vasilev, A. A., *History of the Byzantine Empire*, Wisconsin, London: University of Wisconsin Press, 1980.

Vassiliadis, Petros, 'Orthodox Christianity' in Jacob Neusner (ed.), *God's Rule: The Politics of World Religions*, Washington: Georgetown University Press, 2003.

Veiga, Francisco, *La Mistica del Ultranacionalismo: Historia de la Guardia de Hierron. Rumania, 1919–1941*, Bellaterra: Universitat Autonoma de Barcelona, 1989. In Romanian, *Istoria Gărzii de Fier, 1919–1941, Mistica ultranaționalismului*, Bucharest: Humanitas, 1995.

Velehorschi, Gh., *Colaborarea Bisericii Ortodoxe Române cu Straja Țării: Raport pentru Congresul Național Bisericesc* [Collaboration of the Orthodox Church with the Guardians of the Country: Report for the National Clerical Congress], Bucharest: Tipografia Cărților Bisericești, 1939.

Velicu, Dudu, *Biserica Ortodoxă în perioada sovietizării României. Însemnări zilnice: 1945–1947* [The Orthodox Church during Romania's Sovietization. Dairies: 1945–47], Bucharest: Arhivele Naționale ale României, 2004.

Velicu, Dudu, *Biserica Ortodoxă în perioada sovietizării României. Însemnări zilnice: 1948–1959* [The Orthodox Church during Romania's Sovietization. Diaries: 1948–59], Bucharest: Arhivele Naționale ale României, 2005.

Verdery, Katherine, *National ideology under socialism: Identity and cultural politics in Ceausescu's Romania*, Berkeley: University of California Press, 1991.

Waddams, H. M., *Anglo-Russian Theological Conference, Moscow 1956: A Report of a Theological Conference held between Members of a Delegation from the Russian Orthodox Church and a Delegation from the Church of England. With a Preface by A. M. Ramsey, Archbishop of York*, London: The Faith Press, 1957.

Ware, Timothy, *The Orthodox Church*, London: Penguin Books, 1997.

Wurmbrand, Richard, *Tortured for Christ*, London: Lakeland, 1967.

Index